The Great Power Competition Volume 6

Adib Farhadi · Mark Grzegorzewski ·
Anthony J. Masys
Editors

The Great Power Competition Volume 6

The Rise of China

Springer

Editors
Adib Farhadi
College of Arts and Sciences, University of South Florida
Tampa, FL, USA

Anthony J. Masys
University of South Florida
College of Public Health
Tampa, FL, USA

Mark Grzegorzewski
Department of Security Studies and International Affairs, College of Arts and Sciences
Embry-Riddle Aeronautical University
Daytona Beach, FL, USA

ISBN 978-3-031-70766-7 ISBN 978-3-031-70767-4 (eBook)
https://doi.org/10.1007/978-3-031-70767-4

This Springer imprint is published by the registered company Springer Nature Switzerland AG
The registered company address is: Gewerbestrasse 11, 6330 Cham, Switzerland

Preface

The 2013 launch of China's Belt and Road Initiative (BRI) marked a transformative inflection point in great power competition, signaling China's ambition to rival U.S. influence on the global stage. Designed to provide an alternative to traditional Western aid and development frameworks, the BRI funds large-scale infrastructure projects in developing countries, particularly in resource-rich regions such as Central and South Asia (CASA), the Middle East, and Africa. These projects—spanning deep-sea ports, railways, power plants, and more—enhance economic connectivity and trade while offering China strategic footholds to counter Western influence and secure preferential access to natural resources.

A key factor contributing to the appeal of the BRI is its promise of substantial infrastructure investment coupled with a policy of non-interference in domestic affairs—an attractive proposition for nations wary of Western "conditionalities." To date, the BRI involves approximately 150 countries, representing nearly two-thirds of the global population and over half of the global GDP. Its multidimensional structure includes the Silk Road Economic Belt, Maritime Silk Road, and initiatives like the Digital and Health Silk Roads.

Before the BRI's introduction, the United States largely perceived China as a formidable regional power and a rising trading partner rather than a global competitor. However, the BRI's scope and ambition has reshaped this view, prompting the United States to recalibrate its policies and acknowledge China as a near-peer rival. The initiative not only reflects China's aspiration to reshape the international order but also underscores its transformation into a global power. Projects like the China-Pakistan Economic Corridor (CPEC) and institutions such as the Asian Infrastructure Investment Bank (AIIB) exemplify China's expanded global influence.

China's rise is further bolstered by cultural diplomacy initiatives, such as its international network of Confucius Institutes, which promote Chinese language and culture worldwide. China has also increasingly focused on the role of religion and religious diplomacy in BRI outcomes and risk mitigation, largely through government-sponsored think tanks and institutes. Religious groups, the government, and academia collaborate under the direction of the Chinese state to inform the bulk of BRI religious diplomacy, showcasing a whole of government approach.

The BRI has significantly contributed to China's global rise, positioning it as a central player in economic development and soft power politics. By fostering partnerships across developing nations, the initiative has shifted the dynamics of great power competition and invigorated momentum for the BRICS nations. As the U.S. grapples with this new geopolitical reality, it faces the challenge of adapting its strategies to counterbalance China's growing global influence.

This volume, complementing the seventh Great Power Competition Conference hosted by the University of South Florida, delves into these and other timely themes, offering a comprehensive analysis of China's rise and its implications for modern geopolitics. Addressing policymakers, scholars, and practitioners, it provides interdisciplinary perspectives and actionable insights into the evolving nature of international relations and soft power dynamics. The book aims to illuminate the complexities of great power competition and inform strategies for navigating this shifting global landscape by fostering dialogue and knowledge exchange.

Tampa, USA
December 2024

Adib Farhadi

Acknowledgments

The editors would like to extend their sincerest gratitude to the entire University of South Florida and the National Defense University Near East South Asia Center for Strategic Strategies for their work on the Great Power Competition Conferences Series and the resulting edited volumes.

They would like to express their appreciation to General (Ret.) Frank McKenzie, Executive Director, Global and National Security Institute, Executive Director, Florida Institute for Cybersecurity, and Former Command, United States Central Command, for his continued and enthusiastic support for the initiative.

A most heartfelt thank you to Sarah White for her tireless researching, editing, and coordinating assistance, and to Andrew Roberts for his invaluable graphic design work.

We are very grateful to the contributing authors for making this book possible.

Dr. Adib Farhadi would like to thank his wife, Elaha, children, Adam, Sophia, and mother, Maliha, for their support, patience, and love, not only in regard to this project but also in everything that life brings.

Contents

Contributors

Mark Abdollahian Department of Politics and Economics, Claremont Graduate University, Claremont, CA, USA

Dean Cheng U.S. Institute of Peace, Washington, D.C., USA

G. Douglas Davis MSIR Program, Troy University, Troy, AL, USA

Alice Dell'Era Security Studies and International Affairs, Embry-Riddle Aeronautical University, Daytona Beach, FL, USA

Khaled Eid ACERTAS, Claremont Graduate University, Claremont, CA, USA

Gaspard Estrada Advisory Board Member, Global South Unit, London School of Economics, London, UK

Adib Farhadi College of Arts and Sciences, University of South Florida, Tampa, FL, USA

Mark Grzegorzewski Embry-Riddle Aeronautical University, Daytona Beach, FL, USA

Zhamilia Klycheva TransResearch Consortium (TRC), Claremont, CA, USA

Alvaro Mendez London School of Economics and Political Science (LSE), LSE Global South Unit, London, UK

Peter J. Middlebrook Geopolicity, Dubai, UAE

Lawrence C. Reardon University of New Hampshire, Portsmouth, NH, USA

Leif Rosenberger The Peacekeeping and Stability, Operations Institute (PKSOI), US Army War College, Carlisle, PA, USA

Omar Samad Atlantic Council, Fairfax, VA, USA

Michael O. Slobodchikoff Department of Political Science, Troy University, Troy, AL, USA

Marcus Snow The Peacekeeping and Stability, Operations Institute (PKSOI), US Army War College, Carlisle, PA, USA

Amba Tadaa Geopolicity, Dubai, UAE

Introduction: The Rise of China

Adib Farhadi

Abstract The 2013 launch of the Belt and Road Initiative (BRI) marked a turning point in China's rise as a global power, signaling its intent to challenge U.S. influence. Central to this strategy is the development of large-scale infrastructure and mining projects, coupled with a policy of non-interference in the domestic affairs of partner countries. This pragmatic approach has been pivotal in "winning hearts and minds" across key strategic regions, significantly enhancing China's global influence and drawing parallels to the ancient Silk Road. China's strengthening relations with Afghanistan exemplify this strategy. Following the U.S. withdrawal in 2021, China prioritized economic cooperation with the Taliban-led government, bypassing ideological differences. This strategic engagement is rapidly drawing Afghanistan into China's sphere of influence, reflecting a broader trend of declining U.S. presence in pivotal regions and China's readiness to fill the void to advance its national interests. As China's influence grows in resource rich, strategically vital areas; the United States must recalibrate its foreign policy to address this shift. *The Rise of China* brings together global experts to examine and analyze the implications of China's ascent across economic, political, technological, and security domains, as well as U.S. strategies within the context of great power competition. This introductory chapter highlights China's use of economic soft power to expand its global influence, setting the stage for the subsequent essays that tackle this geopolitical challenge.

Keywords Afghanistan · China · Soft power · Influence · Great power competition · U.S. policy · Middle East · Central Asia-South Asia

---

A. Farhadi (✉)
College of Arts and Sciences, University of South Florida, Tampa, FL, USA
e-mail: Farhadi@usf.edu

A. Farhadi et al. (eds.), *The Great Power Competition Volume 6*,
https://doi.org/10.1007/978-3-031-70767-4_1

1 Introduction

The People's Republic of China (PRC) has been steadily gathering influence on the world stage for the greater part of the 21st century. However, it was not until 2013 that its wider ambitions to become a near-peer economic rival to the U.S. were solidified through the launch of its Belt and Road Initiative (BRI)—a massive interstate infrastructure connectivity network that has propelled China forward as a major player in the modern- day Great Power Competition (GPC) (Farhadi 2023). One of China's key geopolitical strategies and foreign policy objectives is leveraging "soft power" through integrated BRI channels to advance its global influence and national interest.

In keeping with China's global vision, the BRI has galvanized momentum for BRICS, the intergovernmental alliance formed in 2009, which includes Brazil, China, Egypt, Ethiopia, India, Iran, Russia, South Africa, and the United Arab Emirates (BRICS 2024). Focused on enhancing trade and investments, the organization's primary goal is to promote cooperation between member states and emerging economies, while advocating for a multipolar system (BRICS 2024). BRICS collectively represents approximately thirty-five percent of the global GDP (BRICS 2024), challenging Western-dominated financial institutions by establishing alternatives like the multilateral New Development Bank (NDB).

Central also to the wider aims of the PRC is the diminishing of Western liberal democratic governance ideals to weaken U.S. influence and shift the equilibrium of the international order in favor of a multi-polar system (Beckley and Brands 2023). By advancing the integration of critical sectors, China seeks to reconfigure the global landscape to align with its strategic interests (Doshi 2021; Jie and Wallace 2021). This approach is particularly evident in China's expanding presence in key resource rich regions such as the Middle East and Central Asia- South Asia (CASA), where most of China's hydrocarbon imports originate. As noted by General Frank McKenzie at the 2023 GPC Conference, these areas are becoming a focal point of U.S.-China strategic competition, with the potential for Chinese military bases to follow economic endeavors.

Moreover, the BRI, which aims to jump-start economic advancement among partner nations, can be viewed as filling a critical void in the international aid and development landscape. The initiative marks a significant shift from traditional paradigms by emphasizing non-interference policies and the absence of strict "conditionalities" (Hillman 2020). China's approach addresses two major shortcomings of conventional international aid and development: the lack of large-scale infrastructure projects in developing countries and the imposition of Western-centric governance models. By offering essential infrastructure without imposing political or ideological contingencies, the BRI ostensibly grants partner nations greater autonomy and self-governance.

China also deftly leverages religion and culture through diplomatic channels (directly) and policies of non-interference (indirectly), to enhance its influence at the grassroots level and safeguard substantial BRI investments (Raymond 2020).

Through state run religious diplomacy, China engages religious communities and institutions in partner countries, with the objective of building trust and mitigating risks (Jinping 2017; Zhang 2023; Yihua 2012; Yihua 2015). Simultaneously, non-interference in domestic matters can cultivate a sense of autonomy and confidence, particularly among religiously conservative states that may otherwise feel shunned by Western liberal idealism. This approach apparently involves respecting local customs and values, which aids in "winning the hearts and minds" of local populations (Yihua 2015). By creating a favorable environment through this strategic framing, China seeks to preemptively address potential conflicts or tensions during BRI development and operations.

Given China's adept leveraging of religion and culture, even indirectly, to enhance regional influence, two critical questions arise: with its current contingent approach to foreign aid and development, can the United States meet the nuanced challenges of the modern GPC to compete effectively with China in strategically pivotal regions such as Central Asia-South Asia (CASA), the Middle East, and Africa? Are ideologically conditional approaches to foreign partnerships becoming obsolete, as soft power strategies increasingly center on building trust and influence at the grass roots, ideational level?

Navigating these complex geopolitical landscapes requires a nuanced understanding of the role of religion in regions driven as much by enduring cultural and ideological traditions as material necessity (Farhadi 2023; Farhadi 2020). Consequently, the United States now faces the challenge of adopting a more pragmatic approach to strategic competition in CASA and the Middle East, one that takes ground realities into account and can effectively accommodate diverse perspectives without compromising its own ideals. Within this context, religion and religious diplomacy emerge as crucial considerations for expanding U.S. influence (Mandaville 2022). Afghanistan serves as a prime example of how China, despite its own non-religious perspective, leverages non-interference practices in religiously conservative nations to deepen relations and expand regional influence.

China's growing interest in Afghanistan is driven in no small part by the country's vast natural resources, particularly critical minerals such as lithium. The strategic value of these resources extends well beyond China's borders, as they are essential to the broader global economy, especially in the context of U.S. competition with China. According to a 2022 U.S. Department of Defense (DoD) action plan, "China dominates the global advanced battery supply chain, including lithium hydroxide (94 percent), cells (76 percent), electrolyte (76 percent), lithium carbonate (70 percent), anodes (65 percent), and cathodes (53 percent)." However, China's current mineral reserves cannot sustain growing global consumer and military demands indefinitely. China has for some time been exploring other regions such as Afghanistan for additional sources of critical minerals.

Furthermore, Afghanistan serves as a vital land bridge between China and Iran. By developing a land-bridge directly to Iran, China hopes to secure energy resources, particularly oil and gas, and create more efficient trade routes between Central Asia and the Middle East. Afghanistan's central position in Asia makes it a critical piece in this geopolitical puzzle. The China-Pakistan Economic Corridor (CPEC) and the potential expansion of infrastructure networks into Afghanistan can further strengthen Beijing's economic ties with Iran. One of the most notable recent developments was China's announcement in May 2023 to extend the China-Pakistan Economic Corridor (CPEC) to Afghanistan, which will incorporate the country into the BRI. The Taliban's attendance at China's October 2023 Belt and Road Forum in Beijing marked the highest-profile international meeting since the Taliban returned to power in 2021 (Cash 2023).

These actions, coupled with the signing of the 25-year Amu Darya oil contract in January 2023, have substantially strengthened diplomatic relations between China and Afghanistan, deepening economic ties and enhancing mutual cooperation in key sectors. The $540 million deal between Xinjiang Central Asia Petroleum and Gas (CAPEIC) and the Taliban-led administration marked a victory which not only facilitated sorely needed natural resource development but also inched the Taliban-led government closer to international recognition (Yawar 2023).

China continued its diplomatic outreach with the appointment of the first ambassador to Afghanistan under Taliban rule, Zhao Xing, in September 2023, marking a significant step in the development of bilateral relations. Building on this momentum, in June 2024, Chinese Ambassador Geng Shuang emphasized strengthening ties with Afghanistan during his remarks at the UN Security Council Briefing on the Assistance Mission in Afghanistan (UNAMA). He stated, "the Doha meeting will be an opportunity to communicate with the Afghan Government and unleash a new chapter of international engagement. The international community must respect the political reality in that country and push the Government "to gradually become more trusting to the outside world" (United Nations 2024). Geng Shuang also condemned the imposition of unilateral sanctions on Afghanistan, which have severely disrupted its financial systems.

These sentiments were echoed during the quadrilateral meeting of foreign ministers from China, Iran, Pakistan, and Russia at the September 2024 UN General Assembly, where they convened to discuss regional security issues. A follow-up report by China's Ministry of Foreign Affairs in September 2024 states—"They [Ministers] reaffirmed principles of international law, particularly non-interference in its internal affairs, and the right of the Afghan people to independently decide the future of their country in accordance with the principles and purposes of the United Nations Charter... The Ministers pointed out that NATO members should bear the primary responsibility for the current plight in Afghanistan. They should create opportunities for Afghanistan's economic recovery and future development and prosperity, immediately lift unilateral sanctions against Afghanistan, and return Afghanistan's overseas assets for the benefit of the Afghan people." The report highlights China' s resolve to address Afghanistan's challenges through multilateral cooperation while advocating for international accountability.

These recent examples of China's growing influence in Afghanistan should serve as a cautionary tale for U.S. policymakers advocating for deeper isolationist policies. In the aftermath of the chaotic U.S. withdrawal in 2021, China has strategically positioned itself to fill the resulting power vacuum, establishing a foothold in the "heart of Asia." This maneuver enables China to leverage Afghanistan's geostrategic location while influencing the Taliban-led government to secure preferential access to the country's critical mineral resources. China's approach has proven particularly appealing to nations such as Afghanistan and others in the region that resist adopting Western governance models, instead prioritizing the preservation of their cultural traditions and ways of life. By pairing non-interference with investments in critical sectors such as essential infrastructure, healthcare, technology, and trade, China not only fosters economic growth but also builds trust from the ground up (Farhadi, 2023). For many nations wary of external interference in domestic affairs, China's approach presents an appealing framework for building long-term relationships.

In this context, the United States' approach to foreign affairs appears increasingly outdated, as it continues to rely on Cold War-era strategies and institutions—both domestically and internationally—to compete with China in the modern era of great power competition. To effectively counter China's growing influence, the United States should recalibrate its strategy, shifting from an emphasis on imposing ideals and pursuing dominance to fostering soft power cooperation. By prioritizing more inclusive and cooperative soft power initiatives within the framework of strategic competition, the United States can more effectively navigate the complexities of contemporary global power dynamics. Such a shift would enable the United States to maintain its competitive edge and play a more constructive role in shaping the future of great power competition.

To dissect these and other far-reaching implications of China's ascendancy within the context of great power competition, this volume gathers insights from thought leaders, experts, and practitioners worldwide, to examine China's rise through the lenses of security, technology, economics, and diplomacy. The authors' final analyses underscore that in the modern era of the GPC, strategic competition, influence, and soft power are deeply intertwined. This introductory chapter sets the stage for the following essays and touches on China's compelling soft power approach to foreign engagement, involving non-interference alongside economic development, as the hinge point of China's growing influence.

2 Intent and Themes

The primary intent of this edited volume is to provide an interdisciplinary analysis of contemporary global geopolitics, with a focus on a rising China and the evolving nature of soft power and influence within the context of great power competition and national security. Through a collection of essays authored by international thought leaders and experts spanning the fields of geopolitics, geoeconomics, technology, and security studies, *The Rise of China* offers new insights into the complex interplay of influences currently shaping the international competitive landscape. Additionally, it explores the broader repercussions of this shift for global governance and the international order, highlighting potential areas of conflict and cooperation in the 21st century.

Authors examine the global implications of China's expanding influence within the framework of great power competition, with a particular focus on how China leverages the BRI to impact various regions and sectors. The book is structured to cover a range of topics and regions, examining China's expanding influence in Central Asia, the Middle East, Africa, Latin America, and the Indo-Pacific region. Each chapter offers a unique perspective on the implications of China's rise for regional and global dynamics, security concerns, and economic development.

A key theme that emerges is how states advance their geopolitical interests in the competitive global environment by leveraging economic influence as soft power. *The Rise of China* offers a detailed analysis of the economic dimensions of soft power and strategic competition including trade dynamics, investment patterns, and infrastructure development projects. The book also investigates the implications of emerging security threats for global stability and peace, such as cyber warfare, terrorism, and geopolitical rivalries, resulting from shifting power dynamics.

3 Overview of Chapters

3.1 AI, Great Power Competition and the Future Operating Environment

Chapter two examines the critical impact of artificial intelligence (AI) on future operating environments and global power dynamics within the context of geopolitical competition. Author Mark Abdollahian demonstrates how as the United States grapples with the rise of Beijing's economic influence and Russia's assertive geopolitical strategies; AI emerges as a crucial factor. He begins by defining AI, highlighting the associated risks and opportunities across multiple domains. From a complex adaptive systems (CAS) perspective, the author surveys key drivers such as increasing connectivity and uncertainty, which are transforming national and regional security landscapes.

Abdollahian considers how AI and technological advancements are shifting the strategic calculus of great power competition. The author details how these changes enable smaller or rising powers to challenge traditional technological superiority and lead to new sources of innovation, reconfiguration of global value chains, and heightened economic interdependence. Abdollahian suggests this results in a more interconnected, compressed, and complex global environment where nations and actors navigate the conflict-competition-cooperation spectrum.

The chapter offers a critical framework to tackle the increasing complexities of modern operating environments within the context of the GPC. It emphasizes the necessity of adaptive perspectives and a realistic understanding AI's systemic risks and potentials. The author advocates leveraging CAS approaches to decode key drivers of change and suggests that robust, resilient, and adaptable strategies are crucial for future success.

3.2 *The People's Republic of China and the Information Age: Dealing with Informationization and Intelligence-Ization*

Chapter three explores the expanding role that information and communication technologies (ICT) play in the People's Republic of China's (PRC) rise as a global power. Author Dean Cheng demonstrates how the Chinese Communist Party (CCP) strategically leverages ICT advancements to enhance its economic capabilities and bolster its political control. Cheng argues that by integrating these technologies into global supply chains, China not only advances its technological prowess and wider aims but also ensures the CCP's continued dominance.

The author draws on Jackie Fisher's "Dreadnought," to illustrate how the proliferation of ICT has enabled the People's Liberation Army (PLA) to close the technological gap with Western militaries. Cheng contends that with the shifting nature of warfare since the Gulf War, the PLA has made significant strides in enhancing its qualitative capabilities. According to Cheng, the PLA can now compete more effectively with technologically superior adversaries such as the United States, marking a critical shift in the equilibrium of great power competition.

This chapter demonstrates the transformative impact of the ICT revolution and the evolution of "Informationization" and "intelligence-ization," representing a leap beyond traditional technological applications. This new phase will demand innovative approaches and cognitive strategies, fundamentally altering the military and informational landscape of the Great Power Competition.

3.3 *Twilight of the Idols: European Demographics and China's Middle East Prowess*

Chapter four investigates the profound demographic transformations unfolding in the Western world that are "causing its population to simultaneously age and decline." Through detailed historical analysis, G. Douglas Davis charts shifting demographic trends that are prompting both the United States and Europe to reduce their global presence, which in turn creates power vacuums for competitors such as China to fill. Davis argues that nations grappling with such demographic shifts are increasingly forced to re-allocate resources towards social welfare over military expenditures.

Douglas suggests that Beijing is particularly well-placed to benefit from emerging power vacuums left by the United States and Europe in resource rich regions like the Middle East and Africa. The author highlights China's growing ability to provide military aid and defense grants to developing nations in need, thus extending its power and filling the void left by retreating Western powers.

This chapter is crucial for understanding the less explored dynamics of great power competition in the modern era. It illustrates the long range implications of demographic changes in the West and highlights China's expanding role in the global arena across multiple dimensions. The chapter serves as a timely contribution to understanding the intricate interplay between demographic trends, international relations, and the redistribution of global power in the modern GPC.

3.4 Japan, the Gulf, and Sino-American Great Power Competition

Chapter five delves into the evolving nature of Sino-American rivalry, emphasizing its expansion beyond the Indo-Pacific to become a global phenomenon, with a particular focus on the Arab Gulf region. Author Alice Dell'Era argues that China's growing influence in this region will impact not only U.S. interests but also other significant actors, notably Japan, who will adjust its strategies in response to Beijing's regional ambitions.

Dell'Era outlines Japan's current efforts to position itself as a key alternative partner in the Gulf's current energy transition and diversification—a role that Beijing also actively pursues. The author offers a detailed analysis Japan's strategy to strengthen its relations in the region amidst changing global dynamics and assesses the potential for coordinated efforts between the United States and Japan, to protect both nations' interests and better meet the emerging needs of Gulf countries.

The chapter is organized into three sections. The first section provides an overview of China's increasing influence in the Gulf, highlighting the strategic and economic implications of this expansion. The second section focuses on Tokyo's evolving approach, particularly its relations with Saudi Arabia and the UAE. The third and final section explores the opportunities that Japan's initiatives in the Gulf present to the United States in the context of the growing Sino-American rivalry in the region. Through this structured analysis, Dell'Era presents a comprehensive view of shifting power dynamics and the potential for collaborative efforts amid competition.

3.5 Silk Roads and Subtle Blows: China's 'Soft Power' Influence in the Middle East and North Africa

Chapter six explores the strategic shift in China's approach to the Middle East and North Africa (MENA). Author Khaled Eid identifies several contributing factors: the growing need for energy, the push for economic expansion, the quest for greater international diplomatic support for its strategic objectives, and the aim to compete with the waning influence of the United States.

Eid illustrates how these imperatives have led China to develop a robust "soft power" competitive strategy in MENA to supplant U.S. influence and hegemony and entrench its own presence in the region. The author considers China's "development over democracy" model and its appeal to authoritarian regimes in MENA.

Eid suggests that China's approach offers a viable alternative to Western models that often come with demands for political liberalization.

Eid concludes that to effectively counter China's rising "soft power" in MENA, the U.S. must rethink and reprioritize its objectives in the region. The chapter offers several recommendations for creating policies that are more attuned to the current geopolitical realities and the needs of MENA countries. For instance, by focusing on economic cooperation and engaging in more nuanced diplomacy, the United States can strengthen its relations in the region to compete more effectively with China.

3.6 *China's Economic Statecraft: The Ministry of State Security, Rule by Law, and Cyberspace*

Chapter seven analyzes the role of the Ministry of State Security (MSS) in cyber espionage as a strategic "tool for economic statecraft," focusing on its utilization of China's domestic legislation, research, and talent to advance objectives through cyberspace. Author Mark Grzegorzewski investigates the collaboration between the MSS, Ministry of Public Security (MPS), and the People's Liberation Army (PLA) to explain collective efforts in supporting China's cyber espionage campaigns.

Drawing from a variety of official reports, academic studies, and historical analyses, the chapter demonstrates how MSS cyber operations provide China with a significant competitive advantage in its economic development. Grzegorzewski closely examines China's economic objectives, the collaboration between the MSS and MPS in the domestic security sector, external collaborations involving the MSS, PLA, and Chinese contractors, and establishes the MSS's role in achieving China's national economic goals.

In the context of the broader edited volume, the chapter sheds new light on China's approach to gathering economic sway in the global arena. Understanding the MSS's strategic role in fulfilling China's global ambitions is essential for understanding the rise of China as an economic force in the modern GPC.

3.7 *Dragon's Embrace: China's Growing Influence in Central Asia*

Chapter eight examines China's expanding influence in Central Asia, a resource rich region of great strategic interest within the context of the modern GPC. Through thoughtful historical analyses, author Zhamilia Klycheva unveils an intricate web of political, economic, and social entanglements characterizing China's engagement in the region. Klycheva offers a detailed account of China's relations with all five Central Asian countries, highlighting their importance within the Belt and Road Initiative (BRI).

Klycheva outlines the deeper implications of China's growing influence in the region for U.S. interests and the wider GPC, cautioning that while the U.S. still maintains a presence in Central Asia, its influence is waning in comparison to China's expanding BRI footprint. The author demonstrates why the region, abundant in hydrocarbons, minerals, and hydropower, will continue to be crucial to Chinese interests and should be closely monitored to safeguard U.S. interests.

The chapter concludes with a recommendation that the United States should capitalize on the current, fleeting opportunity to counterbalance China's expanding influence in the strategically vital region. Proactive soft power engagement with Central Asian countries is essential for offsetting power imbalances, maintaining regional stability, and protecting U.S. positioning in the GPC.

3.8 *Geopolitical Game-Changer: China's Expanding Role in Latin America's Defense, Aerospace, and Telecommunications*

Chapter nine analyzes the growing influence of the People's Republic of China (PRC) in Latin America and the Caribbean (LAC). Drawing on extensive quantitative and qualitative research, Authors Alvaro Mendez and Gaspard Estrada, assess China's evolving role in key LAC sectors such as defense, aerospace, and telecommunications. The authors explain how Beijing's strategic pursuits in these areas are reshaping regional power dynamics, especially in relation to U.S. interests.

Mendez and Estrada provide a thorough account of Beijing's involvement in LAC, highlighting both the opportunities and challenges it presents. They emphasize that Beijing's growing engagement confronts traditional U.S. dominance in the region, with significant implications for global power dynamics and the future of U.S.-Latin American relations.

The chapter concludes with key policy recommendations for both Latin American countries and the United States. For Latin America, the authors suggest focusing on policies that ensure strategic autonomy while still benefiting from Chinese investments. For the United States, they recommend strengthening economic ties and security partnerships with Latin American countries to counterbalance China's growing influence and reaffirm U.S. intentions as a key ally in the region.

3.9 *The Rise of China and De-Dollarization in a Multi-Polar Digital World*

Chapter ten tackles the profound geopolitical and economic transformations driven by China's emergence as a major global power. Authors Peter J. Middlebrook and Amba Tadaa explore the dynamics of de-dollarization, the process by which China is strategically reducing its reliance on the U.S. dollar. Through in-depth historical analyses Middlebrook and Tadaa illustrate how digital currencies and financial technologies are reshaping the global economic landscape, emphasizing the implications for great-power competition.

Central to the discussion are China's strategic initiatives, such as the Belt and Road Initiative and the introduction of the digital yuan. These efforts are seen as key mechanisms through which China aims to foster a multi-polar financial system, challenging the longstanding dominance of the U.S. dollar in international trade and finance. The chapter provides a thorough examination of the potential economic, political, and technological impacts of these changes, highlighting the growing importance of digital innovation in global finance.

This chapter offers critical insights for policymakers, economists, and scholars, focusing on the implications of China's rise for the future of international monetary systems. The authors' concluding recommendation is that the sustainability of the U.S. dollar's hegemony is contingent on the U.S. addressing internal political fractures, high national debt, and global concerns over the dollar's weaponization. This chapter contributes to a deeper understanding of the shifting power dynamics in global finance and the strategic maneuvers underpinning China's rise.

3.10 Whither the Belt and Road Initiative?

Chapter eleven explores the strategic shifts in Chinese economic policy, particularly within the context of the Belt and Road Initiative (BRI), and its wider implications for great power competition. Author Lawrence C. Reardon argues that China's response to the 2007–2008 Global Economic Crisis evolved into a broader geopolitical strategy aimed at consolidating China's influence globally. The author considers the rationale behind China's deployment of capital and resources, examining both the economic motivations and the strategic missteps that have shaped the BRI.

Reardon explains how the BRI aimed to alleviate the internal pressures of market saturation and production overcapacity by exporting Chinese capital, labor, and expertise abroad. Though a detailed account of BRI investments, the chapter reveals the scale of China's ambitions and the scope of its international economic engagements. The author critically assesses the viability of BRI megaprojects and explores how widespread requests for loan adjustments and forgiveness prompted Xi Jinping to readjust BRI policy in 2021 to focus on smaller, more sustainable projects.

The chapter's analysis is highly relevant for understanding the modern-day great power competition as it sheds new light on China's strategic use of economic policy to expand its global influence. This is essential for grasping great power rivalry in the modern era, in which soft power strategies are as pivotal as military and diplomatic maneuvers.

3.11 *Identifying Risks from US China Economic Decoupling if Trump Wins*

Chapter twelve considers the implications of a renewed trade war under a second Trump administration. Economist Leif Rosenberger addresses the misconception that U.S. manufacturing jobs can be reclaimed from China, suggesting that technological advancements, not trade, are the primary cause of job displacement. Rosenberger examines the evolution of global manufacturing, involving complex supply chains across multiple countries, the democratization of production, and fostering of economic growth in various Asian nations.

The chapter offers a detailed overview of trade dynamics, economic policies, and the competitive landscape. Through in-depth historical analyses, the author demonstrates how protectionism and escalating the trade war would disrupt supply chains, undermining economic interdependence and stability. Rosenberger explains why these actions would harm not only China, but all economies involved in the production process, potentially pushing Asian countries closer to China and away from the United States.

The author concludes that economic decoupling and protectionist policies are counterproductive to U.S. interests, particularly in the context of great power competition. Instead, he advocates for maintaining economic interdependence and shared prosperity as strategic measures to ensure peace and stability. This perspective is crucial for understanding the broader context of great power competition in the modern era, emphasizing the strategic benefits of cooperation over coercion.

3.12 *China's Dominant Economic Influence: IMF and Asia Push Back Against Decoupling*

Chapter thirteen critiques the United States' Indo-Pacific Strategy, formulated in reaction to China's growing economic dominance in the region. Author Leif Rosenberger argues that the United States has failed to acknowledge the key underlying factors behind China's success. Through a series of case studies and historical analyses, Rosenberger demonstrates how China's particular strategy of shared prosperity has strengthened its economic influence and diminished the draw of violent extremism in nations like Bangladesh.

The author suggests this approach stands in stark contrast to the United States, which views China's economic ascent as a threat. Rosenberger argues that the United States worsens the situation by employing economic coercion against China, drawing parallels to the economic policies of post-World War I. The author cites the Biden administration's high tariffs, which have been criticized by the IMF, as examples of such coercion, warning that these measures can threaten global economic stability.

Rosenberger concludes that the United States should adopt a strategy promoting shared prosperity with China to reduce its historical reliance on coercive measures.

The chapter offers essential insights into the complex dynamics of the modern-day Great Power Competition, highlighting alternatives to the current U.S. approach to a rising China.

3.13 China and Afghanistan: Navigating Geoeconomic and Geopolitical Pathways at a Time of Great Power Competition

Chapter fourteen discusses the evolving relationship between China and Afghanistan, particularly following NATO's withdrawal from Afghanistan in 2021. Ambassador Omar Samad contextualizes the historical relationship within the broader dynamics of great power competition and a shifting multipolar world order. Samad examines China's strategic interests in Afghanistan, emphasizing its Belt and Road Initiative (BRI) and how Afghanistan's untapped mineral wealth and strategic location fit into this framework.

The author questions the sustainability of the Taliban's de facto government in balancing relations with major powers and examines the practicalities of China and the U.S.'s assertions that Afghanistan is not a theater for their rivalry. By exploring the historical dynamics of Sino-Afghan relations across multiple dimensions, Samad offers new insights into the future of geopolitical competition in the regional context.

This chapter is particularly relevant for understanding how great powers like China and the United States navigate complex opportunities and challenges in developing regions, reflecting broader trends in international relations and soft power strategies. The chapter highlights Afghanistan's unique role in the modern-day GPC and the geopolitical realities of a region historically situated at the crossroads of global power struggles.

3.14 Dragon Over Trident: Lessons for China from Ukraine

Chapter fifteen examines China's shifting geopolitical stance in light of Western responses to the conflict in Ukraine. Author Michael O. Slobodchikoff argues that China, currently in a short-term position of strength, is now compelled to adopt a more aggressive and less deliberative foreign policy to achieve its strategic objectives in Taiwan. Slobodchikoff notes that diminishing Western interest amid distractions, particularly from the United States and European countries, presents Beijing with a narrow window to act decisively.

Through detailed historical analyses, Slobodchikoff surveys Beijing's strategic calculus, which now involves bracing for immediate economic sanctions and potential aid to Taiwan, with the expectation that international interest and support will eventually wane. He posits that China's demographic challenges and its status as a declining power further pressure Beijing to act swiftly.

This chapter addresses the broader implications of China's maneuvers on great power competition and national security, outlining the urgency that drives Beijing's actions in an increasingly multipolar world. The chapter cautions that policymakers should anticipate a heightened level of assertiveness from China compared to its historical approach.

3.15 African Debt Crises: Causes, the Role of China, and Recommended US Responses

Chapter sixteen dissects the multifaceted causes behind the debt crises in African countries and challenge the prevailing notion that these crises are primarily the result of a Chinese-orchestrated "debt trap." Authors Marcus Snow and Leif Rosenberger argue that while China's economic growth model inspired African nations to borrow heavily for development, the ensuing debt issues are largely due to a combination of transnational external shocks and internal economic mismanagement rather than malicious intent from China.

Snow and Rosenberger identify key external factors that have significantly impacted African economies such as the global pandemic and the war in Ukraine. Additionally, the authors outline and analyze U.S. financial policies, such as the Federal Reserve's interest rate hikes, which have inadvertently exacerbated African debt levels by strengthening the dollar and devaluing local currencies.

The chapter includes a thorough analysis of Africa's capital demands and the roles of both Western and Chinese financial influences. Recommendations highlight the importance of strategic American financial interventions to foster economic resilience and stability in Africa amidst complex challenges. This chapter offers crucial analyses for understanding the evolving power dynamics in Africa in terms of economic influence and development strategies.

4 Conclusion

At its core, the rise of China represents a significant shift in the modern-day Great Power Competition, challenging the long-standing dominance of Western liberal ideals and traditional economic aid and development paradigms. Through extensive BRI-driven large-scale infrastructure projects and economic investments in the mining sector, coupled with a policy of non-interference in the domestic affairs of partner nations, China has effectively expanded its influence across key sectors in regions such as Africa, Central Asia-South Asia (CASA), and the Middle East. This approach—economic engagement paired with stated non-interference—has proven instrumental in consolidating China's role as a near peer U.S. competitor and influencer. Resource-rich Afghanistan serves as a salient example of this strategy, offering a cautionary lesson about the broader implications of isolationist policies, and urging the U.S. to rethink its approach.

The following essays from esteemed global experts offer invaluable insights, examining the implications of China's rise on national security and great power competition from diverse perspectives. Each chapter contributes to a fuller understanding of the complex interplay of forces shaping contemporary geopolitics, with a particular emphasis on China's soft power approach to competition. From the strategic implications of China's expanding presence in resource rich Central Asia to the evolving dynamics of Sino-American rivalry in the Arab Gulf, the volume explores the multifaceted dimensions of China's impact on the current international order.

Key takeaways emphasize that in the modern era of the GPC, the relationship between soft power and influence is profoundly interconnected. Effective soft power engagement is not merely an accessory to hard power but a vital component of geopolitical strategy. Recalibrating foreign policies to address the needs of partner countries at grassroots, institutional, and ideational levels, alongside meaningful investments, is crucial for sustainable competition with China's BRI. This is particularly critical in regions like CASA and the Middle East, where such an approach can assure partner nations that U.S. foreign policy is not solely top-down but is informed by the lived experiences of diverse communities. This can lead to more sustainable partnerships, enhanced regional stability, and a more compelling alternative to the BRI that aligns with the needs of partner countries.

Finally, this volume aims to go beyond theoretical discussions by providing practical insights and recommendations for policymakers, scholars, and practitioners. Its interdisciplinary approach, combined with actionable policy suggestions, seeks to promote informed dialogue and strategic decision-making. By holistically addressing emerging trends in great power competition, the volume contributes widely to the existing literature and offers a comprehensive framework for understanding evolving power dynamics on the global playing field.

References

Beckley and Brands (2023) China's threat to global democracy. J Democracy 34(1):65–79 (paper). Johns Hopkins University Press

BRICS (2024) What is BRICS. https://infobrics.org/page/national-rubrics/

Cash, J (2023) Taliban representative arrives in Beijing to attend Belt and Road Forum. Reuters, October 17. https://www.reuters.com/world/taliban-representative-arrives-beijing-attend-belt-road-forum-2023-10-17/

Doshi R (2021) The long game : China's grand strategy to displace American order. Oxford University Press, New York

Farhadi A (2020) Countering violent extremism by winning hearts and minds. Springer Press, Switzerland

Farhadi A (2023) Lessons learned from Afghanistan: The Heavy price of treating long-term strategic issues with short-term tactical approaches. In: Farhadi A, Masys AJ (eds) Great power competition volume 4: lessons from Afghanistan: America's Longest War. Springer Press (In Press)

Hillman J (2020) The emperor's new road: China and the project of the century. Yale University Press

Jie and Wallace (2021) What is China's Belt and Road Initiative (BRI)? Chatham House. https://www.chathamhouse.org/2021/09/what-chinas-belt-and-road-initiative-bri

Jinping X (2017) Work together to build the silk road economic belt and the 21st century maritime silk road. The belt and road forum for international cooperation. Available at: https://www.chinadaily.com.cn/a/201705/14/WS59bbe0bda310ded8ac188e0b.html

Luft G (2016) It takes a road: china's one belt one road initiative: an American response to the new silk road. Institute for the analysis of global security. chrome-extension://efaidnbmnnnibpcajpcglclefindmkaj, http://www.iags.org/Luft_BRI.pdf

Mandaville P (2022) The growing role of religion in great power rivalry. Bulletin de l'Observatoire international du religieux (39)

Nedopil C (2023) Countries of the belt and road initiative; Shanghai, Green Finance and Development Center, FISF Fudan University. www.greenfdc.org

Peoples Republic of China (2024) Joint Statement Third Quadrilateral Meeting of Foreign Ministers of China, Iran, Pakistan, and Russia. Ministry of Foreign Affairs, September 30. https://www.mfa.gov.cn/mfa_eng/wjbzhd/202409/t20240930_11501285.html

Raymond G (2020) Religion as a tool of influence: Buddhism and China's belt and road initiative in Mainland Southeast Asia. Contemp Southeast Asia 42(3):346–371

United Nations (2024) Doha Meeting on Afghanistan Provides Critical Opportunity to Discuss Women's Rights, Speaker Tells Security Council: Engagement with De Facto Authorities Does Not Mean Legitimization of Taliban, Special Representative Underlines. Meetings and Press Releases, June 21. https://press.un.org/en/2024/sc15740.doc.htm

United States Department of Defense (2022) Securing Defense-Critical Supply Chains An action plan developed in response to President Biden's Executive Order 14017, February. https://media.defense.gov/2022/Feb/24/2002944158/-1/-1/1/DOD-EO-14017-REPORT-SECURING-DEFENSE-CRITICAL-SUPPLY-CHAINS.PDF

Yawar, M.Y. (2023) Afghanistan's Taliban administration in oil extraction deal with Chinese company. Reuters, Januray 5. https://www.reuters.com/business/afghanistans-taliban-administration-oil-extraction-deal-with-chinese-company-2023-01-05/

Yihua XU (2012) Religion and international relations in the age of globalization. J Middle Eastern Islamic Stud (in Asia) 6(4):19–50. https://doi.org/10.1080/19370679.2012.12023212

Yihua XU (2015) Religion and China's public diplomacy in the era of globalization. J Middle Eastern Islamic Stud (in Asia) 9(4):14–35. https://doi.org/10.1080/19370679.2015.12023271

Zhang J (2023) China's religious diplomacy for its belt and road initiative; a review and assessment. Policy Brief #3, Transatlantic policy network on religion and diplomacy. The Berkley center for religion, peace, and world affairs. Georgetown University

AI, Great Power Competition and the Future Operating Environment

Mark Abdollahian

Abstract Artificial intelligence is fundamentally transforming the very fabric of future operating environments. Elements of national power are rapidly changing given advancing technologies, conditions and influences. In terms of great power competition, this forces us to lean ahead into new normals, addressing sustainable US preeminence as both allies and adversaries judge between Chinese financial and economic gains versus US security and political values, all set against the backdrop of aggressive Russian expansion. This chapter first explores different types of artificial intelligence, fundamental drivers and how it is shaping great power competition currently, with a variety of identified risks and opportunities across political, military, economic, social, information and infrastructure domains. Given trends, a horizon scan of underlying operating environment drivers is next performed from a complex adaptive systems perspective. Key drivers include increasing connectivity, compression of timescales, complexity, and uncertainty. These profoundly change the shape, scope and contexts of national security futures with near peer competitors. Complexity frameworks also provide a coherent and comprehensive strategic, cognitive approach to navigating future operating environments through the lens of robust decision making approaches. Finally, this chapter provides some strategic recommendations on how we can intelligently adapt and prepare for change we know is forthcoming.

Keywords Artificial intelligence · Cognitive warfare · Complex adaptive systems · Great power competition · National strategy · Robust decision making

It is not the strongest of the species that survives, nor the most intelligent, but the one most adaptable to change. Although inspired by and often misattributed to Darwin, the actual quote is from Megginson 1963.

M. Abdollahian (✉)
Department of Politics & Economics, Claremont Graduate University, Claremont, CA, USA
e-mail: mark.abdollahian@cgu.edu

A. Farhadi et al. (eds.), *The Great Power Competition Volume 6*,
https://doi.org/10.1007/978-3-031-70767-4_2

1 Introduction

National interests may come and go, but the national interest always remains the same—owning the operating environment. Instead of reacting to crises *du jour*, great power competition calls for adeptly navigating the conflux of shifting human dynamics across the conflict-competition-cooperation spectrum. Here we must separate contemporary national interest events from the national security production function we must achieve and maintain.

While steel and coal were once seen as elements national power, in our fourth industrial revolution, technology, artificial intelligence (AI) and their convergent impact on political, military and economic affairs dominate radical social transformations already occurring. AI impacts national power dynamics by influencing military and non-military aspects of state power alike. These alter the very fabric of societies across individuals, markets, international power hierarchies and geopolitical norms.

Understanding such is paramount for decision makers tasked with shaping the landscapes in which AI development and deployment occurs, ensuring continuation of national security production functions aligned with societal values. All this must be accomplished while simultaneously safeguarding against the risks of unintended consequences. AI already performs some remarkable feats, yet quantum computing and cold fusion overcoming energy limits foreshadow tipping points and a global techno-social phase shift somewhere over the horizon.

As such, many scholars and policy makers alike identify a new era of 'geotech'—the convergence of technology with political, economic and social implications worldwide (Atlantic Council 2021). Data, computers and AI, coupled with new technologies increasingly extends human achievement yet exacerbates social inequality which impact geopolitics across the conflict-competition-cooperation spectrum (Abdollahian and Yang 2022).

> The emergence of generative AI marks a transformational moment that will influence the course of markets and alter the balance of power among nations. Increasingly capable machine intelligence will profoundly impact matters of growth, productivity, competition, national defense and human culture. In this swiftly evolving arena, corporate and political leaders alike are seeking to decipher the implications of this abrupt and powerful wave of innovation, exploring new opportunities and navigating new risks (Cohen and Lee 2023).

Over the last two decades, AI's remarkable progress has accelerated. Figure 1 illustrates AI capability scores across a range of capabilities relative to human performance. These span increasingly difficult cognitive tasks: handwriting, speech and image recognition as well as reading comprehension, language understanding and predictive reasoning. Each task is then scored against an index of human level performance. As clearly seen, not only have most of these AI tasks surpassed human performance, advancements are exponential. Such dramatic change is something humans have little experience with and force us to recalibrate our mental models and timescales.

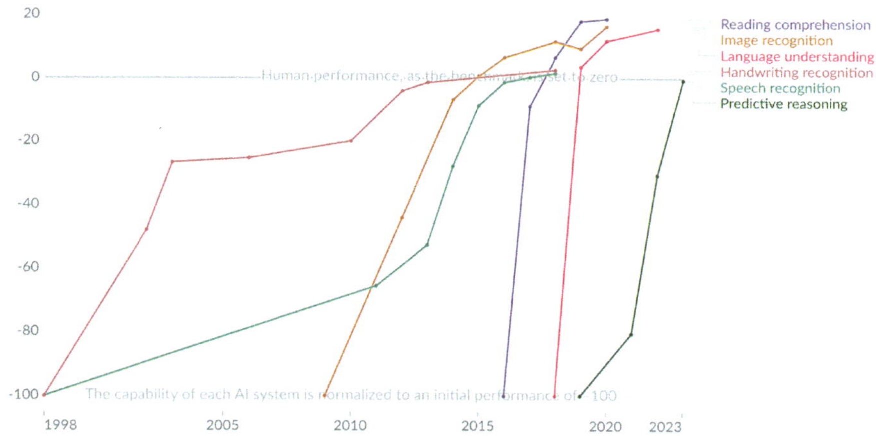

Fig. 1 Test scores of AI systems on various capabilities relative to human performance. *Source* Our World in Data (2024)

Recent analysis by governments, think tanks and academics place AI squarely in the center of great power competition (Atlantic Council 2021). Currently, there is widely varying consensus on who is ahead or winning the AI competition. Some have the US ahead on chip design, market leading foundation and GPT models as well as capital investment. While conversely, China is currently leading on AI research and patents, hardware manufacturing as well as critical minerals and materials. However, both the US and China are largely dependent on foreign chip fabrication in Taiwan and South Korea (Naprys 2023). Thus the US CHIPS Act and China's Big Fund seek to achieve economic independence for their respective domestic semiconductor industries.

Others present a much more pessimistic view, with the US leading in only 11 out of 44 categories advanced technology categories (ASPI 2023). These span advanced information and communication technologies (ICTs); advanced materials and manufacturing; AI technologies; biotechnology, genetic technology and vaccines; defense, space, robotics and transportation, energy and environment; quantum; and finally sensing, timing and navigation. Compared to our Cold War competition with the Soviet Union, where the US possessed superior technology, equipment and human capital, today areas where we do occupy the advantage with near peer competitors are at risk of quickly eroding.

While tallying scores and rankings of country competitors across multiple technological dimensions is necessary for policy, it can be remiss of Heraclitus' warning "there is nothing permanent except change."

> The current age of innovation is marked by an unprecedented pace of change and transformation across the entire spectrum of human activities. Over the last few decades technology has rapidly evolved, while costs of entry for accessing technologies have rapidly decreased.

> Technological innovations will continue to transform human capabilities and provide opportunities to mitigate and adapt to a broad range of challenges from incurable diseases to climate change (NATO 2023).

Our future operating environments are the confluence of political, military, economic, social, information and infrastructure (PMESII) conditions, circumstances, and influences that affect employment of national elements of power. Understanding the status, factors and conditions of the operating enviroment is a prerequisite for all planning activities, operations and strategic initiatives (Parkinson 2019). As such, technology defines our operating environments, both enabling and disrupting, with novel, convergent effects emerging across physical, virtual and the cognitive domains.

> At the same time, the speed of evolving technologies is also broadening attack surfaces, including critical infrastructures and societies' beliefs and values. As a result, new state and non-state behaviours will emerge and additionally, new tools of competition and confrontation will enhance manoeuvres and effects in the non-physical dimensions, to wage cognitive warfare with the objective of achieving physical effects with increased efficiency and impact (NATO 2023).

This forces us to lean ahead into *new normals*, questioning sustainable US preeminence as both allies and adversaries judge between Chinese financial and economic gains versus US security and political values all set against the backdrop of aggressive Russian expansion.

> AI is a cognitive revolution. It will change the ability of militaries to process information on the battlefield and increase their ability to gain information to make sense of it, and likely over time, accelerate the decision tempo of military operations allowing militaries to move faster. (Sharre 2023)

We can run the AI race instinctually and reactively, hoping our 20th century, hard won American experience and optics win the day. Or conversely, we can first map and only then game out the race we must run for smart power projection in the 21st century. Here understanding the map of the underlying dynamics of how AI, great power competition and future operation environments will evolve is paramount—as the rules, runners and racecourse will also change while we are running the race.

Below first is a level set what AI actually is, beyond current hype in AI, large language model (LLM) and generative pre-trained transformers (GPTs). Then the chapter looks into how AI is shaping great power competition currently, with a variety of identified risks and opportunities across PMESII domains. Moving beyond just currently identified symptoms of possible AI and great power interactions, next is a detailed, deep dive into four fundamental drivers of AI advancement and how these will likely change the shape, scope and contexts of national security operating environments. Given AI trends, a horizon scan of underlying operating environment drivers is next performed from a complex adaptive systems (CAS) perspective. Here key drivers include increasing connectivity, compression of timescales, complexity, and resultant uncertainty. This allows for a comprehensive and coherent strategic approach to navigating future operating environments through the lens of robust decision making approaches. Finally, the chapter provides some strategic recommendations on how we can intelligently adapt and prepare for change we know is forthcoming.

2 AI Fundamentals

AI can broadly be defined as "a variety of information processing techniques and technologies used to perform a goal-oriented task and the means to reason in the pursuit of that task" (DOD 2018). Activities encompass *perceptions* to interpret various sensory and data inputs; *knowledge* representation of data, objects, events and concepts; human *language* understanding and interactions; *reasoning* based upon data and simulated experience; and finally *problem solving* goal directed behavior using optimization, heuristic searches and decision making. As such, AI is a multidisciplinary field, spanning computer science, mathematics and the cognitive sciences to name only a few.

AI is usually categorized along two dimensions: by *capabilities* such as narrow, general and super (Russell and Norvig 2003; Hutter 2022; Kurzweil 2005) as well as by *functions* such as reactive, limited memory, theory of mind and self aware (IBM 2023). We already rely on narrow AI in our everyday lives, including voice recognition, recommender systems and self-driving cars. However, super AI or artificial general intelligence (AGI) remains to be realized. This is in despite recent GPT and LLM advancements from a multitude of teams worldwide that some claim indicate the nascence for generalizable intelligence (Baum 2021; Dupree 2023).

Narrow AI is engineered to execute specific tasks with a high degree of proficiency but without the broader understanding or adaptability seen in human intelligence. This type of AI operates primarily within the reactive and limited memory functionalities. Reactive AI systems, such as traditional rule-based expert systems, including IBM's original Deep Blue chess playing computer, stock trading bots, spam filters, facial recognition, missile guidance and electronic warfare systems respond to a set of fixed inputs with predictable outputs but do not learn from past interactions. A step above reactive AI are limited memory systems, including credit scoring, interactive voice and customer chatbots, drone navigation, cyber defense, and battlefield simulations such as DARPA's Air Combat Evolution (ACE) programs that leverage past interactions, data and experiences to make better decisions in even more capable systems.

In contrast, general AI still remains to be realized. However, it would possess capabilities to perform any intellectual task that a human can do, equipped with the ability to learn and apply both fluid and crystalized knowledge (Catell 1963). This form of AI would begin to integrate functionalities associated with the theory of mind and self-aware stages. A theory of mind AI would understand and predict others' emotions and intentions, facilitating a deeper interaction with humans such as envisoned by Google's LaMDA research initiatives, advanced companion robots, mental health chatbots, intelligent MISO/PSYOPS platforms as well as advanced, autonomous and attritable combat systems (Clark 2023). Moreover, self-aware AI would possess consciousness and self-perception, enabling it to make complex decisions in unpredictable environments such as personal coaches, autonomous legal advisors, space exploration robots, strategic AI advisors and autonomous field commanders.

Although currently still over the horizon, super AI could surpass the broadest and most sophisticated human cognitive abilities. While human intelligence and agency encompass goal directed behavior dependent upon perceptions, cognition and choices as well as our subsequent ability to influence outcomes (Otto et al. 2020), super AI would possess *hyper-agentic* behavior; exceeding human intelligence across all tasks, incorporating exponentially scaled enhancements being self aware. Science fiction examples rife in popular culture include autonomous warfighting robots, global security monitoring systems, meta-creative AI and scientific research discovery of new physics (Runco 2015; Blades 2021; Chen et al. 2022).

The development of super AI could lead to scenarios where AI systems make decisions with far-reaching implications, understanding and manipulating PMESII domains and environments with superhuman efficiency. The potential capabilities of super AI raise profound ethical and safety issues, necessitating preemptive international discourse and stringent regulatory frameworks that could also hamper US great power competitiveness (Meacham 2023). This level of AI challenges current governance structures and calls for a reevaluation of how societies engage with and control transformative technologies not seen since the dawn of the nuclear era.

3 AI and Great Power Competition

AI is significantly transforming the dynamics of great power competition by enhancing national security capabilities while creating complex, new geopolitical challenges across our operating environments. Technology such as stealth, social media and LLMs define both problem and solution sets in our fourth industrial revolution (World Economic Forum 2020). AI technologies influence both great power military and economic competitions as well as deeply pervading the fabric of all our other political, social, information and infrastructure operating environments.

These geotech and AI transformations will contribute to step changes for individuals, firms, markets and societies, fundamentally reshaping national security dynamics, international relations and economics (Atlantic Council 2021; Abdollahian and Yang 2022) with some specific examples below.

- AI as an enabling technology influences the pace and nature of international competition, potentially altering first-mover advantages, eroding deterrence and reshaping military strategies (Horowitz 2018).
- AI's integration into military and cybersecurity enhances national security capabilities, augmenting conventional, cyber, and nuclear threats while significantly complicating the security landscape. This necessitates strategic recalibrations, particularly among great power competitors (Schmidt 2022).

- AI technologies fundamentally alter great power competitions by influencing the creation and exercise of national power, whether kinetic military or economic financial hard power, diplomatic soft power or full spectrum smart power (Nye 2013), demanding a reevaluation of strategies against authoritarian governments as well as questioning the resilience of democracies (Daniels and Chang 2021).
- The rise of AI could potentially lead to multi-super AI competition scenarios, challenging the dominance of single AI entities and respective nation states subsequently impacting global power structures (Wu 2021).

As generative AI and LLMs have proliferated in the last few years, so have calls both for and against the development and risks from a wide range of individuals, academics, firms and governments. More recently, AI *foundation* models have emerged (Stanford 2024), such as OpenAI's GPT models that are "trained on broad data (generally using self-supervision at scale) that can be adapted (e.g., fine-tuned) to a wide range of downstream tasks" (Bommasani et al. 2023).

Such models hold vast promises for increasing individuals' productivity and corporate efficiencies as well as driving better healthcare outcomes, financial markets, warfighting and public policy across multiple PMESII domains. Specifically for defense, these include disinformation, deception, offensive cyber capabilities, surveillance, logistics and chemical-biological weapons engineering to name just a few.

Obviously like most other technologies, AI is dual use for both civilian and military purposes. Thus recently a new term of art has surfaced, *frontier* models, which are highly advanced foundation models with the potential to "possess dangerous capabilities sufficient to pose severe risks to public safety" (Anderljung et al. 2023) such as general theory of mind and self aware AI. Given transformative dual use, such frontier models recently prompted a plethora of both industry experts' (Roose 2023) and governmental warnings.

> Artificial Intelligence (AI) holds extraordinary potential for both promise and peril. Responsible AI use has the potential to help solve urgent challenges while making our world more prosperous, productive, innovative, and secure. At the same time, irresponsible use could exacerbate societal harms such as fraud, discrimination, bias, and disinformation; displace and disempower workers; stifle competition; and pose risks to national security. Harnessing AI for good and realizing its myriad benefits requires mitigating its substantial risks. This endeavor demands a society-wide effort that includes government, the private sector, academia, and civil society (US Government 2023).

As such, advances in AI, human computer interactions (HCI) and big data present equal risks and opportunities across our operating environments. Each type of AI brings distinct capabilities and associated functionalities, presenting unique national security opportunities and challenges. Already comprehensive horizon scanning of potential AI impacts are continuously performed across academia, the private sector and governments globally (Atlantic Council 2021; Stanford 2024; Cohen and Lee 2023; NATO 2023). Rather than duplicating excellent community wide efforts, Table 1 samples only some of today's key risks as well as opportunities of AI surveyed across the operating environment. While obviously not complete, it is illustrative

Table 1 Some key AI risks and opportunities across domains

	POLITICAL	MILITARY	ECONOMIC	SOCIAL	INFORMATION	INFRASTRUCTURE
RISKS	Political Interference & Dissolving Democracy	AI-enabled Arms Race Escalation	Technological Hegemony & Economic Coercion	Social Polarization	Disinformation & Information Warfare	Critical Infrastructure Vulnerabilities
	Coercion through AI Dominance	Autonomous Weapons Proliferation	Job Displacement & Social Unrest	Surveillance State Expansion & Social Control	Censorship & Digital Authoritarianism	Supply Chain Disruptions & Cyber Intrusions
	Surveillance State Competition	Cyber/Cognitive Warfare Dominance	Monopoly Power & Market Control	Disinformation & Propaganda	Algorithmic Bias & Manipulation	Environmental Impacts & Energy Consumption
	Cyber Warfare Dominance	Dual-Use Technology Exploitation	Cyber Economic Espionage	Algorithmic Discrimination & Social Injustice	Privacy Erosion & Data Exploitation	Smart City Surveillance & Social Control
	Ethical Dilemmas in Governance	Ethical Dilemmas in AI-driven Warfare	Technological Fragmentation & Standards Wars	Privacy Erosion & Data Exploitation	Cybersecurity Threats & Vulnerabilities	Digital Divide & Inequitable Access
OPPORTUNTIES	Digital Democracy	Deterrence & Strategic Stability	Innovation & Technological Competitiveness	AI-enabled Social Services & Healthcare	Fact-Checking & Information Verification	Smart Infrastructure Optimization
	Diplomacy & Soft Power Projection	Precision & Efficiency in Military Operations	Entrepreneurship & Start-Up Ecosystems	Community Empowerment & Grassroots Initiatives	Content Moderation & Safety Measures	Transportation Efficiency & Mobility Solutions
	International Norms Development	Risk Reduction & Decision Support	Economic Diversification & Resilience	Education & Lifelong Learning	Privacy Preservation & Data Security	Renewable Energy Integration & Grid Optimization
	Conflict Prevention & Resolution	Arms Control & Confidence-Building Measures	Education & Workforce Development	Mental Health & Well-being Support	Cybersecurity Cooperation	Sustainable Urban Planning & Development
	Transparency & Trust-building Measures	Ethical Guidelines & Norms Development	International Collaboration & Trade	Social Impact Investing & Corporate Social Responsibility	Critical Thinking Education	Disaster Preparedness & Emergency Response

of the extent and scale potential of AI impact in the future operating environment relevant to varying national security contexts. Note that the risks and opportunities presented here are simplified, isolated single nodal events, not interconnected or complex accumulative risks with higher order, unintended ripple effects we know can occur (Jeworrek and Sachs 2011; AFDP 2016).

4 Decoding AI Drivers

While the above operating environment risks and opportunities are insightful, they are still symptomatic, currently identified potential impacts, with a plurality of others which we might be missing. Here looking at the core, underlying drivers of AI advancements can be more insightful in preparing for future operating environments.

AI progress is based upon four key fundamental resources: *data, models, compute* and *energy*. Enough data for AI models to learn from and then be validated against; accurate algorithmic models that generate, predict or classify based upon that data; compute power to actually train such models in efficient timeframes, and finally energy for the vast amounts of computational power needed to train models often with billions of parameters. Each of these components plays a crucial role in the development and deployment of AI technologies, which in turn explicitly influence our future operating environments.

To understand how these drivers contribute, interact and accelerate AI development, an example is useful. GPT-3 was groundbreaking when released in 2020, with

175 B parameters using a neural network transformer model architecture trained on 570 GB of publicly available information (PAI) data using 1,024 A100 GPUs. This probably cost around \$5 M in compute time and 1.27GW of electricity and could have trained GPT-3 in as little as 34 days (McGuinness 2023). In comparison, GPT4 was released in 2023 as a significantly scaled up and optimized mixture of experts (MOE) model, with 1.8 T parameters trained on approximately 78 TB of data with an estimated compute cost of \$63 M using 25,000 A100s GPUs over 100 days of training (Patterson et al. 2023; Yalalov and Myakin 2023). Given an order of magnitude increase between GPT-3 and GPT-4 capabilities, cognitively we must recalibrate our timescales on AI developments. Even currently narrow, in silico AI learning far outpaces real world, in vivio human learning timescales.

Data is the foundational element for training AI systems, making it a strategic geopolitical asset. In great power competition, the control over and access to large datasets provide competitive advantages. Governments are increasingly recognizing data as a critical national resource, leading to policies aimed at securing data from foreign access and manipulation. For example, the EU's General Data Protection Regulation (GDPR) is not only a framework for data protection but also a digital geopolitical tool, influencing international data flows and the strategic operations of multinational corporations (Bradford 2020).

Conversely, given the last decade's big data revolution (McAfee and Brynjolfsson 2012), the vast amounts of open source, publicly available data will likely continue to accelerate. This leads to digital and data ubiquity (Iansiti and Lakhani 2014), where access to data to train AI models will be democratized and proliferated simultaneously, accessible to anyone anywhere. Data sources for new AI innovation and disruption will only scale and proliferate across individuals, firms, economies and states (Abdollahian 2021). However, this will also drive up the value of private information, proprietary data and models in specific use cases. This may be for long or short periods of time depending on protectionist policies, before becoming pressured to be open source, incentivized by individuals, researchers and developers pursuing public goods and social objectives.

Models based on machine learning and deep learning are central to modern AI systems. Their ability to possess memory allows them to learn from vast amounts of data and improve over time. In the sphere of great power competition, the development of sophisticated AI models can enhance national capabilities in areas such as cyber defense, autonomous weapons systems, and intelligence analysis. The US, China and Russia among others, invest heavily in AI research to develop models that can provide a strategic edge, such as processing satellite imagery faster than human analysts could.

However, most AI algorithms and models are open source, with online repositories composing training data sets, models and applications. GitHub is currently the world's largest source code host, with over 100 million developers and 420 million code repositories (Github 2024). This includes some current GPTs, self-play reinforcement learning of Alpha Go Zero which in 2017 beat 2000 years of collective human strategy (Silver et al. 2017), and convolutional, generative adversarial network

(GAN) foundation models that can power deepfake to cheapfake disinformation campaigns (Hameleers 2024).

Compute power is essential for training complex AI models. Enhanced compute power allows for faster and more efficient training of narrow and general AI models, which can be crucial for real-time applications in defense and cybersecurity. Countries like China have made significant investments in quantum computing and supercomputing to bolster their compute capabilities, aiming to surpass US computing power in strategic domains. The race for superior compute capabilities is often likened to the new arms race.

As compute resources significantly increase, the economics of costs decrease, driven by advances in semiconductor technology. The cost of Alpha Go Zero in 2016 was $25 M, trained in under 80 h to beat human grandmasters. Today that probably can be achieved at 1/100th the cost. While a narrow AI application, it demonstrated how AI can beat collective human wisdom, all self-taught, with no human intervention or prior data. The GPT-3 versus GPT-4 example above is also illustrative of the relative low costs for increasing AI capabilities significantly. With enough compute resources and general AI capability, a "Thucydides-Sun Tzu-Washington-Clausewitz-MacArthur" AI capability would be existentially consequential indeed.

And as costs decrease, accessibility increases via the cloud for anyone anywhere with internet access. Spending on 'pay as you go' public cloud services, such as Microsoft Azure, Google Cloud or Amazon Web Services, increased 467% since 2017 to an estimated $678 B for 2024 worldwide (Statistica 2024). Given decentralization trends (Johnson 2021), many of the current AI monopolies of nation states and businesses with billion dollar balance sheets will be pressured by highly specialized research teams globally. Agile, well capitalized and unbureaucratic start-ups, new market entrants worldwide and investment funds will affect even more disruptive change.

Such dynamics will disintermediate markets and existing value chains, creating new winners and losers of innovation, leapfrogging established players and nation states (Abdollahian and Yang 2022). Furthermore, the recent dawn of AI *copilots*, for documents or presentations, and low code to no code skills necessary for application development, will only quicken. All these are packaged for anyone from individuals to nation states to use, modify for specific use cases or create subsequent derivative works, eroding existing great power knowledge arbitrages.

While generative AI and LLMs currently lead headlines, quantum computing will accelerate us into a fifth industrial revolution sometime on the horizon. Recently, Google's prototype Sycamore quantum computer managed to outperform in *seconds* a task that would take the world's largest classical supercomputer *47 years* to accomplish (Neild 2023). New cyber strategies are also emerging. Pre quantum computing 'harvest now, decrypt later' (HNDL) activities occur as we prepare for *Q-day*: a post quantum cyber decryption regime of zero privacy and security in the absence of quantum-resistant cryptographic algorithms (NSA 2021; NIST 2022). If deployed covertly by adversaries, national security effects would be devastating

(Parker 2023), consequentially outstripping the Allies deciphering German Enigma communications during World War 2.

Often overlooked, the energy requirements for running advanced AI systems are substantial. As AI technologies become more integrated into national security and economic systems, energy consumption impacts sustainability and operational costs. The race to build more sustainable and energy-efficient AI systems becomes a component of national strategy. Innovative approaches to reduce the AI energy, through the use of more energy efficient neural network designs or green data centers, are being explored and implemented by leading powers (Strubell et al. 2019).

However, in both a post-quantum and cold fusion world, where the rules of energy economics will be re-written (Conrad and Rondeau 2020), human kind will most certainly experience the next industrial revolution. With sustainable energy to drive astonishing compute power, many believe both the necessary and sufficient conditions for general or super AI can emerge (Kurzweil 2005; Afifi-Sabet 2024; Dilmegani 2024). Foundational general AI could scale and connect with narrower, more precise, mixture of expert (MOE) AI models with exponential data and parameter tuning capabilities.

5 AI Futures

The convergence of data, models, compute and energy trends with quantum and/or cold fusion would result in an epistemological step change: AI performing continuous, self-reinforced collective learning across multiple domains for general, if not super AI.

> The technology's development will shape its geopolitical effects. New capabilities will lead to novel use cases and broader impact. We expect generative AI models to become increasingly multimodal, seamlessly reckoning with text, images, audio and video. Focus will also shift toward performance and value engineering, making this technology less costly and more sustainable. The incorporation of persistent memory will enable better long-range planning and true agentic scenarios. Perhaps the timeliest question is whether generative AI will follow a scale up trajectory, with models increasing in size and likely dominated by large commercial enterprises, or whether they will scale down, favoring a broader set of providers and open source approaches (Cohen and Lee 2023).

So is the proverbial genie is out of the bottle? Not quite yet, but we are getting closer to the tipping point. While possibly hard to believe as beyond the realm of our individual experience, we do have precedence for such tipping points and system phase shifts. This is no different from both evolutionary trajectories and the history of human development. Over 530 MYA, the Cambrian explosion marked the transition from simple, single cellular organisms to highly specialized, multicellular life via blood, brain and bone cell specialization networking. Diversity exploded, accelerating different organisms, orders and production functions, propelling new strategies, predator–prey ecological dynamics and complex food webs (Abdollahian 2021; Zhang and Shu 2021).

From our own historical perspective, human beings harnessed the power of collective learning through written language (Cristian 2018)—the ability to transfer information across time, space and each other, not condemned to reinventing or rediscovering knowledge individually each time. With the advent of the 15th century printing press, knowledge transmission proliferated, disintermediating traditional power brokers such as religious and sovereign authorities, spawning the Renaissance and Enlightenment eras of great social, political, economic and military change (Provvidera 2022). More recently with the creation of the internet, instant global connectivity and social media, the proliferation of change scales dramatically through coordinated social action transcending nation state boundaries. Given these trends, strategy and human dynamics will continue to co-evolve with technology to define new defense and security challenges from multiple domains (Abdollahian 2021).

6 A Horizon Scan of the Operating Environment

At a future quantum-cold fusion tipping point, the AI genie is definitely out of the bottle; whether we may or may not be inside the genie's bottle then is another question. This leads us to next understand the impact of AI across our future operating environments before this occurs. National security has evolved to become an extraordinarily complex, dynamic, and adaptive system in which new actors, both state and non-state, interact (NATO SHAPE and HQ SACT 2008.) Both the US' Joint Intelligence Preparation of the Operational Environment (JIPOE) and Joint Chiefs Human Aspects of Military Operations (JC-HAMO) explicitly recognize the need for system-of-systems perspectives of relationships and interdependencies among actors, adversaries and allies alike in a holistic way to anticipate and shape behavior, capabilities and interactions (Antúnez 2021).

Going beyond already identified risks and opportunities above, a complex adaptive systems (CAS) framework coupled with a horizon scan on AI's impact on great power competition illuminates likely future outcomes across PMESII domains and our future operating environment. A CAS is a specialized type of open system where diverse, interconnected components are layered across multi-level structures which can exhibit dynamic behaviors and adaptability in response to changes in the environment or internal conditions. CAS frameworks are central to understanding our world around us, from the natural sciences (Gell-Mann 1995; Holland 1995, 1998; Kaufman 2000) to the social sciences (McKelvy 1999a, b; Miller and Page 2007; Mitchell 2009). The theoretical underpinnings of complex adaptive systems derive from work across multiple disciplines, including physics, biology, economics, and computer science.

A CAS is usually defined by three key characteristics: complexity, self-organization and adaptivity (Arthur 1990, 1995, 2021; Epstein and Axtel 1996; Mitleton-Kelly 2003; Page 2015). From a complexity perspective, these systems have multiple interacting components, whose aggregate activity is nonlinear and not easily predictable from the properties of the individual elements. This complexity arises from the intricate interconnections and interdependencies among the system's components, often nested across micro-meso-macro scale activities, which can include agents and groups with goal directed and strategic behavior, processes, or information flows. These systems exhibit novel, emergent behavior from agent interactions, leading to new order and structures without external direction. Self-organization is a hallmark of complex systems, indicating a capacity for spontaneous adaptation and pattern formation. Agents with goal directed behavior learn from experience and adapt their strategies based on interactions with other agents and the environment. This adaptation is driven by co-evolutionary behavior, allowing the system as a whole to evolve over time. Figure 2 below outlines some key complexity components embedded across PMESII operating environments.

Over the last two decades, CAS approaches have been increasingly applied to analyze national security, foreign policy and PMESII operating environments to overcome limits of traditional analysis (Abdollahian et al. 2006; Abdollahian et al.

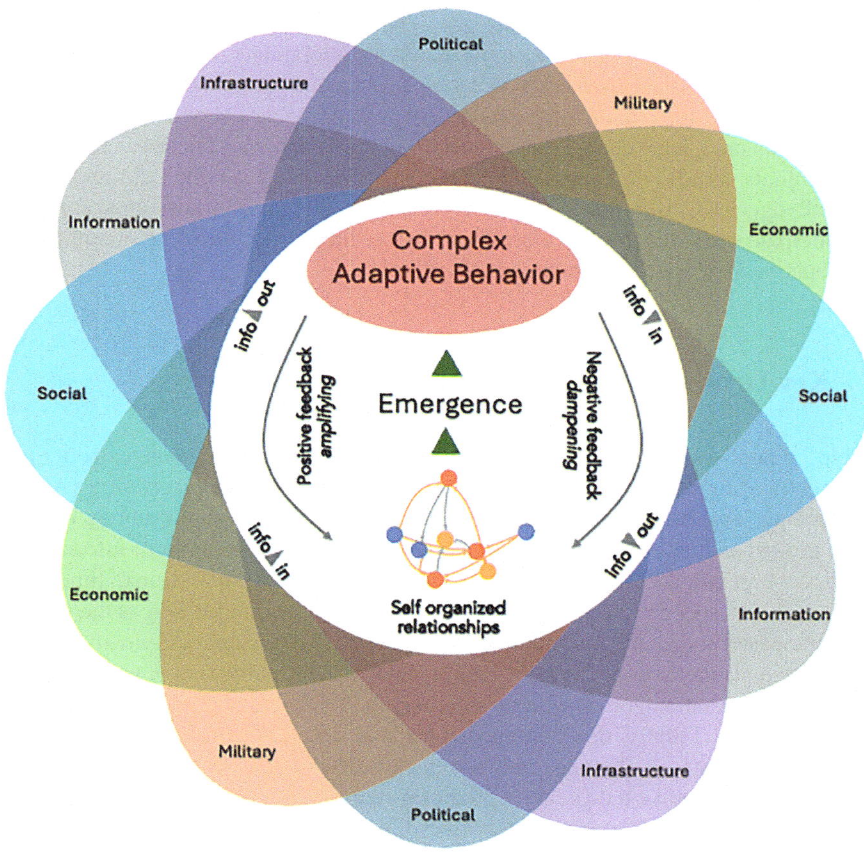

Fig. 2 Components of complex adaptive systems. Adapted from Wikipedia

2010; Cioffa-Revellia and O'Brien 2007; Green 2011; Abdollahian et al. 2013, Yang et al. 2016; Neal et al. 2017; Pike 2019; Abdollahian and Yang 2022). A CAS lens "can help analysts see the extent to which a given system incorporates CAS principles and understand what parts of the CAS space are more or less represented within a single system or across a set of systems" (Lingel et al. 2021).

> Failing to account for the system's unique groups at multiple levels implicitly causes an underestimation of the system's resilience, since there are no layers of subcomponents. Identifying the unique groups then requires an understanding of each group's interdependencies (e.g., income, relationships), which reveals how the groups survive and compete within their system. This analysis provides the possibility of assessing how altering interdependencies may weaken some groups, strengthen others, or cause sub-group(s) to realign. [Joint Intelligence Preparation of the Operating Environment] accounts for the main elements of complex systems by viewing them as networks, but it misses the essence of these dynamic systems whose behavior is determined by the flow of resources through the layers of groups (Pike 2019).

With CAS lenses in place, a horizon scan on AI and technology's impact on great power competition is illustrative. Horizon scanning is a systematic process of gathering and analyzing information about trends, developments, emerging issues, and potential opportunities or threats that could impact an organization, industry, or society in the future (Inayatullah 2007). Various approaches center on determining 'what is constant, what changes, and what constantly changes.' These can be weak or early signals, trends, wild cards and persistent problems in specific risks and threats to challenge past assumptions (Cuhls et al. 2015). The goal of horizon scanning is to anticipate such changes and disruptions before they occur, allowing organizations to prepare and adapt their strategies accordingly.

7 Key Drivers of Operating Environment Change

Table 2 diagnoses four key CAS technology drivers underpinning change in our operating environments: *connectivity*, *compression*, *complexity* and *uncertainty* (Abdollahian and Yang 2022). In a CAS, increasing connectivity networks among individuals, groups and nations yields faster and widespread dissemination of information, resources and influence. Increasing connectivity leads to a compression of timescales as interactions speed up with higher order effects rippling quickly across the system. This increased speed and interconnectivity contribute to greater complexity, as relationships and interactions between multi-scale, system components generate nonlinear, unexpected outcomes with novel emergent behavior, new groupings and hierarchical orders. Finally, this complexity leads to greater uncertainty in predicting future system states, as dynamic interactions, feedbacks and sensitive dependence on initial conditions make it difficult to predict system evolution precisely. This coupled sequence illustrates how tightly interwoven CAS dynamics are and underscores the challenges faced by analysts and decision makers alike across the community.

Table 2 CAS operating environment drivers across domains

OPERATING ENVIRONMENT DRIVERS	POLITICAL	MILITARY	ECONOMIC	SOCIAL	INFORMATION	INFRASTRUCTURE
Connectivity	Enhanced global interdependence & diplomatic channels	Multidomain warfare & global surveillance capabilities	Integration of global markets & rise of digital economies	Democratization of information leading to social mobilization	Explosion of digital communication channels	Expansion of global internet infrastructure & smart technologies
Compression	Accelerated policy responses & diplomatic interactions	Rapid AI deployment & reaction capabilities	Speed of financial transactions & market movements	Instant global communication affecting societal trends	Real-time information dissemination & consumption	Accelerating development & deployment of AI enabled technological solutions
Complexity	Complex multilateral relationships & 'whole of government' policy challenges	Integrated conflict-competition-cooperation operations, hybrid, cognitive &	Complex global supply chains & AI enabled financial systems	Diversity of social identities & AI mediated exchanges	Multifaceted media landscapes & AI information sources	Integrated IOT & smart AI systems
Uncertainty	Unpredictable shifts in global power dynamics	Uncertain security environments & emerging threats	Market volatility & unpredictable economic trends	Changing social norms & unpredictable societal activity	Rapidly evolving narratives & information reliability issues	Future-proofing against unknown technological advancements

Today we are already confronted with great power competition outcomes given increased connectivity, compression, complexity and uncertainty. Table 3 details just a few current great power competition PMESII operating environment concerns spanning the entire conflict-competition-cooperation spectrum. However, like the AI risks and opportunities in Table 1 above, they are only some of the expressed symptoms of underlying key CAS operating environment drivers.

> Simultaneously, new technologies, fragilities in the public health sector, and environmental changes are more frequent, often have global impact and are harder to forecast. One need only look at the Gaza crisis—triggered by a highly capable non-state terrorist group in Hamas, fueled in part by a regionally ambitious Iran, and exacerbated by narratives encouraged by China and Russia to undermine the United States on the global stage—to see how a regional crisis can have widespread spillover effects and complicate international cooperation on other pressing issues (ODNI 2024).

Obviously technology significantly increases the connectivity of the global economy, societies, and political systems, both geo-physically with infrastructure, transportation and communication advances as well as virtually through the internet. We have also witnessed a geometric increase in the number of new actors globally and the volume of interstate flows. The proliferation of the internet, smartphones, and social media platforms facilitates an unprecedented level of global interaction, democratizing access to information and enabling instantaneous communication across the world.

Rapid technology advancements accelerate and compress time scales, with the development of COVID-19 vaccines in months instead of decades. Spurring economic growth globally, trade openness grew from 10% post war to over 55% today (World Bank 2023). This interdependence accelerated economic development time scales, diminishing extreme global poverty from 38% in 1990 to 9% today (World Bank 2023). And the promise of better futures birthed many political leaders' promises of sustainable growth narratives (Abdollahian et al. 2012). However, rapid but uneven progress also exacerbates inequalities and civil conflict (Collier and Hoeffler 2004; Schneider 2014). This can be viewed either as globalization's downside

Table 3 Some current great power competition outcomes across domains

OPERATING ENVIRONMENT OUTCOMES	POLITICAL	MILITARY	ECONOMIC	SOCIAL	INFORMATION	INFRASTRUCTURE
Connectivity	Complex negotiations & multiple dimensions of US-China relations, involving tariffs, intellectual property, & technology transfer	Russia's Syria & Ukraine involvent, affecting US strategic interests in Europe and Middle East	China's Belt & Road Initiative expanding economic influence across Asia, Africa, & Europe	Exchanges & educational programs fostering cross-cultural understanding, despite geopolitical tensions	Interference in elections highlights the use of information tactics to influence foreign politics	Expansion technology by Chinese firms like Huawei, influencing global telecommunications infrastructure
Compression	Rapid US sanctions on Russian & Chinese officials in response to human rights, security or economics issues, showing quick policy shifts	US & Russia's modernization of nuclear arsenals to maintain strategic parity and quick deployment capabilities	Flash crashes in stock markets due to AI algorithms, where US-China trade news can cause rapid market reactions	Social media's role in the rapid global spread of movements, affecting perceptions in the US, China & Russia	Real-time global news coverage & social media reactions during domestic protests, impacting international viewpoints	Quick adaptation and rollout of remote working technologies to COVID-19, changing workplace infrastructure
Complexity	Mutually depedent US-China economic ties cause trade tensions & fuel domestic politicization	Hybrid warfare tactics by Russia in Ukraine, combining conventional military operations with cyberattacks & misinformation	Intertwining of global supply chains & technology sectors, complicating economic relations between the US, China & allies	Growing influence of Chinese diaspora and Confucius Institutes, which the US and others see as tools for cultural and social influence	Complex landscape of global media influenced by state-backed outlets, promoting national narratives abroad	Development of smart cities in China with integrated surveillance systems, presenting new standards for infrastructure that contrast with Western models
Uncertainty	US-China tensions in the South China Sea, with unpredictable shifts in military posturing impacting regional stability	Uncertainty in global security environment due to Russia's unpredictable military interventions in Syria & Ukraine	Uncertain impacts of US-China trade tensions on global economic stability, affecting markets, investments & alliances	Domestic social protests and unrest, driven by uncertainties in governance and fears of increasing foreign control	Misinformation campaigns, where the US, China, & Russia disseminate conflicting information, complicating policy responses	Cybersecurity threats impacting critical infrastructure, with state-sponsored attacks creating uncertainties in security protocols

risk exposure or the source of upside sustainable growth gains for many economies. Equally it can be viewed as political security for many governments (Fearon 2018; Abdollahian 2021; Abdollahian et al. 2021). Such developments challenge existing institutional norms and frameworks, where traditional cycles of policy and diplomacy may no longer be sufficient to address the speed of technological and societal change.

The emergence of complex digital platforms and ecosystems, such as those led by Apple and Alphabet, redefine traditional economic and business models, creating new opportunities for value creation while also disrupting established industries (Abdollahian and Yang 2022). Complexity also encompasses global challenges whose solutions are often dictated by local constraints. Current US policy initiatives and options in Ukraine are limited by domestic political considerations and fiscal constraints. US-China economic interdependence is being pulled by US agricultural exports and American's appetite for cheap Chinese tech exports while pushed by domestic political narratives and national security considerations on both sides. These global versus local constraints across PMESII domains indicate coupling of scale activities in much more complex systems than often discussed.

The convergence of AI, emerging powers, shifting geopolitical alliances, and new cognitive strategies across the conflict-competition-cooperation spectrum increase uncertainty in domains often previously taken for granted. In 2024 the world's three largest democracies, US, India and Indonesia with approximately 41% of global population, will participate in national elections (Bucholz 2024). Today, democratic elections increasingly present potential vast attack surfaces for disinformation and manipulation threat vectors from both domestic and foreign malign actors alike.

Conversely, autocracies empowered with AI become more politically capable via self-reinforcing controls, spawning ‘AI-tocracies’ (Beraja et al. 2023). Increasingly complex systems also induce nonlinear, far from equilibrium behavior, with dynamic opening and closing of political and military decision spaces. Here, variegated US policy responses in both uncertain Ukraine and Israel-Gaza conflict contexts remain illustrative.

In terms of US strategic engagement with China, many allies and adversaries hedge in the cognitive domain. At the heart of today’s great power competition is the choice between American security and political values leadership versus Chinese economic growth and market expansion. From an economic perspective, US FDI outflows to the world averaged 25% of all global activity in the 1990s, today it is less than half of that. Conversely, China FDI outflows averaged less than 1% in the 1990s, today it is almost 18% of all global activities (UNCTAD 2023).

Where we do lead in technology, healthcare, financial markets and self-actualization of human attainment, our potentially winning strategic narratives can be drowned out by the noise of domestic polarization and fake news. More often than not, emanating from ourselves. Pundits on all sides fight the same battle: balancing decoupling and reshoring political security with global economic growth, climate, energy, and existential AI challenges which necessitate international collective action.

The US and China are the world’s top AI competitors, but they are also the top AI collaborators. Their technology competition will see geopolitical priorities drive economic decision-making, including export controls, sanctions, tariffs, industrial policy, investment screening, and other measures deployed to increase both absolute and relative advantages. Geopolitical swing states will have an increasingly meaningful role in shaping the AI-enabled future. In particular the EU, UK, UAE, Israel, Japan, the Netherlands, South Korea, Taiwan, and India will remain key. Moreover, these players may form new innovation blocs, such as BRICS, creating alliances and partnerships with more dominant states or cooperating with each other to pursue common goals (Goldman Sachs 2023).

Where we go from here invokes Mile’s Law from bureaucratic politics of “where you stand depends on where you sit” (Miles 1978). Firms and nation states will choose based upon their respective interests, constraints and resources given increasing future operating environment connectivity, compression, complexity and uncertainty.

> The world that emerges from this tumultuous period will be shaped by whoever offers the most persuasive arguments for how the world should be governed, how societies should be organized, and which systems are most effective at advancing economic growth and providing benefits for more people, and by the powers—both state and non-state—that are most able and willing to act on solutions to transnational issues and regional crises (ODNI 2024).

Given that the world has seismically shifted from US preeminence at the end of the Cold War to emerging great power rivalries reverberating across regions, what can we do to cognitively navigate the 21st century?

8 Towards Robust and Adaptive Cognitive Strategy

In order to answer this call to action, we need to aggressively outmaneuver and overmatch near peer competitors in the cognitive domain. While various definitions exist, at its core, cognitive warfare focuses on human dynamics: influencing the perceptions, understanding and decision calculus of individuals, groups and populations across the entire conflict-competition-cooperation spectrum in order to achieve strategic objectives (Du Cluzel 2020; Pappalardo 2022a, b, c).

> Cognitive warfare is distinct from information warfare in that information warfare "focuses on controlling the flow of information", while cognitive warfare "aims to control the responses of targets to the presented information" (Hung and Hung 2022) **...** Cognitive operations "can be used before, during and after kinetic actions, while remaining outside current international definitions of what constitutes an act of war" (Claverie and Du Cluzel 2022). These operations will also become far more prevalent and effective with the increasingly rapid development of emerging technologies such as AI or neuro-technologies, and with the increasing integration of these technologies into daily life (Rickli et al. 2023).

Current military operations are essentially population centric, conducted to influence actors' will and decisions in the operating environment compatible with mission objectives (Antúnez 2021). Nuanced and constantly changing operating environment characteristics make it necessary to transform capabilities to adapt to the human environment. They also force employment of 'whole of government' responses across the conflict-competition-cooperation continuum. As such, understanding the underlying currents of connectivity, compression, complexity and uncertainty is fundamental for cognitive maneuvering. In dynamic human environments, we are reminded by Clausewitz for "a philosophy of strategy that contains the seeds of its constant rejuvenation—a way to chart strategy in an unstable environment" (Clausewitz 1831).

Here work over the last few decades on robust decision making (RDM) provides the seeds for strategic success. RDM represents a paradigm shift, extending along the continuum of conventional prediction-based decision analysis towards flexible, adaptive frameworks under conditions of deep uncertainty (Lempert 2002). It is predicated on the notion that as complexity and uncertainty dramatically increase, probabilistic forecasts from predictive models and plans morph into fuzzier states of possible alternative futures stretching beyond the limits of predictability (Abdollahian 2021; Abdollahian et al. 2021). RDM has been applied across a diverse set of strategy challenges, including policy, national security and climate change (World Resources Institute 2011; Marchau et al. 2019).

RDM leverages computational modeling and simulation advancements to facilitate better decision-making under conditions of profound uncertainty. This methodology integrates various analytical tools and processes, including assumption-based planning, decision and scenario analysis, exploratory modeling and more recently strategic foresight to rigorously evaluate policy strategies across a broad spectrum of anticipated, possible and plausible alternative futures (Coates et al. 2010; Lempert 2019; NATO, 2023).

Strategies are categorized and evaluated against a diverse array of plausible futures, leading to vulnerability analysis identifying potential weaknesses (Lempert et al. 2006). Through trade-off analysis, decision-makers can explore and weigh different strategic options, ultimately formulating robust, adaptive strategies that promise effectiveness across various future scenarios (Lempert et al. 2006). However RDM comes at a cost. Just like the stock market, aggressive investing in riskier, more uncertain assets can lead to vast rewards while more cautious investing accepts lower returns but with less risk assumed.

Figure 3 illustrates a notional classification and characterization framework different RDM strategies across phenomena with increasing uncertainty and complexity. Of course many other decision and game theoretic strategy types and nuances also exist, such as minmax, maximin, no regret, counter, pure, mixed etc. Below a time dimension is explicitly removed, as predictability of heterogenous phenomena across multiple domains obviously varies greatly over different time horizons. Also, strategy classification under high complexity and deep uncertainty can be debated. However, in the current context of AI's impact on our operational environments this framework will suffice.

Core strategies satisfy the fundamental production function of any organization or business' mission, their *raison d'être*. Within the core set are both *optimal* strategies that seek to maximize lines of effort gains and *defensive* strategies that protect outcomes. As complexity increases but uncertainty remains low, decision makers

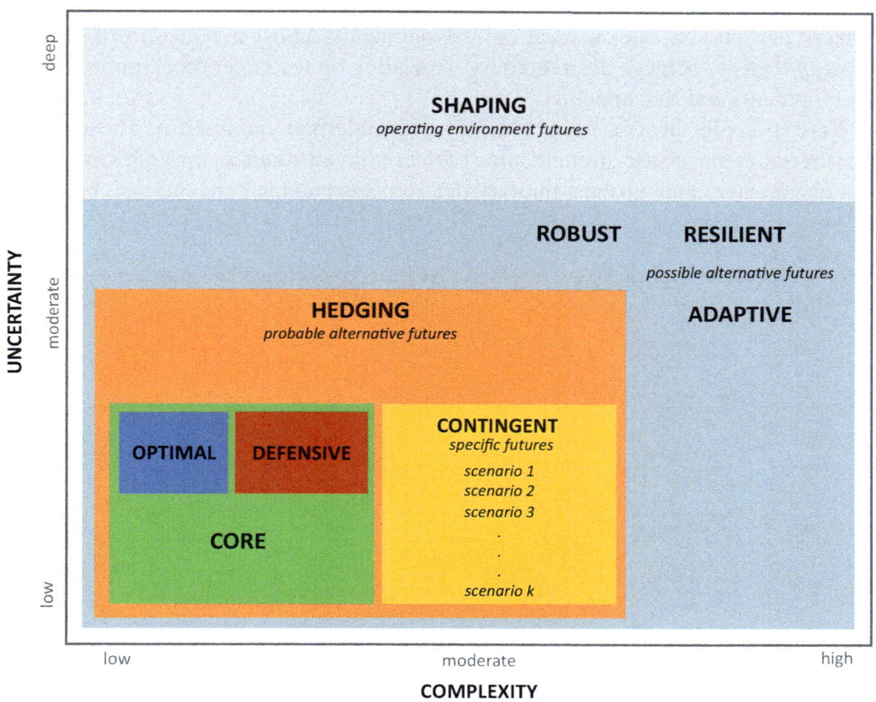

Fig. 3 Strategy types across uncertainty and complexity

usually can identify a portfolio of 'known unknowns.' These are scenarios of probable, anticipated futures where *contingent* strategies and plans arise. Given moderate levels of uncertainty, where scenarios begin to multiply geometrically, *hedging* strategies insure against a portfolio of probable alternative futures to achieve desired outcomes at some cost of potential gains.

When moving to even higher uncertainty and complexity, several different strategy types exist. *Robust* strategies perform well against a range of both possible and plausible alternative futures, where wicked problems, 'unknown unknowns' and spoilers cannot be identified. In this realm are also *adaptive* strategies, which allow flexibility to change due to new circumstances by not precluding future options. *Resilient* strategies here perform well in spite of unforeseen shocks and quickly recover to deliver goals. Going beyond into deep uncertainty regardless of complexity, *shaping* strategies help change the fabric of environmental operating conditions to be more in line with strategic goals.

Table 4 applies some of this strategy framework for AI's transformative impact on great power competition. Today we already see a variety of these strategy initiatives, albeit in piecemeal fashion. Tailored to both domestic and international cognitive objectives, this framework provides a comprehensive, coherent and integrated strategy set to meet challenges and capture opportunities. Each of these strategies provide a roadmap for navigating AI and great power competition challenges and opportunities across different probable and possible alternative futures. Decision makers can choose one, several or a dynamically adjust a portfolio of multiple strategy types to achieve desired goals, depending on resources, constraints, timing, developments and risk appetite.

Core strategies sit at the nexus of national and international stability. These emphasize the maintenance and strengthening of international alliances through robust political engagement and a robust military deterrence spanning conventional, cyber and

Table 4 Robust decision making strategies for AI great power competition across domains

STRATEGY TYPE	POLITICAL	MILITARY	ECONOMIC	SOCIAL	INFORMATION	INFRASTRUCTURE
Core	Maintain & strengthen international alliances	Maintain a strong conventional & nuclear deterrent across multiple domains	Drive global economic development, free trade & protect intellectual property	Promote democratic values & human rights	Strengthen information credibility through independent, verifiable media	Invest in advanced, resilient infrastructure
Defensive	Prepare for diplomatic isolation by enhancing national unity	Strengthen defenses against hybrid, cyber & cognitive warfare	Implement protective tariffs to shield critical industries	Counter foreign influence operations through education	Protect critical information infrastructure from espionage	Enhance redundancy of critical infrastructure for continuity
Optimal	Leverage diplomatic influence to broker international agreements	Develop cutting-edge military technologies like AI and hypersonics	Enhance competitiveness through innovation in AI & high-tech industries	Leverage economic diplomacy to improve international relations	Lead in the development of global cyber norms & AI use	Lead global standards in infrastructure safety & sustainability
Hedging	Balance engagement with rival powers while preparing for conflicts	Prepare for multi-theater operations across different environments	Diversify economic partnerships to reduce reliance on strategic competitors	Support social cohesion by addressing domestic inequalities	Prepare for information warfare with defensive & offensive capabilities	Develop alternative energy sources to reduce geopolitical dependencies
Robust	Invest in global institutions that support international law and norms	Build flexible & rapid military deployment capabilities	Build resilient supply chains to withstand disruptions	Enhance social services to improve societal resilience against shocks	Invest in AI/quantum technologies that secure data & communications	Build smart, connected infrastructure integrating security features
Adaptive	Continuously assess & respond to global political shifts	Continuously train & adapt military tactics to evolving threats	Adjust economic policies based on global economic trends	Monitor and respond to social trends, integrating public opinion into policies	Update cyber policies in response to new threats & technological developments	Implement real-time monitoring systems for infrastructure management
Environmental Shaping	Influence international organizations	Build & maintain strong military alliances	Lead global economic development, practice & standards	Influence global social norms through international collaborations	Set international standards for AI, compute & data protection	Influence global infrastructure & development standards

nuclear capabilities. Economically, these strategies focus on driving global development, promoting free trade, and protecting intellectual property given current operating environment conditions. Socially, it underscores the promotion of democratic values and human rights, supported by a strong informational infrastructure that ensures media credibility and resilience in national infrastructures to sustain advanced societal needs.

Defensive strategies safeguard national sovereignty, minimizing attack surfaces and mitigating downside losses. Under defensive strategies, emphasis is on fortifying national unity to potentially withstand diplomatic isolation while enhancing protection against multifaceted asymmetric, hybrid, cyber, and cognitive threats. Economically, protective tariffs shield vital industries, accompanied by social strategies to counter foreign malign influence through educational, cyber and electoral initiatives. Strategies here also call for stringent protection of critical information infrastructures and enhancing the redundancy of crucial national infrastructures.

Optimal strategies in AI and great power competition leverage national advantages across the conflict-competition-cooperation spectrum, capturing upside opportunities and enhancing capability vectors. Political initiatives influence and broker significant international agreements while military efforts continue to develop cutting-edge technologies such as AI, quantum, robotics and hypersonics to deter conflict. These strategies enhance economic competitiveness through continuous AI and technology innovation for market growth, while using economic engagement and development diplomacy to foster peace, security and cooperation. Optimal strategies also set global cyber norms and promote the responsible use of AI for misinformation inoculation and multiplier effects from infrastructure investment returns.

Hedging strategies prepare for a myriad of possible uncertainties, across a multiplicity of contingencies. Here such strategies balance competition-cooperation engagement with rival powers and preparations for potential conflicts, emphasizing readiness for multi-theater, multidomain military operations. Underlying such is a diversification of economic partnerships to reduce strategic dependencies. Social policies support cohesion by addressing domestic inequalities, complemented by robust preparations for defensive and offensive information warfare capabilities.

Moving into the realm of higher complexity and uncertainty, robust and resilient strategies are evaluated across possible AI and great power competition alternative futures. Investments in global institutions that support international law, institutions and re-establish norms characterize robustness. These focus on building flexible and rapid military deployment capabilities and durable economic value chains that can withstand both supply and demand side shocks. Domestic social services are enhanced to insure vibrant civic participation with significant investments in AI and quantum technologies to secure data, communications and smart infrastructure.

Adaptive strategies provide responsive and on-the-fly dynamic adjustments, similar to financial real options analysis. Adaptive strategies highlight the importance of continuous political assessment and diplomatic responsiveness to operating environment shifts while still leaving options on the table. Military tactics are dynamically adjusted to meet evolving threats while economic policies are continuously recalibrated based on global economic trends. Social strategies are tuned to monitor and integrate public opinion into responsive policy making, with proactive cyber policies minimizing regulatory lag to tackle new threats and technological advancements.

Finally moving out to deep uncertainty across various levels of complexity, environmental shaping strategies influence global standards embedded in the fabric of future operating environments. These strategies influence international norms, institutions and organizations to align with national objectives, maintaining strong military alliances for economic development. This strategic approach influences global social norms through international collaborations and sets precedence for AI, computing, and data protection standards.

9 Conclusions

As seen, AI and technological shifts alter the strategic calculus of great power competition in several ways. First, they erode great power preeminence by providing individuals, groups, companies, smaller or rising powers with the means to challenge technological superiority traditionally held by established powers. Secondly, they lead to new sources of innovation, reconfiguration of global value chains and increased economic dependency, as nations strive to harness technological capabilities for economic growth and strategic autonomy. Finally, these factors contribute to a more connected, compressed, complex and uncertain complex global operating environment where individuals, actors and nations navigate the cognitive domain across the conflict-competition-cooperation spectrum.

> The coming decade—the "GeoTech Decade"—must address the sophisticated but potentially fragile systems that now connect people and nations, and incorporate resiliency as a necessary foundational pillar of modern life. Additionally, the rapidity of machines to make sense of large datasets and the speed of worldwide communications networks means that any event can escalate and cascade quickly across regions and borders— with the potential to further entrench economic inequities, widen disparities in access to adequate healthcare, as well as to hasten increased exploitation of the natural environment. The coming years also will present new avenues for criminals and terrorists to do harm; authoritarian nations to monitor, control, and oppress their people; and diplomatic disputes to escalate to armed conflict not just on land, sea, and in the air, but also in space and cyberspace. (Atlantic Council 2021)

While a daunting challenge, the framework presented here demonstrates the task is not unsurmountable. Leveraging both experience and insights, we can alter our perspectives seeing the forest through the trees by diagnosing our operating environments and decoding key drivers of change. CAS approaches offer such a robust framework for understanding and beneficially shaping interactions and adaptability of agents across PMESII domains.

Here navigating an accelerating complexity of modern operating environments is crucial. Whether we recognize it or not, intuition is the product of our own hard won, individual experiences. This results from our deeply human, agentic, goal directed behavioral experiences learning in multiple contexts. However, the operating environment is the emergent sum of all humanity's experiences, continuously evolving and adapting, far beyond the bounds of any one individual's intuition, experience and geo-physical agency.

Today the pace and tempo of events happens faster than any individual can cognitively process. Here we already rely of intuition as well as crowdsourcing information, common operational pictures, and collective wisdom in augmenting human intelligence. For decision makers, leveraging insights from alternative perspectives and methods provide a deeper understanding of systemic risks and potentials to also offset cognitive biases. RDM strategies empower success in an increasingly uncertain world. The goal is not to anticipate every challenge, but be robust, resilient and adaptable to the change we know is coming.

This forces us to recalibrate not only timescales but existentially question our worldview optics, policy initiatives and risk tolerance for uncertainty. We cannot use the same lenses and policy levers from a 20th century world that does not exist anymore. Individuals, firms, allies and adversaries globally balance the tradeoff between domestic demand for political, social, and cultural security against a global knowledge economy's supply of sustainable and potentially inclusive growth. How will we cognitively compete?

Whether in Tehran or Tampa, most people care about some form of Maslow's hierarchy of needs: air, water, food, security, cashflow … and often only then do they have the luxury to pursue self expression, democracy and human attainment (Inglehart and Welzel 2003; Abdollahian et al. 2012). Global economic development over the last seventy years dramatically improved living standards and the human condition. This has led to increasing happiness and subjective well-being for individuals (Helliwell et al. 2021; Abdollahian et al. 2024) as well as transitions to democratic and post-materialistic values (Inglehart and Welzel 2003) with better societal outcomes (Sachs et al. 2021).

While obviously not disagreeing with democratic values, engagement with populations, allies and adversaries shape our narratives. Messaging, channels and context sculpt how they are accepted by audiences and subsequently countered by peer competitors. Successful narrative campaigns, backed up by deliverable actions in the Sahel could be non-starters in Saudi Arabia. Across multiple contexts, how can we beneficially shape individuals, firms, and national leaders' choices sets across the conflict–competition–cooperation spectrum?

While decision making under certainty is trivial, the real challenge is decision making under risk and uncertainty. However most individuals are uncomfortable with uncertainty. Humans have been evolutionarily optimized for managing uncertainty with society predicated on the foundations to protect us from natural risks and minimize political, economic and social uncertainties. Yet individuals are more sensitive to losses, becoming risk averse compared to acquiring potential gains and being risk acceptant (Kahneman and Tversky 1979, 1992). For AI and great power competition, we might need to rebalance our own risk appetite: from being risk averse, chasing downside losses of a retrospective, eroding *status quo* to becoming more risk acceptant, capturing upside gains in creating the *new normal* as sources of innovation emerge. Herein rests the distinction between management and administration compared to leadership and vision. Ultimately our own risk appetite might determine the great power competition winner.

Where all of this leaves us still sits squarely in the human domain. What choices we make regarding AI and great power competition are still our purview, for now. Decoding complexity and understanding AI and operating environment drivers are necessary for great power competition. For hundreds of years, we have relied on technology and machines to augment human capabilities every day for sensing, navigating and traversing physical environments. At the dawn of a new era, we must now leverage AI for intelligence augmentation, to navigate the next century. Inspired by Darwin, our very success is dependent upon embracing complexity, uncertainty and continuous change to master the cognitive art of great power competition.

References

Abdollahian M (2021) Surfing the conflux: technology information environments and great power competition convergence. In: Farhadi A et al (eds) The great power competition, vol. 1. Springer, Berlin

Abdollahian M et al (2012) The dynamics of cultural change: the human development perspective. Int Stud Q 56(3):1–17

Abdollahian M et al (2013) Human development dynamics: an agent-based simulation of macro social systems and individual heterogeneous evolutionary games. Complex Adapt Syst Model 1:18

Abdollahian M, Yang Z (2022) Geotech: converging technology markets and politics in international political economy. In: Deese D (ed) A research agenda for international political economy. Edward Elgar

Abdollahian M et al (2006) Senturion: a predictive political simulation model. In: Defense and technology, vol. 32. National Defense University

Abdollahian M et al (2010) Politics and Power. In: Kott A, Citrenbaum G (eds) Estimating impact: a handbook of computational methods and models for anticipating economic social political and security effects. Springer, Berlin

Abdollahian M et al (2021) Technology growth and inequality: an agent-based model of micro transactional behaviors and meso technology networks for macroeconomic growth. In: Nunes IL (ed) Advances in Human Factors and System Interactions, vol 265. Springer Nature, Cham, pp 39–48

Abdollahian M et al (2024) Happiness and the state: going beyond GDP with political performance measures for societal subjective well-being. Soc Sci Humanit Open 10

Afifi-Sabet K (2024) AI singularity may come in 2027 with artificial ‘super intelligence’ sooner than we think, says top scientist. Live Science

Air Force Doctrine Publication (2016) AFDP 3–0 operations and planning: an effects-based approach to planning

Anderljung M et al (2023) Frontier AI regulation: managing emerging risks to public safety. arXiv: 2307.03718

Antúnez JC (2021) Understanding the operational environment: the human dimension. Global Strategy Reports—Política de Defensa

Arthur BW (1990) Positive feedbacks in the economy. Scientific American

Arthur BW (1995) Increasing returns and path dependence in the economy. Michigan

Arthur BW (2021) Foundations of complexity economics. Nature 3, February

Atlantic Council (2021) Report of the commission on the geopolitical impacts of new technologies and data
Australian Strategic Policy Institute (2023) ASPI's Critical Technology Tracker: Top 5 Country Visual Snapshot
Baum S (2021) A survey of artificial general intelligence projects for ethics, risk, and policy. Global Catastrophic Risk Institute Working Paper 20
Beraja M et al (2023) AI-tocracy. Q J Econ 138(3). https://doi.org/10.1093/qje/qjad012
Blades R (2021) AI generates hypotheses human scientists have not thought of. Scientific American
Bommasani R et al (2023) On the opportunities and risks of foundation models. arXiv:2108.07258
Bradford A (2020) The brussels effect: how the European union rules the world. Oxford University Press
Bucholz K (2024) The super election year. Statistica, Jan 19. https://www.statista.com/chart/31604/countrieswhere-a-national-election-is-was-held-in-2024/
Cattell RB (1963) Theory of fluid and crystallized intelligence: a critical experiment. J Educ Psychol 54:1–22
Central Intelligence Agency (2023) Annual Threat Assessment Of The U.S. Intelligence Community
Chen B et al (2022) Automated discovery of fundamental variables hidden in experimental data. Nat Comput Sci
Christian D (2018) Origin story: a big history of everything. Little Brown and Company, New York
Cioffa-Revellia C, O'Brien SP (2007) Computational analysis in US foreign and defense policy. In: The first international conference on computational cultural dynamics, pp 26–36
Clark J (2023) Hicks underscores U.S. innovation in unveiling strategy to counter China's Military Buildup. DOD News
Clausewitz CV (1831) On war
Claverie B, Du Cluzel F (2022) Cognitive warfare: the advent of the concept of 'Cognitics' in the field of warfare. In: Claverie B et al. (eds) Cognitive warfare: the future of cognitive dominance. HAL Open Science
Coates J et al (2010) Strategic foresight issue: introduction. Technol Forecast Soc Chang 77(9):1423–1425. https://doi.org/10.1016/j.techfore.2010.08.001
Cohen J, Lee G (2023) The generative world order: AI, geopolitics, and power. Goldman Sachs
Collier P, Hoeffler A (2004) Greed and grievance in civil war. Oxf Econ Pap 56(4):563–595
Conrad J, Rondeau D (2020) Natural resource economics: analysis theory and applications. Cambridge University Press
Cuhls K et al (2015) Models of horizon scanning: how to integrate horizon scanning into European research and innovation policies. Fraunhofer Institute for Systems and Innovation Research, European Commission
Daniels M, Chang B (2021) National power after AI
Department of Defense (2018) Summary of the 2018 DoD artificial intelligence strategy–harnessing AI to advance our security and prosperity
Dilmegani C (2024) When will singularity happen? 1700 expert opinions of AGI [2024]. AIMultiple
Du Cluzel F (2020) Cognitive warfare. NATO Innovation Hub
Dupree MH (2023) Microsoft researchers claim GPT-4 Is showing 'Sparks' of AGI. Futurism
Epstein JM, Axtel R (1996) Growing artificial societies: social science from the bottom up. Brookings Institution Press
Fearon JD (2018) Cooperation, conflict, and the costs of anarchy. Int Organ 72(3):523–559
Gell-Mann M (1995) What is complexity? Complexity J
GitHub (2024) Number of repositories on GitHub. GitHub. https://github.com/about
Green KL (2011) Complex adaptive systems in military analysis. Institute for Defense Analysis
Hameleers M (2024) Cheap versus deep manipulation: the effects of cheapfakes versus deepfakes in a political setting. Int J Public Opin Res 36(1)
Helliwell J et al (2021) World happiness report
Holland J (1995) Hidden order: how adaptation builds complexity. Addison Wesley
Holland J (1998) Emergence: from chaos to order addison. Wesley

Horowitz MC (2018) Artificial intelligence, international competition, and the balance of power. Texas Natl Secur Rev 1(3)
Hung TC, Hung TW (2022) How China's cognitive warfare works: a frontline perspective on Taiwan's anti-disinformation wars. J Global Secur Stud 7(4)
Hutter M (2022) Universal artificial intelligence: sequential decisions based on algorithmic probability. In: Texts in theoretical computer science. Springer, Berlin
Iansiti M, Lakhani K (2014) Digital ubiquity: how connections, sensors, and data are revolutionizing business. Harvard Business Review
IBM (2023) Understanding the different types of artificial intelligence. https://www.ibm.com/think/topics/artificialintelligence-types
Inayatullah S (2007) Questioning the future: methods and tools for organizational and societal transformation, 3rd edn. Tamkang University Press.
Inglehart RF, Welzel C (2003) The theory of human development: a cross-cultural analysis. Eur J Polit Res 42(3):341–379
Jeworrek T, Sachs R (2011) Complex accumulation risks–a challenges for the insurance industry. Munich RE
Johnson K (2021) Disintermediation and decentralization in financial markets. The Regulatory Review
Kauffman S (2000) Investigations. Oxford University Press
Kahneman D, Tversky A (1979) Prospect theory: an analysis of decision under risk. Econometrica 47(2):263–291
Kahneman D, Tversky A (1992) Advances in prospect theory: cumulative representation of uncertainty. J Risk Uncertain 5(4):297–323
Kiela D et al (2023) Plotting progress in AI. Our World in Data. Accessed from https://ourworldindata.org/brief-history-of-ai
Kurzweil R (2005) Long live AI. Forbes
Lee KF (2017) The real threat of artificial intelligence. New York Times
Lempert RJ (2002) A new decision sciences for complex systems. Proc Natl Acad Sci USA 99(Suppl 3):7309–7313. https://doi.org/10.1073/pnas.082081699
Lempert RJ (2019) Robust Decision Making (RDM). In: Marchau et al (eds) decision making under deep uncertainty: from theory to practice. Springer. https://doi.org/10.1007/978-3-030-05252-2_2
Lempert RJ et al (2006) A general, analytic method for generating robust strategies and narrative scenarios. Manag Sci 52(4):514–528
Lempert R et al (2007) Managing the risk of uncertain threshold responses: comparison of robust, optimum, and precautionary approaches. Risk Anal 27(4):1009–1026
Lingel S et al (2021) Leveraging complexity in great-power competition and warfare. Volume I, an initial exploration of how complex adaptive systems thinking can frame opportunities and challenges. RAND
Marchau VAWJ et al (2019) Decision making under deep uncertainty from theory to practice. Springer, Berlin
McAfee A, Brynjolfsson E (2012) Big data: the management revolution. Harvard Bus Rev
McGuinness P (2023) GPT-4: 1.8T parameter mixture-of-experts model trained on 13T tokens and optimized for inference. Accessed from https://patmcguinness.substack.com/p/gpt-4-details-revealed
McKelvey B (1999a) Self-organization, complexity catastrophe, and microstate models at the edge of chaos. In: Baum JAC, McKelvey B (eds) Variations in organization science: In Honor of Donald T. Campbell. Sage
McKelvey B (1999b) Visionary leadership vs distributed intelligence: strategy, microcoevolution, complexity. In: Proceedings of EIASM workshop. Brussels
Meacham S (2023) A race to extinction: how great power competition is making artificial intelligence existentially dangerous. Harvard Int Rev 8

Megginson LC (1963) Lessons from Europe for American business. Southwestern Soc. Sci. Q. 44(1)
Miles R (1978) The origin and meaning of Miles' law. Public Adm Rev 38(5):399–403
Miller J, Page S (2007) Complex adaptive systems: an introduction to computational models of social life. Princeton University Press
Mitchell M (2009) Complexity: a guided tour. Oxford University Press
Mitleton-Kelly E (2003) Ten principles of complexity and enabling infrastructures in complex systems and evolutionary perspectives on organizations: the application of complexity theory to organizations. Elsevier
Naprys E (2023) China versus US: who's winning the race for AI supremacy. Cybernews
National Institute of Standards and Technology (2022) NIST announces first four quantum-resistant cryptographic algorithms
National Security Agency, Quantum Computing and Post-Quantum Cryptography (2021). https://media.defense.gov/2021/Aug/04/2002821837/-1/-1/1/Quantum_FAQs_20210804.PDF
NATO (2023) Strategic foresight analysis. Brussels
Neild D (2023)Google quantum computer Is '47 Years' Faster Than #1 supercomputer. Physics
Neal PD et al (2017) Spectral analysis of network effects in a complex adaptive system of human and nature dynamics. In: Proceeding of summer simulation conference. The Society for Modeling and Simulation International.
Nicolis G, Prigogine I (1989) Exploring complexity W.H. Freeman
Nye JS (2013) Hard, soft, and smart power. The Oxford Handbook of Modern Diplomacy, Oxford
ODNI (2024) Annual threat assessment of the US intelligence community. Office of the Director of National Intelligence, Feb 5
Otto I et al (2020) Human agency in the anthropocene. Ecol Econ 167
Pappalardo D (2022) Win the war before the war? A French perspective on cognitive warfare. War on the Rocks
Parker E (2023) When a quantum computer is able to break our encryption, it won't be a secret. Rand
Parkinson J (2019) Is fluidity the key to effective multi-domain operations? In: Joint air & space power conference
Patterson D et al (2023) Carbon emissions and large neural network training. Accessed from https://arxiv.org/ftp/arxiv/papers/2104/2104.10350.pdf
Pike T (2019) Analysis and artificial intelligence in integrated campaigning. Strategic Multilateral Assessment
Provvidera T (2022) Printing and publishing in the renaissance. In: Sgarbi M (eds) Encyclopedia of renaissance philosophy. Springer, Cham. https://doi.org/10.1007/978-3-319-02848-4_682-1
Rickli JM et al (2023) Cognitive warfare and the governance of subversion in the 21st century. Geneva Center for Security Policy
Roose K, Poses AI (2023) Risk of extinction. Industry leaders warn. The New York Times
Runco MA (2015) Meta-creativity: being creative about creativity. Creat Res J 27(3):295–298
Russell SJ, Norvig P (2003) Artificial intelligence: a modern approach, 2nd edn. Prentice Hall
Sachs JD et al (2021) Sustainable development report 2021
Scharre P (2023) Artificial intelligence and great power competition. Council on Foreign Relations
Schmidt E (2022) AI, great power competition & national security. Daedalus 151:288–298
Schneider G (2014) Peace through globalization and capitalism? prospects of two liberal propositions. J Peace Res 51(2):173–183
Silver D et al (2017) Mastering the game of Go without human knowledge. Nature 550(7676):354–359. https://doi.org/10.1038/nature24270
Stanford University (2024) Artificial intelligence index report 2023
Statistica (2024) Public cloud services end-user spending worldwide from 2017 to 2024. https://www.statista.com/statistics/273818/global-revenue-generated-with-cloud-computing-since-2009/

Strubell E et al (2019) Energy and policy considerations for deep learning in NLP. https://doi.org/10.48550/arXiv.1906.02243
UK Government Office for Science (2023) Future Risks of Frontier AI. UK Government
UNCTAD (2023) World investment report 2023. United Nations
US Executive Order 14110 (2023) Safe, Secure and Trustworthy Development and Use of Artificial Intelligence
US Government (2023) Executive order 14110 safe, secure and trustworthy development and use of artificial intelligence
World Bank (2023) Annual report 2023, World Bank Group.
World Economic Forum (2020) The global risks report 2020 15th edn. World Economic Forum. https://www.weforum.org/reports/the-global-risks-report-2020
World Resources Institute (2011) World resources report 2010–2011: decision making in a changing climate. UNDP
Wu J (2021) Possibilities and implications of the multi-AI competition (2021). arXiv:2209.00509
Wurm D (2022) Cognitive domain: hybrid and future forms of prosecuting conflicts in the cognitive domain. Austrian Federal Ministry of Defence
Yalalov D, Myakin D (2023) GPT-4's leaked details shed light on its massive scale and impressive architecture. https://mpost.io/gpt-4s-leaked-details-shed-light-on-its-massive-scale-and-impressive-architecture/
Yang Z et al (2016) Social spatial heterogeneity and system entrainment in modeling human and nature dynamics. In: Zhang L et al (eds) Theory, methodology, tools and applications for modeling and simulation of complex systems. Springer, Berlin
Zelli F et al (2020) Global governance in complex times: exploring new concepts and theories on institutional complexity. University of Bamberg Press
Zhang X, Shu D (2021) Current understanding on the Cambrian Explosion: questions and answers. PalZ 95:641–660. https://doi.org/10.1007/s12542-021-00568-5

The People's Republic of China and the Information Age: Dealing with Informationization and Intelligence-Ization

Dean Cheng

Abstract Chinese Communist Party (CCP) leaders assess that the revolution in information and communications technology (ICT) has ushered in the Information Age. As ICT steadily advances, the PRC has first sought to informationize, and now to intelligence-ize the nation, including its economy and military. Informationization facilitated the movement of information among people and organizations. Intelligence-ization is the movement of information among machines and systems without human intervention. Key technologies associated with intelligence-ization include not only artificial intelligence, but big data, networking, and digitization. At the same time, the CCP seeks "information security," which in the Chinese context means the ability to control the flow and monitor the content of information. This is as aimed in part at the domestic Chinese information space, to ensure the continued rule of the CCP. In addition, the CCP is engaging in "informational mercantilism," where the PRC seeks access to a range of foreign information, including economic and national security data, while denying outsiders access to China's. Both informationization and intelligence-ization are also key parts of the ongoing Chinese military modernization effort. For the PLA, and the Chinese leadership in general, the combination of efforts goes beyond applying information technologies to traditional tasks and missions. Instead, entirely new technologies and processes are being generated, requiring new thinking, including tactics, techniques, and procedures.

Keywords PLA · Informationization · Intelligence-ization · Cyber · Joint operations · Cyber security

---

D. Cheng (✉)
U.S. Institute of Peace, Washington, D.C., USA
e-mail: sunzi36@yahoo.com

A. Farhadi et al. (eds.), *The Great Power Competition Volume 6*,
https://doi.org/10.1007/978-3-031-70767-4_3

1 Introduction

When British First Sea Lord Jackie Fisher authorized the construction of the HMS *Dreadnought*, the first all-big-gun battleship, he wrought a global revolution in naval power. In one stroke, he made every other battleship in existence obsolete. But in doing so, this included every British battleship. In essence, the *Dreadnought* revolution reset global naval power, with all states essentially starting from the same point.

In many ways, the People's Republic of China (PRC) and its ruling Chinese Communist Party (CCP) see the world as undergoing a comparable set of fundamental changes due to the proliferation and rapid development of information and communications technology (ICT). Since the 1990s, the growing role of ICT has altered the currency of power, leading to the evolution from the Industrial Age to the Information Age. Where broad industrial capacity was the centerpiece of the Industrial Age, the ability to generate, move, analyze, and exploit information is now the key determinant of national and international power.

For the CCP, this resetting of power provides China with the ability to catch up with the West. If the PRC can generate information, access it, and exploit it more rapidly than other nations, then it can become a major, developed power on a par with other major powers. If it can develop the industries necessary to manufacture information systems and build the associated infrastructure, it can integrate itself into the global information networks at a physical as well as informational level. To this end, the PRC has openly pursued a policy of "informationization."

The Information Age itself, however, has also evolved as ICT has steadily improved. Whereas this new age began with the proliferation of microelectronics and computers, the advent of 5G and 6G networks, artificial intelligence, and the Internet of Things is seen as marking a new stage of information exploitation. Information technology now permeates all aspects of society and economics, such that one cannot escape its impact. As important, nations and systems that fall behind in the development of both hardware and software, who cannot meet their own national demands will be strategically vulnerable. "Intelligence-ization," the exploitation of these new aspects of information technology, has become as important as "informationization" for nations and systems to remain competitive.

2 Informationization of Chinese Society

The growing importance of information technology, beginning in the 1970s but blossoming in the 1990s and early 2000s led to the promotion of "informationization (xinxihua; 信息化)." From the Chinese perspective,

> Informationization is a comprehensive system of systems, where the broad use of information technology is the guide, where information resources are the core, where information networks are the foundation, where information industry is the support, where information

> talent is a key factor, where laws, policies, and standards are the safeguard. (State Council Information Office 2002)

For the CCP, the explosion of information technology was initially seen as facilitating systematic national economic development. The 1993 establishment of the State Economic Informationization Joint Council was aimed at helping the state gather more and better economic information. CCP leaders soon recognized, however, that "informationization" meant more than simply applying better information technology to economic planning. Instead, it meant the application of ICT to all aspects of society and government, the development of information-related industries, but also defending against the free flow of information in order to safeguard CCP rule. The PRC would join the global information web—but on its own terms.

The Chinese Ninth Five Year Plan (1996–2000) emphasized the construction of telecommunications infrastructure across the PRC. This included developing digital mobile communications equipment and program-controlled switchboards, as well as the extensive laying of fiber optic cables. The first step towards informationizing the PRC would be networking the nation.

In addition, the CCP introduced a series of "Golden projects," which would create the software infrastructure to allow Chinese netizens to plug in. These included:

- Golden Bridge (*jinqiao*; 金桥). An information infrastructure to facilitate the movement of economic information
- Golden Card (*jinka*; 金卡). A nation-wide payment system promoting the use of credit and debit cards in what had been a cash-driven economy
- Golden Tax (*jinshui*; 金税). Computerization of the nation's tax system, to reduce fraud and tax dodging while simplifying tax payments (Qiang 2006).

Perhaps the most well known and extensive part of these "Golden projects" was "Golden Shield (*jindun*; 金盾). While China was interested in joining the global telecommunications network, it nonetheless sought to control what could be accessed. Even as China was taking its first steps into connectivity, research was underway to ensure that those connections were firmly under the control and supervision of the CCP and its censors. The Golden Shield project works alongside "the Great Firewall of China," and together these elements constitute an essential means of defending the PRC from unauthorized information proliferation, from without—and within.

Subsequently, informationization was repeatedly emphasized by Chinese leaders as a key part of improving Chinese "comprehensive national power (*zonghe guojia liliang*; 综合国家力量)." General Secretary Jiang Zemin in 2002 emphasized the Chinese path to industrialization and economic modernization would rely more on the information sector. Jiang noted that information technology was the "logical choice," if Chinese industrialization was to accelerate, especially since informationization would promote other benefits, including raising the overall level of scientific and technical awareness, reducing resource consumption, and developing Chinese human resources. Therefore, "we must give priority to the development of the information industry and apply IT in all areas of economic and social development" (Jiang 2022).

His successor, Hu Jintao, established the "National Strategy for Informationization Development, 2006–2020," laying out further plans for expanding and deepening the role of information technology in Chinese development. This strategy called for expanding the level of informationization throughout China's economy and society, including promoting the application of information technology in healthcare and government operations, as well as making China a globally competitive in the production of information related technologies.

3 Xi Jinping and Informationization

These efforts were redoubled under Xi Jinping, who pushed for not only continued growth in Chinese application and exploitation of information technologies, but also for self-sufficiency in key related technologies and production. This dual emphasis is reflected in the "Made in China 2025" program. Set forth in 2015, "Made in China 2025" borrows from the German concept of "Industry 4.0," the application of advanced information technologies to manufacturing (Kennedy 2025). The Chinese plan goes further, however, identifying ten industrial sectors where the PRC would strive to move up the value chain in production and reduce reliance on imports, while improving the efficiency and sophistication of Chinese production capabilities. Among the emphasized industrial sectors are new generation information technology including chip manufacturing; aerospace; advanced robotics; and advanced materials, but also agricultural machinery, railway systems, and ocean transport (Sutter 2023).

China's push for "informationization" entails both assuming a central role in the development and manufacturing of key components for information systems, but also applying advanced information technology to traditional industrial sectors, including "smokestack industries" such as shipbuilding, railways, and agricultural machinery. In this manner, the entire Chinese economic base will be improved, benefiting from the improvement in information flow and feedback.

Even as the CCP has been pushing the more widespread application of information technologies, and integrating China's economy into foreign supply chains, it has also recognized that many of the threats to PRC national security, including the party's grip on power, have also become informationized.

The spread of information technology means that potential adversaries have unprecedented access to each others' national economy, as well as the broader population and the top decision-makers. Just as the bomber and long-range missile allows an opponent to directly strike a nation without having to first break through ground or naval defenses, information technology similarly outflanks traditional military and security measures. The proliferation of information technology into society and economics makes them vulnerable to a range of new pressures and threats.

These threats extend beyond information networks (e.g., vulnerability to denial-of-service attacks) and component computers (e.g., computer viruses, malware). Instead, the very information itself can constitute a threat, if, for example, its content influences key decision-makers, saps popular support for a conflict, or erodes the will of the military to fight. Similarly, foreign access to various types of information can provide insights into not only Chinese military systems, but key research fields and agendas, popular opinion, decision-making weaknesses, and leadership and popular preferences. Consequently, China's interpretation of its informational national interests has expanded, in step with the expanding impact of information writ large on China.

Consequently, since Xi Jinping took power in 2012, Chinese efforts to secure its information space have redoubled. One line of effort has been the promulgation of a series of national security-related laws, many of which include significant provisions for controlling information flow.

National Security Law of the People's Republic of China (2015). This law defines the PRC's concept of national security, and makes clear that the CCP has adopted a broad, sweeping set of definitions. National security encompasses not only military, economic, and food security, but cultural security, scientific and technological security, and information security (National Security Law of the People's Republic of China 2015). As one American legal assessment of the National Security law notes, the legislation "imposes broad obligations on [Chinese] citizens and corporations to assist and cooperate with the government," including the provision of "data, information, and technological support or assistance to security agencies, law enforcement agencies, and the military" (Covington and Burling 2015).

Cybersecurity Law (2016). This law, passed in 2016 and in force since 2017, lays out a range of rules for cyber activity (Cybersecurity Law 2016). One of its most important aspects is establishing sovereignty over information networks and cyberspace, by making clear that Beijing can and will establish rules for information activities that it will enforce. In essence, the Cybersecurity Law establishes that there is a ***Chinese*** Internet, for all netizens operating within the borders of the PRC. Various provisions of the law include requiring data localization, making clear the responsibility of Chinese Internet service providers (ISP) and network operators for maintaining information security and cooperating with Chinese security services, and banning the export of "important data" deemed to be of national security interest (Wagner 2017). In particular, flow of data from sectors deemed to be "critical information infrastructure" is heavily restricted. The Cybersecurity Law also mandates that sales of foreign software and hardware be made contingent upon certification, which in turn potentially allows Chinese officials to demand source codes and encryption techniques (Alsabah 2017). Only state-approved encryption technologies are allowed in any case.

National Intelligence Law of the People's Republic of China (2017). This law covers intelligence and counterespionage activities (PRC National Intelligence Law 2018). Under the law, businesses operating in the PRC can be required to hand over information to the Ministry of State Security; they can also be required to conceal the release of this information. Notably, the law applies not only to companies incorporated in the PRC, but potentially to foreign businesses operating in the country as well. Similarly, PRC companies operating abroad can, under the law, be required to share foreign data that they have acquired. A 2023 amendment to the related Counter-Espionage Law further expands potential state authorities. For example, the Chinese definition of "espionage" shifted from state secrets and intelligence (as is typical in many countries) to all "documents, data, materials or items related to national security and interests", (McCarthy and Gan 2023) without specifying specific parameters for how these terms are defined.

Data Security Law (DSL 2021). This law expands the regime established under the 2016 Cybersecurity Law (Digichina 2021). While the Cybersecurity Law prohibited operators of "critical information infrastructures" from transferring "important data" and personal data outside of China, the DSL goes further, limiting any transfers of "important data" outside of China, regardless of whether it is related to "critical information infrastructure" (Sidley and Austin 2022). Furthermore, transfers of any data outside of China are contingent upon a security assessment and review process. In addition, any provision of data to foreign law enforcement agencies requires prior approval by "competent authorities," which is left ambiguous (Chen 2021).

Personal Information Protection Law of the People's Republic of China (PIPL 2022). The PIPL is the first Chinese law specifically focusing on personal information security (Creemers and Webster 2021). It requires foreign companies to store Chinese citizens' data in the PRC (as also required by other PRC laws), as well as enjoins them from cooperating with foreign law enforcement in accessing said data without the specific permission and participation of Chinese law enforcement representatives. The law provides for PRC extraterritoriality; that is, companies that process personal information of Chinese citizens abroad are subject to the PIPL (China's Personal Information Protection Law 2022).

In combination, these laws forge a comprehensive set of controls on information access and flow, covering any aspect that might touch upon the CCP's very broad definition of "national security." Indeed, the CCP's concept of "information security" is primarily about limiting the flow of data both within and out of the PRC. This is in sharp contrast to Western concerns about information security, much of which is focused on safeguarding the privacy of citizens' or corporations' data. This network of laws does not meaningfully limit the Chinese state's ability to surveil either; indeed, it provides a legal justification for intrusive inspection and surveillance of both Chinese citizens and corporations, and foreign companies that might interact with them. In a world where information is the currency of power, the ability to constrain foreign access to Chinese information can generate significant impacts.

As important, Beijing's broad approach as reflected in these laws is one of "informational mercantilism," wherein the PRC can access foreign information (including intellectual property, academic and scientific literature, and personal and corporate data), while restricting foreign access to China's. This arguably reduces Chinese vulnerabilities, gives China certain competitive advantages in both domestic and foreign markets, and limits foreign influence.

Chinese actions under these laws in recent years has spotlighted the extensive reach that Chinese authorities can now exploit. Under the authorities granted to them by these laws, Chinese security officials have questioned employees at Bain & Co., an American consulting firm, and detained employees of the Mintz Group, a firm specializing in due diligence investigations (Yu and Strumpf 2023). These actions were justified by the information control provisions of various Chinese laws, including those relating to intelligence, national security, and anti-espionage.

Consequently, some companies have chosen to alter their business practices, voluntarily avoiding Chinese information in order to reduce their legal exposure. Wind Information Company, a Chinese firm which collects data on a variety of enterprises and their executives and legal representatives, including shareholders, investment activities, and operating risks, chose to limit access to its corporate registry database for clients outside the PRC, presumably for fear that it might violate Chinese information security regulations (China Restricts Overseas Access to Key Corporate Information 2023). The international law firm Dentons broke ties with its Chinese partner (McDonald 2023). It is believed it did so to avoid suspicions that it might have access to information on Chinese citizens.

Such self-censorship meets China's interest in curtailing foreign access to Chinese information, with the added benefit that it cannot be directly attributed to Chinese actions.

Chinese efforts at preserving their information security have not been limited to legal restrictions, however. In order to limit foreign ability to access and influence the Chinese population, the PRC has also put in place a variety of filters and barriers on the Chinese Internet, many incorporated within the "Great Firewall of China." The Great Firewall employs a variety of software to detect and halt the flow of ideas and information deemed inappropriate or dangerous by the CCP. Authoritative name servers and domain name servers in the PRC are run by Chinese companies, or the state itself, so their cooperation is both legally mandated and generally freely provided. Similarly, China's main Internet Service Providers, China Mobile, China Telecom, and China Unicom are all state-owned enterprises, and are an integral part of the Great Firewall's filtering process (The Great Firewall of China: Obstacles to Monitoring Performance n.d.).

An"on-path system," the Great firewall shunts inbound and outbound data packets that are flowing into and out of China through the three main international exchange points, China's connection to the broader global Internet. These packets are sent to what is believed to be separate clusters of government-run computers, reassembling the data to some extent, so that it can examine their destinations and even their content. If the IP address, URL, or constituent packets are found to be problematic, the Great Firewall can then block access, misdirect responses, or in some cases even filter out specific web pages and images.

Notably, the Great Firewall of China does not rely upon human actions. There is no army of censors monitoring the traffic passing through the international exchange points. Instead, the Great Firewall of China is a display of Chinese prowess in artificial intelligence and machine learning, employing automated inspection and responses. As important, the Great Firewall responds to changes in terms and content. When CCP authorities cracked down on images of "Winnie the Pooh" (often used in reference to Xi Jinping), the Great Firewall promptly began to filter out such images.

4 Growing Emphasis on "Intelligence-Ization"

The steady development of information technologies, including advances in big data, cloud computing, 5G networking, and artificial intelligence, has led to increasingly abilities to exploit information that had not been available at the start of Chinese informationization efforts. These are enablers of the process of "intelligence-ization" (Intelligence-ization: Explaining Its Development and New Capacities 2020). Chinese thinking about "intelligence-ization" goes back to at least 2006, when the Informationization Development Strategy included references to "intelligence-ization (*zhineng hua*; 智能化)."

From the Chinese perspective, "informationization" was the application of ICT to existing processes, systems, and activities. Advances in ICT created the conditions for linking people and businesses with each other, allowing them to make more effective use of information that had hitherto not been exploited. Individual, corporate, government, and military activities could all become more efficient and effective with the incorporation of these new technologies. But the additional advances since the early 2000s have meant that information is not just augmenting or enhancing capabilities, but creating entirely new opportunities (http://media.people.com.cn/n1/2019/0301/c40606-30951460.html). The result is "intelligence-ization," building atop "informationization."

The Cyberspace Administration of China has detailed how various technologies are playing a role in the "intelligence-ization" process.

Digitization (*shuzi hua*; 数字化) and *data-ization* (*shuju hua*; 数据化). More and more activities, things, and actions, and the consequent information, can be expressed in coded, digital form. That digitized data in turn can be more easily manipulated, assembled into databases, and therefore analyzed.

Big data. The accumulation of massive databases and the ability to manipulate them allows analysts to link together what had previously been disparate fragments of information. These links in turn allow analysts to identify patterns which might otherwise be undetected, which in turn can be exploited.

Network-ization. The expansion of databases in turn leads to the linking of ever more networks, Information is therefore more rapidly collected and disseminated, and can be more efficiently exploited (http://www.cac.gov.cn/2019-03/01/c_1124178478.html).

For Chinese analysts, the combination of these elements, as well as the general proliferation of big data and networks, coupled with improved processing has created the conditions for "intelligence-ization." If "informationization" is the broad process of applying information technology to all aspects of social, economic, and military affairs, "intelligence-ization" is the most current phase of this process, reflecting advances in ICT that had not necessarily been envisioned thirty years ago. Indeed, several Chinese analysts have noted that "intelligence-ization" is not only about the growth of artificial intelligence, but the evolution of the overall information environment to allow the growth and interplay of these various new technologies.

One Chinese analysis notes, for example, that intelligence-ization extends information exchanges from people-to-people and business-to-business to linkages between people and items and between items and items. The resulting "Internet of Things (IoT)" is as much part of intelligence-ization as artificial intelligence and machine learning. Enabled by the growth of sensor technology, wireless communications technology, including 5G networks, and data management, there can now be constant interchange of data among various sources. This, in turn, will lead to improved recognition, positioning, tracking, surveillance, and management, integrated into equipment and facilities and a routine part of their operation (http://www.cac.gov.cn/2019-03/01/c_1124178478.html).

The IoT will eventually involve the linking of various smart systems together, exploiting their built-in abilities to assess, sense, and learn, whether in terms of intelligent manufacturing or intelligent cities.

5 Military Significance of Informationization

From the CCP perspective, this growing importance of information technology inevitably influences the nature of warfare. There is always an interplay between military affairs and the larger economy and society; thus, as the currency of power changes, this affects all aspects of human interaction and society. Mechanized military forces are a reflection of the Industrial Age, including both industrial economics and an industrialized society. Correspondingly, an informationized society will create an informationized military, while an informationized military can only be produced by an informationized society and economy. In short, from the CCP's perspective, informationized societies and economies lead to informationized wars, which in turn require informationized militaries to fight them successfully. In the Chinese view, the People's Liberation Army (PLA) and broader security establishment must be prepared for "informationized warfare (*xinxihua zhanzheng*; 信息化战争)."

From the PLA's observations of other people's wars of the 1990 and 2000s, it concluded that future wars would primarily entail joint operations, for which the PLA must therefore be prepared. Such operations, however, would have to go beyond simply placing forces from different services in the same area of operations, with provisions for communications and some planning among them. Instead, they would involve activities across land, sea, air, outer space, and the electromagnetic and informational domains. Moreover, they would be "unified (yitihua; 一体化) joint operations," requiring four major improvements (Cui and Wang 2004).

- The creation of a "unified operational theory (*zuozhan lilun yitihua*; 作战理论一体化)," in which a common operational theory was created to allow different services to share common operational concepts. Without a unified, or consolidated, set of regulations (*tongyi de tiaoling*; 统一的条令), then it would not be possible to have different services' formations act in a unified, or consolidated, manner (*tongyi xingdong*; 统一行动).

- A unified operational theory, in turn, requires a "unified command (*zhihui yitihua*; 指挥一体化)" to implement it. It is no longer sufficient to coordinate at the highest levels, and then rely on the services to implement operations along their own lines. Instead, the PLA would have to establish an appropriate command system (*zhihui tizhi*; 指挥体制) for all the various forces, capable of executing planning across service lines.
- The unified command system can only manage joint operations through the creation of a "unified C3I system." This system is necessary, so that all participating forces can be linked together not only at the top, but farther down the chain of command.
- Only such a unified command structure and C3I system would allow the unification of knowledge (*zhishi yitihua*; 知识一体化)," so that there is common situational awareness of the situation, but also mutual understanding of each service's perspectives and approaches to warfare.

The massive organizational overhaul of the PLA at the end of 2015, and subsequent organizational and equipment modernization efforts, have codified many of these improvements. The General Staff Department of the Central Military Commission has been reoriented into the Joint Staff Department, to underscore that joint operations are now the standard for planning by the entire PLA. The seven military regions, long dominated by the ground forces, have been reorganized into five military theater commands. These commands are joint, rather than ground force dominated, and several are now headed by commanders drawn from the PLA Navy and PLA Air Force.

Meanwhile, the creation of the PLA Strategic Support Force (PLASSF), which controls PLA network warfare, electronic warfare, and space warfare forces means that there is a service which is responsible for operations in the information, electromagnetic, and space domains, supporting the more traditional services. This supports the idea that, for the PLA, joint operations occur across multiple domains, and not just across different services, and addressed the shortcoming of "jointness in form but not in actual capability (*xing ju er neng bu ju*; 形聚而能不聚)" (Li 2014).

Such a structure in turn entails dedicated human, organizational, hardware, and software elements. As Chinese analyses note, not only must there be a more unified joint campaign command organizational structure, but the personnel staffing it must be accustomed to thinking in more joint terms. Unified joint operations require command and control perspectives that have a holistic view of the mission and the participating forces, rather than starting from the point of view of a particular service or branch. Successful joint operations require expertise in not only air, land, and sea operations, but also special operations, space operations, and operations in the electromagnetic domain. Similarly, the planning for command, control, intelligence, communications, and defensive counter-measures should reflect this comprehensive, unified approach (Zhang 2012). Above all, successful joint operations require the ability to rapidly and accurately manage massive amounts of information, including compilation, analysis, and dissemination.

Indeed, a major challenge for the commanders and staff of joint operations is to incorporate information from all sources, including not only military ones (of all the services), but also local military commands, Party officials, and civilian assets. The information systems must allow for smooth sharing of information, which requires compatibility in both hardware and software terms. Thus, there must be many different communications channels, able to support the transfer of many different forms of information (e.g., video, audio, data), in various formats (Xu 2013). There must also be a common set of procedures that spans all the services, so that personnel assigned to this unified joint campaign command structure have a common understanding of staffing processes.

According to Chinese writings, successful future joint operations on the informationized battlefield is built upon several key pillars.

- Real-time information.
- Accurate data.
- Collection of many different kinds of information for many different users.
- Intelligent information processing.
- Reliable communications.

All of these efforts will ideally generate a common situational picture, one that can receive inputs from all the participating forces and their various platforms (including their weapons). Each input, in turn, will be shared near simultaneously by all the participating forces. By generating such shared situational awareness, exploiting all the available information sources, Chinese analysts expect a more rapid cycling of information, allowing commanders' decisions to be more rapidly disseminated to the units, leading to a more flexible, rapid, tailored response. Enemy vulnerabilities will be rapidly identified, allowing for prompt exploitation by all available forces. Command will be in real-time, and operations will be highly adaptive. Rather than a linear progression, operations will be able to proceed in parallel, across the depth and breadth of a theater, with precise attacks paralyzing an adversary, rather than relying upon brute force to bludgeon them into submission (Chinese Military Encyclopedia 2nd Edition Editorial Committee 2007).

At the same time, this common situational picture would allow commanders to better track not only adversary forces, but also friendly units. This latter aspect is especially important, given the involvement of forces drawn from all the different services, who would be operating across multiple domains. As one Chinese analysis observed, even Sun Tzu had written that only by knowing oneself as well as the adversary can one hope to be ever victorious (Zou and Cha 2011).This would be even more true in the Information Age.

From the Chinese perspective, a clear demonstration of what such information sharing can achieve was provided by the American-led coalition's operations against Iraq in the 2003 Iraq war. Because the coalition forces had superior C4ISR capabilities, they could forge a truly joint operational approach to the conflict, with smooth communications among the various forces. This was a significant advance over what had been undertaken in the Gulf War, a decade previously, where coalition ground forces had some difficulties coordinating with naval and air forces (He 2004).

Much of the PLA's modernization of the past three decades has therefore been aimed at improving the level of information collection, dissemination, and exploitation. The creation of the PLASSF is only facet of this broad effort. PLA Navy combatants, for example, are outfitted with a variety of modern sensors. The PLA Air Force has not only been acquiring new combat aircraft, but also airborne early warning and airborne electronic warfare platforms. Nearly 15 years ago, the Office of Naval Intelligence was already observing that the PLA Air Force and PLA Naval Aviation were incorporating datalinks and electronic attack methods into their training (Office of Naval Intelligence 2009). The 2022 DOD report on Chinese military capabilities notes that the PLA continues to emphasize C4I modernization, while also improving its electronic and cyber warfare capabilities to counter adversaries' command and control systems (https://navyleaguehonolulu.org/maritime-security/ewExternalFiles/2022-military-and-security-developments-involving-the-peoples-republic-of-china.pdf).

6 Military Significance of Intelligence-Ization

Just as the evolution of ICT has led to intelligence-ization as the next phase of informationization of the broader society and economy, the effects of that same evolution are felt in military operations. PLA analysts have concluded that modernization must now include not only informationization, in terms of the widespread deployment and exploitation of information systems, but also intelligence-ization of military equipment, organization, and training.

In this regard, "intelligence-ization" of the PLA is not simply a matter of applying artificial intelligence/machine learning, any more than "intelligence-ization" of the broader society and economy is a function of applying those technologies. Indeed, one Chinese article makes this specific point, noting that "Military Intelligence is not 'military activities + AI (*junshi zhineng bu shi junshi + AI*; 军事智能不是 '军事 + AI')'" (Liu 2021). Instead, according to this researcher, artificial intelligence is only one of a suite of technologies, as well as processes and procedures, necessary to effect the intelligence-ization of the PLA.

Other Chinese analyses support this view. At a conference on improving command and control, one presenter noted that key technologies associated with intelligence-ization include human–machine interfacing, intelligent manufacturing, and swarm technology (Wu 2020). It also includes the incorporation of stealth technologies and techniques. As important, intelligence-ization involves different applications of these technologies. This ranges from incorporating machine intelligence and advanced sensors to individual equipment, to coordination between manned and unmanned systems, to improvements in system-of-systems. It also includes public opinion warfare and other aspects of political warfare and cognitive warfare (Wu 2020).

Intelligence-ization involves not only technologies, but the human element as well. Another PLA analysis observes that how the PLA officer is educated must also evolve if the PLA as a whole is to intelligence-ize. The intelligence-ized officer must be conversant with man–machine interface technologies and capacities. They must have an understanding of cognition and cognitive warfare, network warfare, as well as more traditional military operations. They must be educated in thinking along system-of-system lines. To this end, the professional military education system of the PLA must also be adjusted and reformed, if it is to produce officers fit to serve and lead an intelligence-ized force (Tang 2021).

As yet another PLA analysis suggests, for the PLA, intelligence-ization entails merging the physical elements, the informational and sensor elements, and the cognitive into military activities, while applying the latest advances in information technology. This spans not only combat operations, but logistics, training, and military education. Where "informationized warfare" focused on joint operations through networking of forces, "intelligence-ized warfare" is cognition-centric warfare (*renzhi zhongxin zhan*; 认知中心战), with intelligence (including suitably trained personnel as well as artificial intelligence decision aids) managing and controlling firepower, mobility, and information capabilities (Intelligence-ization Brings New Changes to Conflict 2021).

Through intelligence-ization, the PLA hopes to employ the next stage of advances in ICT in order to both fully exploit the enormous amounts of data now available as well as harnessing the range of new systems-of-systems and technologies that these advances have also created. At a minimum, the use of artificial intelligence and machine learning can help manage the terabytes of data available to the modern commander from the myriad of sensors, from tactical ground systems and those integrated into weapons to persistent space- and near-space-based systems.

But advances in ICT can do more than just manage data flow. As one PLA analysis notes, the proper, prompt analysis of big data, for example, can potentially reveal patterns that have otherwise not been noticed, helping reduce the "fog of war." The same analysis suggests that by relying on algorithms, human biases can potentially be reduced, allowing for more objective assessments of data to produce better situational analysis (Li 2019).

Recent combat experience with drones in Nagorno-Karabakh and Ukrainian battlefields highlight how advanced manufacturing, improvements in sensors, and better software allows unmanned systems ranging from table-top and cardboard drones to HALE (high altitude, long endurance) and MALE (medium altitude, long endurance) remotely piloted vehicles to engage in a range of combat activities from long-term surveillance to tactical attacks against armored vehicles, ships, airfields, and headquarter facilities. A *PLA Daily* article suggests that future unmanned systems will be able to undertake network attack and defense, electronic combat, multi-source sensing, battle damage assessment, tracking, surveillance, and other such tasks (Intelligence-ization Brings New Changes to Conflict 2021).

7 Conclusion

For the CCP, the growth of information technologies has been a boon in strengthening the PRC. Integrated into the world's supply chains, the PRC has been able to harness its advantages not only to produce more advanced technologies, but also to exploit those same technologies to help preserve the CCP's power. The advances in ICT that are leading to intelligence-ization are providing the CCP with additional tools for both strengthening the Chinese economy and maintaining its grip over the Chinese people.

In military terms, like Jackie Fisher's *Dreadnought*, the proliferation of ICT, and the new phase of intelligence-ization, has leveled the playing field globally, allowing the PLA to catch up with other militaries, including those of the West. This is a significant shift in the strategic situation, as the PLA has long been less technologically sophisticated than its primary potential opponents, and had to hope that it could employ sheer numbers to offset this disadvantage.

Since the first Gulf War (1990–1991), however, the PLA has recognized that it must improve qualitative capacity if it is to successfully fight future conflicts, which will be marked by a far higher reliance on advanced technologies. By the late 1990s, information technologies were recognized as the foremost of the new technologies that the PLA needed to incorporate. The rapid evolution of ICT, however, has provided the PRC an opportunity to leapfrog older technologies, much like cellular phone networks have allowed countries to bypass copper and even fiber infrastructure and yet achieve widespread connectivity for its populace.

The rapid transformation offered by the ongoing ICT revolution, however, has led to further advances, encapsulated in the Chinese concept of "intelligence-ization." This new phase goes beyond applying information technologies to traditional tasks and missions, and instead generates entirely new technologies (and demands new tactics, techniques, and procedures). It also requires new ways of thinking, not only in terms of technologies and tactics, but the entire informational landscape—hence the emphasis on "cognition (*renzhi*; 认知)," and "cognitive warfare." Under these circumstances, the PLA's lack of combat experience is not necessarily as much a hindrance, as the far-reaching impact of informationization, and even more of intelligence-ization, is potentially leading to a radically different context, if not a paradigmatic shift.

References

Alsabah N (2017) China's cyber regulations: a headache for foreign companies. MERICS Comment. https://merics.org/en/comment/chinas-cyber-regulations-headache-foreign-companies

Chen A (2021) A Close Reading of China's Data Security Law In Effect September 1, 2021, China Briefing. https://www.china-briefing.com/news/a-close-reading-of-chinas-data-security-law-in-effect-sept-1-2021/

China Restricts Overseas Access to Key Corporate Information (2023) Bloomberg News. https://www.bloomberg.com/news/articles/2023-05-03/china-restricts-overseas-access-to-corporate-registry-databases#xj4y7vzkg
China's Personal Information Protection Law (PIPL) (2022) Bloomberg Law. https://pro.bloomberglaw.com/brief/china-personal-information-protection-law-pipl-faqs/
Chinese Military Encyclopedia 2nd Edition Editorial Committee (2007) PLA Encyclopedia, 2nd Edition, Campaigns. China Encyclopedia Publishing House, Beijing, PRC, p 127
Covington and Burling (2015) China Enacts New National Security Law. https://www.cov.com/~/media/files/corporate/publications/2015/06/china_passes_new_national_security_law.pdf
Creemers R, Webster G (2021) Translation: the personal information protection law of the People's Republic of China, Effective November 1, 2021. Digichina. https://digichina.stanford.edu/work/translation-personal-information-protection-law-of-the-peoples-republic-of-china-effective-nov-1-2021/
Cui S, Wang J (2004) Advancing military transformation with Chinese characteristics, strengthening integrated-style joint operations. People's Liberation Army Daily
Cybersecurity Law (2016) China Law Translate. https://www.chinalawtranslate.com/en/2016-cybersecurity-law/
Digichina (2021) Translation: data security law of the People's Republic of China (Effective September 1, 2021). Digichina. https://digichina.stanford.edu/work/translation-data-security-law-of-the-peoples-republic-of-china/
He Z (2004) Experts assess the Iraq war. Military Science Publishing House, Beijing, PRC, p 146
http://media.people.com.cn/n1/2019/0301/c40606-30951460.html
http://www.cac.gov.cn/2019-03/01/c_1124178478.htm
https://navyleaguehonolulu.org/maritime-security/ewExternalFiles/2022-military-and-security-developments-involving-the-peoples-republic-of-china.pdf
Intelligence-ization brings new changes to conflict (2021) People's Liberation Army Daily. http://military.people.com.cn/n1/2021/0107/c1011-31992241.html
Intelligence-ization: Explaining Its Development and New Capacities (2020) People's Daily. http://www.xinhuanet.com/comments/2020-07/13/c_1126229075.htm
Jiang Z (2022) Work Report to the 16th Party Congress. Xinhua. http://www.china.org.cn/english/features/49007.htm
Kennedy S (2025) Made in China 2025. Center for Strategic and International Studies, Washington, DC. https://www.csis.org/analysis/made-china-2025
Li D (2014) Reconsideration of the mechanisms for winning informationized warfare. China Military Science (#6, 2014), pp 80–81
Li M (2019) What is the mechanism for victory in intelligence-ized warfare?" People's Liberation Army Daily. http://www.mod.gov.cn/gfbw/jmsd/4834525.html?big=fan
Liu W (2021) What are the bottlenecks and key links of military intelligence-ization: questions and research. In: Front lines of research. http://www.rmlt.com.cn/2021/0815/621863.shtml
McDonald J (2023) A global law firm separates from its Chinese Partner, Citing Cybersecurity and Data Rules, AP. https://apnews.com/article/china-cybersecurity-law-dentons-data-privacy-investment-0caeb8e6b38a9fd06066e5076c83b152
McCarthy S, Gan N (2023) China has widened its already sweeping counter-espionage law. Experts say foreign businesses should be worried. CNN. https://www.cnn.com/2023/04/27/china/china-counter-espionage-law-revision-intl-hnk/index.html
National Security Law of the People's Republic of China (2015) China Law Translate. https://www.chinalawtranslate.com/en/2015nsl/
Office of Naval Intelligence (2009) A modern navy with Chinese Characteristics. Office of Naval Inteliigence, Suitland, MD, p 36. https://irp.fas.org/agency/oni/pla-navy.pdf
PRC National Intelligence Law (as Amended in 2018) (2017) China Law Translate. https://www.chinalawtranslate.com/en/national-intelligence-law-of-the-p-r-c-2017/
Qiang C (2006) China's Information Revolution. World Bank Publications, Washington, DC, p. 93; and GUO Liang, Under the Golden 'Shine': China's Effort to Bridge Government and Citizens.

Chinese Academy of Social Sciences, Beijing, PRC, pp 4–6. http://unpan1.un.org/intradoc/groups/public/documents/un-dpadm/unpan042815.pdf
Sidley and Austin (2022) Understanding China's data regulatory regime: what are important data? and can they be transferred outside of China? https://datamatters.sidley.com/2022/04/11/understanding-chinas-data-regulatory-regime-what-are-important-data-and-can-they-be-transferred-outside-of-china/
State Council Information Office (2002) Tenth five year plan for national economic and social development, informationization key point special plans. http://www.cia.org.cn/information/information_01_xxhgh_3.htm
Sutter K (2023) Made in China 2025' industrial policies: issues for congress, IF10964. Congressional Research Service, Washington, DC. https://sgp.fas.org/crs/row/IF10964.pdf
Tang W (2021) Higher Level Military Education for the Intelligence-ized Warfare Era. People's Liberation Army Daily. http://www.mod.gov.cn/education/2021-05/13/content_4885203.htm
The great firewall of China: obstacles to monitoring performance, dotcom-monitor. https://www.dotcom-monitor.com/blog/the-great-firewall-of-china-obstacles-to-monitoring-performance/
Wagner J (2017) China's cybersecurity law: what you need to know. The Diplomat. https://thediplomat.com/2017/06/chinas-cybersecurity-law-what-you-need-to-know/&cd=11&hl=en&ct=clnk&gl=us
Wu M (2020) The five main directions of military intelligence-ization. In: China command and control conference. http://www.sohu.com/a/409610381_358040
Xu G (2013) Research on our military's information operations strength construction. Military Science Publishing House, Beijing, PRC, pp 76–77
Yu E, Strumpf D (2023) US Companies in China worry due diligence will end in spy dramas. Wall Street J. https://www.wsj.com/articles/foreign-firms-in-china-squeezed-by-u-s-disclosure-demands-beijings-security-clampdown-658ef8f?mod=itp_wsj&ru=yahoo
Zhang P (2012) Joint campaign command teaching materials. Military Science Publishing House, Beijing, PRC, p 33.
Zou Z, Cha R (2011) Command information capabilities research, based on systems combat between information systems. Oceans Publishing House, Beijing, PRC, p 57

Twilight of the Idols: European Demographics and China's Middle East Prowess

G. Douglas Davis

Abstract The Western world is undergoing a demographic transformation never seen in recorded history as falling birthrates and improved life expectancy cause its population to simultaneously age and decline. These demographic changes are causing both the United States and Europe to contract their global presence and slowly extract itself from the Middle East and other regions. This creates power vacuums that adversaries will be able to exploit. The demographic change will weaken Europe by seeing its working age population fall and, at the same time, its retired population will grow and create significant pressures on the national budget. European states will face strong pressures to divert funds to preserve its social welfare system that will cause it to lower military appropriations. Europe will be forced to make difficult adjustments that will weaken its military prowess and cause it to pull back from international commitments. China is increasing its military strength and is replacing Europe as America's primary strategic concern. As Europe adjusts to its new reality it will be forced to move away from the Middle East and Africa. Beijing is best situated to benefit from Western demographic changes as China can fill the vacuums that emerge when European powers pull out. Beijing is able to provide military aide and defense grants to potential client states searching for help.

Western demographic changes will create strategic necessity for Europe and the United States to gradually step away from the Middle East and China is primed to be the main beneficiary. The West is undergoing a demographic transition unlike any other in human history and all the long-term consequences are impossible to predict. Some elements are possible to foresee and they will change the great power relationships present in the Middle East. As the West pulls back, China is the great power is the best situated country to be able to set into the region and increase its regional influence and access to the region's vast energy resources. Its strength is that it does not need to actively oppose the West with military force, but can use its hybrid tactics to delegitimize the EU and US while waiting for it to pull back. For

G. D. Davis (✉)
MSIR Program, Troy University, Troy, AL, USA
e-mail: gddavis@troy.edu

A. Farhadi et al. (eds.), *The Great Power Competition Volume 6*,
https://doi.org/10.1007/978-3-031-70767-4_4

the short-term, time favors Beijing and it is likely to exert greater influence in the region. The West's demographic challenge will force it to retract and bring about a change in the global balance of power. This will be an unstable period globally and conflicts are likely to emerge as states compete to fill the vacuum where the West pulls back.

There is some parallel between the contemporary experience and Nietzsche's last major work, *Twilight of the Idols, or, How to Philosophize with a Hammer.*[1] This work criticizes Europe's turn to moralism at the same time that it's religious base has eroded and Nietzsche focuses on the contradiction between the focus on public morality and the nihilism based on secularism that was behind it. In other words, there was discord between the public proclamations for morality and the difficult reality that there was no basis for the normative claims. The philosophical base when examined was empty. This, unfortunately, symbolizes the difficulty confronting Europe today. Its plan for the world was to continue to institutionalize governmental relationships so that conflict can be solved through mutual consent in the institution assigned to deal with the disagreement. The problem is that when one looks deeply at the European proposal and one explores it deeply, there is a similar emptiness. One does not need to go too far to see that public proclamation for democracy and the institutional constraints on European governments that prevent them from responding to popular demands. When one considers China's rise within the Middle East—and other global regions—Beijing's emergence is due to the decline of the West. Perhaps the greatest cause of Europe's falling international prowess is its demographic reality.

One major problem is that the changing security situation globally indicates that the European progect to manage the international system has failed. Some attritribute this to the cultural assumptions[2] and others to the way liberalism was institutionalized.[3] In *On World Order*[4] Henry Kissinger notes that there is a basic confusion within European capitals and institutions that confuse the EU formation with contributing a coherent foreign policy and contributing to the newly emerging global order. The very intense attention devoted to building the multinational governing institutions are seen as Europe's contribution to forming the global order. Kissinger notes that Europe is abandoning the international system it worked to construct and its withdrawal is itself a threat to world stability. If the global realm were a theater production, it would resemble an old actor who confuses their performance resume with contemporary relevance. The EU is failing to participate as a character in the global play underway to determine the new global architecture and it confuses its success in building complex international governing institutions as a model for states that have stronger economies, stronger militaries, and greater international influence. The demographic transformation will further make Europe focus internally and become a weaker protagonist and become an observer to the greatest international conflicts confronting it. The outcome will be that it will withdraw from other regions and other great powers will move into the regions. Thus, Beijing will have more room to navigate in the Middle East.

1 The Demographic Crisis

The Western world is undergoing a demographic change never seen in recorded history produced by falling birthrates and improved life expectancy. There are many European states where the population is simultaneously declining and aging. While these transitions are also present in the United States, but at a lower level, the European demographic reality is more immediate threat to Europe's military strength in both absolute and relative terms. There are two difficult tensions that will weaken the European security architecture: first, aging populations require governments to devote additional resources to maintain their social welfare system. Since the shrinking labor pool lowers the social tax base, funds will need to be diverted from defense to pensions and health care. Second, low birth rates have significantly lowered the young adult population from which societies draw personnel to serve in the military and defense agencies. European states will see their military prowess deteriorate as its financial resources and military age population falls.

Considering this within NATO, the Atlantic alliance was born when the West was dominant globally. Its nations had the most advanced economies, technology, and military capacity at its foundation, but these advantages, where they still exist, are marginal and weaker than ever. It is forgotten that the West was also dominant demographically and this gave it the ability to deploy armed forces to geographically distant places for long times and not hurt the domestic economy. NATO was strategically strong in the areas that would allow it to maintain its great power status. When the Atlantic alliance was signed, NATO states were responsible for 60% of the world's economy, but today when there are more member states, it constitutes only 38%.[5] The aggregate population of NATO was 13.7% of global population at its foundation and, although there are around twice as many member countries, it constitutes 12.5% today, which is a smaller percentage. The UN data forecasts expect this number to fall to 10.3% by 2050.[6] The overall statistics miss an important dimension, the United States is expected to grow significantly while Europe will decline. While the US population grows, Europe will fall from 12% of world population to around 7% by 2050. The UN population forecasts show that while the United States population is expected to expand, the European population will fall.

While the UN statistics are conservative and provide the forecasts used in this work, the actual population change is likely to be worse. The scholars who construct the predictions use models that have taken decades to refine; if the world did not change, there accuracy would continuously improve. The assumptions used to generate the forecasts are representative of a different time and are unlikely to be accurate in the future.[7] The estimates employ two problematic assumptions: The models employ unrealistically high estimates for both birthrates and immigration. The actual decline is likely to be greater than the UN models forecast. Another reason to be skeptical is that demographic forecasts may be good in the short run, but historically they have proven to be very poor.[8] Future predictions are notoriously poor and it is very possible that Europe's demographic decline will be much greater than what is portrayed in the UN models. The population drop for Europe could be worse than the projections indicate which means would further weaken its security institutions.

Europe deserves closer analysis because the demographic changes will require governments to make substantial changes that will see its contribution to military operational competence decline substantively. Europe's population would have dropped more dramatically but there is some good news, life expectancy has been increasing for over a century and this is the primary reason for population growth.[9] The bad news is that birthrates have been below replacement level for decades. The problems generated by not having enough babied to sustain society take decades to become manifest, but they are now emerging and will get worse in the future. Western European state's birthrates have been too low to sustain their population since 1975[10] and around 2000 another critical threshold was reached: their overall population began to decline.[11] The transformation confronting Europe is not new and has been building up for decades. Since demographic data is widely-available, this decline was easy to foresee ahead of time. Since there is an approximately twenty-year time delay before low birth rates create a labor shortage, it was politically expedient to ignore the issue. Those governing would be out-of-office when the negative effects emerge and actually facing the problem would require them to invest a substantial part of their political capital to oppose the popular sentiment for population control. No leader was willing to make the sacrifice and, since the consequences would emerge after they left office, there was no political cost to ignoring the problem. It was left for future leaders.[12] We are now entering the time when Europe the negative consequences will begin to emerge and this will affect its international standing, security, and commitment to its international allies.

Demographic decline and aging constitutes a new and crucial reality that the alliance needs to confront. When NATO was founded Europe was strategically important given its geographic proximity to the communist world and its large population and market. While NATO's institutions may appear to be consistent to its original structure, the world in which it operates today is fundamentally different. The demographic transition will weaken NATO's absolute and relative military aptitude and cause Europe to further decline in global importance.[13]

2 Aging, Welfare, and Military Spending

European NATO member states have made promises to increase their military and defense spending so that their armed forces will have greater prowess. The assurances may have been made with good intentions, but it will be impossible to honor this commitment given their low birthrates and aging populations. The armed services will have a more difficult time going forward maintaining their personnel as the military age population declines. Universities and businesses have a need to attract talented students and workers and this will weaken the ability of European armies to maintain adequate troop levels. This will create multiple problems for the conventional military forces as they will have less capacity to contribute to international operations. The military will have less ability to conduct foreign operations or deploy when they face a manpower shortage. The will affect NATO's overall prowess as the conventional military capacity provided by European powers is in decline and going forward it will be more difficult for the alliance to conduct large operations.

European military prowess will decline due to the decrease in its military age population, but their technological advantage and advanced weapons stock will also fall. The growing complexity of advanced weapon systems means that it is becoming increasingly difficult to integrate the technological advances into the apparatuses. These systems are scientifically sophisticated and expensive to maintain, manage, and deploy. Advanced weapons have become more expensive as per-unit-costs grow faster than inflation which automatically lowers the production numbers unless additional funds are applied. When states merely maintain their budget in real terms, they get fewer units. Today, even when states increase their appropriation budget, their fleets have been shrinking.[14] The problem confronts all NATO member states, including the United States, which means that their advanced fleet size is declining. In particular, Washington has fewer ships and planes although its budget has increases significantly in real terms. The same inflationary pressures are also confronting European states who will see their military capacity further decline in the coming decades.

While Washington was able to increase the funds available to acquire additional advanced weapons, European states have smaller budgets and lack the domestic political support to even maintain their current weapons systems. The inflation problem means that even in the unlikely event that European states are able to maintain their current defense spending (in real terms), their military capacity would still decline. The political demand to support the social welfare system is great and the demographic changes that are emerging within European states mean that budgetary pressures will increase. The challenge is how to maintain healthcare systems, pensions, and elderly care when the laborer pool, financing the system, declines. Governments will not be able to maintain their current defense spending levels and funds devoted to military research and procurement today will be diverted to their welfare system. The technological competence European states now enjoy will decline and European NATO member states will decline in both absolute and relative terms as they have smaller defense budgets, small advanced weapons stocks, and fewer soldiers. The demographic transition will weaken NATO's operational capacity to the extent that it depends on its European members.

European militaries will have less operational capacity and fewer tools as its population ages and its social welfare system becomes more expensive. The outcome is that European states will become more dependent on the United States contrary to its promise at the Wales Summit where all members promised to increase or maintain their defense spending. The timing is problematic for Brussels as Washington is focused on China's rise and is working to protect its Asian allies from Beijing's military and economic strength. One particular problem for Europe is that China's advanced missile systems make it a significant challenge to the U.S. military in Asia. Some analysts argue that this weapons system gives Beijing military supremacy in East Asia, but others argue they will gain this status by 2025. China's growing military and its efforts to develop blue water navy means that it will become more important to the United States going forward. The major problem for NATO member states is that Europe is no longer Washington's primary security focus. NATO is growing more dependent on Washington when the institution is more marginal to U.S. interests. If Europe will have less operational capacity and is unlikely to be able to contribute to American security, NATO is a greater drain on Washington that offers few benefits. Is it likely that the United States will continue to support allies that are unwilling to defend themselves?

NATO has survived many crises and perhaps the forthcoming European military decline will constitute another challenge that it can survive. The challenges confronting the institution are fundamentally different from the past. Historically, Europe could take for granted that the United States would support NATO even when Washington was disproportionally burdened with expenses because, as Robert Kromer expressed in congressional testimony in 1982, "the Europeans know that we need them as much as they need us." He was responding to why Europe never responded to threats that Washington would pull troops out of Europe. Washington had a strategic imperative to protect Europe from communism and this meant that NATO partner states could cut military spending and reallocate funds to their social welfare system knowing that the United States would continue to protect them. While Europe was strategically important to Washington, NATO allies could effectively depend on the United States to subsidize their defense capacity. While NATO was effectively able to survive past crises, it was always because the American people continued to see value in providing protection to our European allies. Are the circumstances that allowed the United States to support the alliance even when Europeans failed to contribute their share to common defense still present?

3 Aging and Military Prowess

The aging population presents new challenges are unlike any in recorded history and this will require states to make significant cuts to their military budgets. Europe, South Korea, and Japan are historical anomalies because their social welfare systems are already burdened with a large elderly population and the changes in the coming decades, they will have societies older than any ever known in recorded history.[15] Every European NATO member state is undergoing this demographic transformation and their society is aging significantly and this presents challenges to their social welfare system.

European societies are more vulnerable to funding problems emerging from low birth rates and an aging population than the U.S. Europeans have accepted lower wages and paid higher taxes than their American counterparts and the state has assumed the responsibility to provide pensions and care for the elderly population. European NATO members are facing an unprecedented challenge in trying to preserve their costly social welfare system while their labor pool and overall population declines. The retirement options available to the typical American works, such as a private pension, 401 k plan, or other private investment are not options for most Europeans.[16] Often their income is too low as individuals live paycheck to paycheck and there is naïve confidence that the government will take care of them so there is no need to make additional arrangements. Continental Europe features coordinated market economies[17] that compound the difficulties emerging through aging societies because once the government stepped in to become the main social welfare provider, this eliminated other private and religious associations that were caring for the poor and elderly. Once this task was assumed, a growing percentage of its population has become dependent on the state and there is no other actor that can provide elderly care in the event that the government fails. As the elderly population grows, the government has no alternative but preserve its programs. This will require it to raise taxes or divert funds by cutting spending. European states[18] have long prioritized social welfare spending over defense and when tax revenues decline as their workforce becomes smaller, the easiest path for governments is to reapportion funding from defense and foreign aid to its social welfare system. As the European population ages and the number of workers declines, governments will have to make difficult decisions that will require military spending to fall and the armed services will have fewer resources.

An aging population will introduce structural imperatives and while some problems can be foreseen, since this is unprecedented historically there will be many consequences that are unknown. One possibility is that elderly societies will be increasingly at risk for seasonal illnesses or pandemics. The coronavirus was not a problem for most young to early middle age adults, but it constitutes a serious illness for older populations. The ordinary flu also takes thousands of lives every year and the majority of those are older citizens who have pre-existing health conditions. It is normal for a person's health to grow more vulnerable as they age and develop a medical history. As a society ages even ordinary illnesses present more difficulties and this can have an impact on the economy, openness, and even basic freedom within society. While one can predict that health care will be more expensive and challenging as society ages, the one element that cannot be foreseen is a contagion, an outbreak of a serious illness. European NATO partners will be more vulnerable going forward to ordinary diseases, such as the flu, but their susceptibility toward a serious pandemic increases even more. While it is too early to predict the social changes that the coronavirus will bring, the changes are likely to be most necessary in Europe. While it is too early to see the outcome, it poses a risk to the economy which could further reduce the military budget.

The military and security institutions will have fewer resources as society ages because overall tax base will decline as welfare costs rise. The social-economic structure changes when fewer babies are born and the labor pool shrinks. Since fewer people are present, the aggregate demand for goods declines and this leads to a contraction in the overall market size. This leads to reductions in personal and institutional income that work to reduce tax revenue. Ruchir Sharma[19] reports that for every one percent decline in the population growth rate, this eventually leads to a one percent decline in the economic growth rate. The German labor pool is smaller today, as well as its overall population, than it was a decade ago and its population is aging. The local market will adjust to lower demand for goods and less overall consumption.[20] As society ages and the working population falls, European national economies will likewise decline in relative and absolute terms.

The demographic transformation presents an unprecedented challenge to global capitalism because there is no working economic model that assumes that markets shrink over time. The conventional wisdom presumes that aggregate demand increases over time so that expanded exchange of goods leads to an increase in social wealth. Firms need to continuously integrate new technology to remain competitive and to grow their production capacity.[21] If fewer goods are needed over time, this will constitute difficult adjustments for firms and, if their technological and production improvements were secured through borrowing, enterprises will have difficulty making debt payments and bankruptcy will become more common. This is particularly problematic for Europe because even though its young adult population is low, youth unemployment remains high in several regions. Even though the labor pool has fallen over time, European firms have not been able to absorb the potential workers already present. Labor market inflexibility is not likely to go away if there is an expectation that aggregate demand is likely to fall over time. As a result, the budget shortfall projections that measure the potential worker pool have unrealistic assumptions regarding labor participation and underestimate the actual decline in tax revenue. There are economic reasons that the budget shortfalls could be worse than expected and this would significantly hurt defense funding.[22]

The aging European population will have a more difficult time conducting military operations and will lose conventional capacity and their military-technology status. The research necessary to continuously refine and improve advanced weapons requires a constant budget and European states will need to find creative means to fund this research. The increased social welfare costs associated with aging populations will require states to make difficult decisions on which budgets should be cut. European states have smaller budgets than Washington already and lack the domestic support to authorize additional funds even to maintain their current defense. If this is not possible, how can one expect them to get support to preserve military research appropriations? Since the per-unit cost is increasing, merely maintain a budget is not enough and even if Europe were able to do this, its strategic prowess would still decline. Strategically, it would be more beneficial to increase spending to maintain their existing weapons supply.

4 Domestic Tension

The European response to the demographic crisis has been to ignore its primary cause, the decline in its birthrates, and to turn to foreign populations to meet their labor force needs.[23] While this is a nice idea in theory, the developing countries that have traditionally been the origination point for European immigration are also experiencing falling birthrates. Western societies cannot count on an immigrant supply in the forthcoming years.[24] There are areas where birthrates continue to be high that could be utilized for immigration. The problem is that these regions have the worst governments, pedagogical systems, and infrastructures. As immigrants arrive from these areas, there will be significant costs associated with educating and training them so that they can contribute to the European labor market. While immigration does constitute a solution, it could also work to further worsen state budgets if the labor market remains inflexible. Youth unemployment is already a problem for individuals born and educated within Europe, the job prospects for individuals who lack the language skills and have fewer years of formal education are even worse.[25] Even foreigners who strongly desire to work and make positive contributions to society will have a difficult time until the labor market becomes flexible and able to absorb existing workers. Thus, this proposed solution sounds good on paper and is politically attractive, but will not work as expected because of structural problems within European labor market.

While most literature examining the relationship between demographic change and conflict finds that specific scenarios where population growth combined with instability can lead to conflict. Since Europe's population is aging and declining, this risk is not present on the continent. There are multiple scenarios where the demographic transition can instigate tensions. The first friction point is the turn toward immigrants as means to supplement its low young adult population and to preserve its social welfare system. Since immigration is offered as a potential option to alleviate budgetary pressure and to maintain pensions and the social welfare state, this solution could also become a source of tension. Will the immigrant population passively accept paying higher taxes to preserve the welfare of elderly populations from a different ethnic group? Huntington's 1968 *Political Order in Changing Societies* shows how economic transition within ethnically homogenous societies can be a source of conflict when a society is advancing because new empowered groups will pressure government for change while those the status-quo benefits will work to preserve the system. These tensions will be greater within a multiethnic state. Absurdly high tax rates that are perceived to disproportionally benefit other ethnic groups are sure to generate opposition from those bearing the financial loss.

Some experts focus on immigration as a method to prevent a pension crisis because new arrivals will join the labor market and generate revenue that can be used to pay taxes. This does not address labor market inflexibility, which makes it difficult for recent college graduates and the educated workers to find jobs. It is unlikely that new arrivals that are less educated on average than the native population will be able to make substantial contributions to the labor market when linguistically-proficient and highly educated job seekers are not. Labor market inflexibility will not be helped through immigration. In the disciplinary literature, youth bulges can generate conflict when the economy does not grow quickly enough to integrate them into the economy. Although Europe does not have a growing population, it has long suffered from high unemployment and low wages for young workers. New arrivals will experience frustration and not find a labor market that can integrate them. A problematic labor market is a source of instability and possible tension.

The second friction point concerns the sociological consequences arising from immigrants who are less educated, poor, and predominately male. Most migrants come to Europe from the Middle East, Africa, or South-Central Asia and their sociological characteristics, namely young, unschooled males, ill-equipped for the modern work-force, are the factors that predict crime, violence, and domestic unrest. While the migrants come from lands where a different dominant religion, language, and culture are present, these differences do not predict violence or crime. Nevertheless, the typical migrant possesses the factors that are positively associated with delinquency and criminality.

Since the arrivals are predominantly men and the reality of male-dominated poor immigrants could become a source of violence. In some cities, these arrivals are crowded into ghetto-like urban areas where they are segregated from the local population and can constitute additional domestic security threats. These ethnic suburbs are more male (there is male–female imbalance), poorer, and less educated than other areas and this generates resentment and can lead to radicalization, crime, or violence. The immigrants are mostly confined to urban areas where they suffer hardships and indignity due to their economic weakness and social isolation. The problem is that immigrants are likely to create an even bigger burden on the security and intelligence institutions and require additional resources at a time when their budgets are in decline. Governments will have to devote additional resources domestically while their overall security resources decline. The rising immigrant populations will further reduce European states capacity to act internationally.

The migrants come from places that are culturally, religiously, and ethnically distinct and their arrival in Europe can generate a backlash and create problems for governments and security forces. The state will encounter difficulties because the newcomers can lead to a rise in nationalistic sentiment and the immigrants themselves will be likely to engage in unlawful activities. Local police and security organs will need to increase their intelligence gathering and surveillance of the arriving immigrants and the domestic groups who have the potential to use violence against the immigrants. The problem is that there is a potential for conflict from both sides: the immigrant populations are likely to generate violence and criminal activity and the receiving nation is likely to react negatively to new arrivals who demand cultural changes within society. This is especially problematic when host governments practice censorship over the issue and prevent any democratic discourse that could create a popular consensus to accept the refugees. In addition to the rise in domestic tension, there are additional economic factors that are contingent on Europe maintaining the appearance of safety.

Since many European states are economically dependent on tourism, one outcome is that greater resources will need to be devoted to domestic security. The state will need to intervene to create barriers that protect tourists and actively segregate visitors from immigrant communities. The industrial sector requires that tourists are protected from crime as this is essential to maintaining its attractiveness to future travelers. This will consume additional intelligence, police, and security resources to protect the tourist industry and to preserve state's reputation for safety. The dependency on tourism is a national security issue and it is imperative that a state preserve its safety. Violence or social tensions that become public can generate significant economic costs if tourism is negatively affected. One consequence is that European states will need to devote more resources to protect its cultural sites and the visiting tourists and this means that governments will have fewer resources to deploy internationally. EU member states will need to devote additional security resources to preserve domestic stability and enhanced intelligence sharing will be essential so that it can prevent all threats that can affect its industries.

All European states have some dependency on their tourist sector and this shared vulnerability increases the incentives for intelligence sharing between governments. If the situation becomes problematic within one country, it affects the aggregate number of tourists visiting its neighbors. Governments have additional reasons to coordinate their intelligence gathering and security enforcement so that the free movement across the Schengen area does not jeopardize public safety. Since the United States is also working to keep potential troublemakers under observation, the daily diplomatic and military contact NATO provides can work to further coordinate these efforts.

While selective immigration may work to allow European states to preserve their pension and social welfare systems, this is also likely to require additional resources to be devoted to domestic security. The budgetary changes required to preserve their entitlements require these states to cut overall defense spending while additional resources will be needed domestically. This means that this solution will further weaken European state's ability to make security contributions to the United States or NATO going forward. The Atlantic alliance will be even more unequally dependent on Washington as Europe's solution to preserve its social welfare system will require them to devote more security resources domestically.

NATO survived the Cold War and continues to accept new members, but Europe no longer has the same strategic importance to the United States. China's meteoric rise and the East Asia's economic growth have changed the global balance of power in ways that have required Washington to rethink its strategic orientation. In 2012, the Obama Administration announced its Pivot to Asia which began to remove U.S. forces from Europe and the Middle East so that they could be redeployed in Asia. European NATO states finally came to recognize that the global incentives that kept U.S. forces on the continent for decades are weaker than at any time since NATO's foundation. One response was that European partners made concessions to Obama at the 2014 Wales Summit where they promised to spend two percent of their GDP on defense by 2024. To further strengthen the promise, twenty percent of their defense spending (or 0.4% of its GDP) would be applied to major equipment or research and development. European allies hoped that this assurance would deflect further criticism from Washington, but it has not. While many states may have had the intention of honoring their promise, but there are no penalties for failing to keep your word. Europe's demographic aging will make it less likely, if not impossible, that it will be able to keep this promise.

Most Americans wish to see the United States maintain its commitment to NATO even if they are growing more frustrated with our European allies. When Donald Trump won the 2016 U.S. presidential election, he campaigned against the Atlantic alliance and this sentiment broadly resonated with the American public as a 2015 Pew research polls that showed that only 49% of Americans supported NATO.[26]

Those numbers have since rebounded and today the majority of Americans support the alliance, but European member states should be aware that U.S. support may have an expiration date if they cannot improve and sustain their military spending. American military guarantees, particularly the nuclear shield Washington provides NATO, has allowed our allies to neglect their militaries and to spend more on their domestic economy. NATO is a political union and cannot be taken for granted. It is likely to remain strong only while all parties receive benefits from preserving it. European allies need to remain aware that its strategic importance is falling and that Washington needs to re-focus on Asia. As such, the time when the American commitment to NATO could be taken for granted is past; NATO's long-term survival depends on its member's political will to maintain it.

It is unreasonable to preserve military research appropriations when the armed services themselves have aircrafts that cannot fly, paints broomsticks black to participate in NATO exercises because they lack sufficient weapons for their already scarce personnel. It is difficult to see what actual benefit would take place by increasing its advanced research budget. If the technological and scientific studies were undertaken, the budgetary overstretch would prevent the weapons from being produced in sufficient numbers to make a meaningful difference. Also, these weapons are expensive and difficult to maintain and after production it is doubtful that they would be able to keep them operational. Even if there is an effort to use arm exports as a means to sustain their defense industry, they would face competition from American, Russian, Chinese, and Israeli producers.[27] To further complicate matters, the per-unit cost of advanced weapons is growing faster than inflation and this means that benefits the weapons provide declines even when additional funds are applied. To maintain their technological prowess, European states would need their sales profits to be sufficient not only to replace the budgetary loss, but it would need to generate funds to keep pace with inflation. Maintaining their research and production appropriation is not enough, additional spending is necessary. This is a very difficult prospect and it is more likely that European militaries, whose advanced weapons stock is already ineffective and feeble, will fall further behind. Secondly, their government will spend less on weapons acquisition and thereby have less capacity in the future. Even though the European NATO partners pledges to increase their military spending, their aging societies will make this financially and politically difficult, if not impossible. This is another dimension in how aging societies work to weaken the European contribution to NATO's military prowess. The traditionally technologically-proficient European militaries will become a nothing more than a memory. It will be difficult for them to adjust their expectations for international influence and lower them to their actual geopolitical position.

While advanced weapons have become increasingly expensive, this is also true of conventional operations. The 2003 Iraq War provides an important case showing that managing foreign conflicts is more difficult and expensive. The complexity it entails requires considerable planning and organization. The cost of maintaining two brigades in Iraq was the equivalent to the New Zealand's entire GDP. Since European conventional forces already face a decline in their military age population, their budget cuts will also make some foreign operations impossible. The geography where European NATO partners can operate will decline and fall over time. As societies age, additional resources will need to diverted from the military and defense institutions, which will reduce their conventional operating area. It will also grow increasingly difficult for European partners to meaningfully contribute to out-of-theatre campaigns. Europe will become more dependent on Washington and should see its conventional forces decline in both relative and absolute terms.

UK and France are the only continental armies that have the ability to carry out advanced operations independently from the U.S., but both are expected to decline in the coming years and become less distinguishable from the other Western European armed forces.

The negative consequences generated by aging will be significant for all NATO members but the effect will be most intense within Europe. The United States will constitute an ever greater portion of NATO's overall capacity and Washington will receive less military help from the alliance.

The aging of US allies is a global problem for Washington as the strongest U.S. Pacific partners are in the midst of an even greater aging crisis. South Korea and Japan are undergoing an accelerated demographic transformation as their birthrates are even lower and both have a higher elderly percentage than European states. While there is no broad security institution, like NATO, in Asia, the U.S. maintains individual partnerships where it offers protection to individual countries where it a special relationship exists. Seoul and Tokyo are very dependent on Washington's military protection and their proximity to China give them increasing strategic importance and a greater dependency on the United States.[28] China's increasing military strength, its intelligence network, and persistent diplomacy, and its ability to advance its interests in secrecy mean that our Asian partners become increasingly vulnerable over time. China's main objective is to gain military supremacy within East Asia which also leaves the U.S. military in Asia increasingly exposed to dangers. While China's military power is growing, the relative advantage Beijing maintains over Seoul and Tokyo continue to grow. This superiority will allow Beijing to intervene more in distal regions—creating the opportunity for it to deepen its presence in the Middle East.

The security risks confronting Europe and the United States converged over much of the post-World War II era and this meant that Washington was concerned with protecting our NATO partners from the Soviet Union and later Russian threats. Today, Russia remains the primary defense focus for European allies in East and Central Europe, but it is no longer the primary American strategic focus. Russia poses a specific risk to the Baltics, Poland, Romania, and states in its geographic proximity, but Washington needs to contend with the growing East Asia's growing geopolitical importance.[29] Europe is less strategically important to American interests than at any point since the WWII. The shared strategic worldview is no longer present and the strategic imperative that kept the U.S. in Europe for over half a century no longer exists. When the Cold War ended, Washington continued to focus on Russia and this kept the United States engaged in Europe even though the global security environment was changing.[30] The Obama Administration took the first steps in removing American forces from Europe when it undertook the pivot to Asia.

Going forward, Washington's conventional force presence within Europe is likely to continue to decline. This will occur as Europe's military capacity also grows weaker as it ages and devotes greater resources to maintain its social welfare state. The Atlantic alliance confronts a new strategic reality protecting Europe is no longer the primary strategic focus for Washington. This will be a difficult adjustment, but necessary if NATO can continue to endure into the indefinite future.

Small states bordering Russia, such as the Baltics, wish to maintain U.S. troops domestically because this makes it more difficult for Moscow to invade. If a military campaign directly involves the American military, this would mean that Russia would suffer a significant retaliation that would mitigate any strategic gain an invasion would yield. The U.S. military could offer this protection with a base and a relatively small numbers of troops. The United States needs to find an appropriately small number that offers this strategic balance and does not need massive forces. This constitutes a problem for the forward-deployed troops who would be further isolated from major support and supply systems as the U.S. repositions itself in Asia, or, more likely, it would require a substantial American presence to maintain them. The U.S. troops would act as a shield, in a way similar to insurgent groups who fire from civilian areas who use innocent peers as shields. Their presence would not increase the Baltic state's defensive strength, but it make a Russian attack on Estonia, Latvia, or Lithuania less likely because this would guarantee American retaliation. Perhaps, this would constitute a more strategic threat to Moscow than NATO itself. While this is an option favored by the Baltic States and Poland, this is a less optimal solution than having Europe increase its overall prowess to change the defensive calculation.

If the U.S. were to adopt this solution this could serve as an excuse for the small states within Europe to remain content with their current defense and not work to increase their security. While the Baltic States, Poland, and other states in the proximity of Russia would continue to maintain their defense spending at or slightly above NATO's minimum spending threshold, this limit is not adequate to increase their deterrence. Lithuania, Latvia, and Estonia will remain vulnerable and dependent on allied protection even if they maintain their spending levels. American military presence in the Baltics would increase the deterrence against Russia, but it would maintain an unhealthy dependency on Washington. The real problem is that Europe does not have any credible military threats that have enough strength to protect small states. If the U.S. were to pursue this path, it would certainly strengthen our NATO allies' security by increasing the cost of attacking them, but there would be little domestic benefit to Washington. The other negative is that American troops would again be placed in a vulnerable position to protect NATO[31] partners while European nations fail to act as a protagonist to protect themselves.

Europe's global prestige has fallen and there is a disparity between how the EU presents and sees itself and the rest of the world. While the EU maintains that it a normative global power, its influence is falling and its status is much lower internationally than is realized within Europe.[32] In truth, Europe has been a secondary power following World War II and had to rely on the United States to defend it. It is more difficult to conceal Europe's actual decline and growing international irrelevance. To people outside the continent, Europe has less status than ever in modern history and this creates a problem for its status as a normative power. There is no plan to counteract or respond to European decline and going forward, European countries will have less international influence.

The Middle East is likely to see Western powers—both the United States and Europe—pull out and this creates an opportunity for foreign adversaries to enter the arena and provide security in return for access to energy resources.[33] Beijing is best situated to move into the vacated areas and it is likely to become more dominant, not because of its prowess, but due to the internal demographic Western decline.

Notes

1. Nietzsche, F. (2003). Twilight of the idols and The Anti-Christ. United Kingdom: Penguin Publishing Group.
2. Davis, G. Doug, and Michael O. Slobodchikoff. *Cultural Imperialism and the Decline of the Liberal Order: Russian and Western Soft Power in Eastern Europe*. Russian, Eurasian, and Eastern European Politics. Lanham, Maryland: Lexington Books, 2019.
3. Gray, J. (1992). Against the New Liberalism-Rawls, Dworkin and the Emptying Of Political Life. *TLS-The Times Literary Supplement*, (4657), 13-15.
4. See Kissinger, Henry. 2014. World Order: Reflections on the Character of Nations and the Course of History. London: Allen Lane.
5. Ceccorulli, Fassi, and Lucarelli provide an interesting overview of NATO's relative decline in terms of population and economy in their 2017 study: Ceccorulli, Michela, Enrico Fassi, and Sonia Lucarelli. "NATO's demographic paradox." *Global Change, Peace and Security* 29, no. 3 (2017): 249–271.
6. Ceccorulli, Fassi, and Lucarelli 2017.
7. Bricker and Ibbitson (2019) point out, "The UN is employing a faulty model based on assumptions that worked in the past but may not apply in the future." See Bricker, Darrell Jay, and John Ibbitson. 2019. *Empty Planet: The Shock of Global Population Decline*. New York: Crown Publishers.
8. Morland, Paul. 2019. *The Human Tide: How Population Shaped the Modern World*. First US edition. New York: PublicAffairs, an imprint of Perseus Books, LLC, a subsidiary of Hachette Book Group, Inc.
9. See Eberstadt (2014).
10. See Chesnais (1998).
11. See Lutz et al. (2003).
12. Margaret Thatcher was an exception as she tried to repeatedly raise the demographic aging issue at G-7 meetings. Peter Peterson (1999) notes she was rebuffed by other heads-of-state who ignored the subject and replied, "Of course aging is a profound challenge, but it doesn't hit until early in the next century—after my watch [52]." See Peterson, Peter G. 1999. "Gray Dawn: The Global Aging Crisis." *Foreign Affairs* 78, no. 1: 42–55.
13. Slobodchikoff et al. (2021).
14. Freedman (2017) provides some inflationary data on U.S. advanced weapons procurement. He notes that although the Navy's budget increased by 22% in real terms from 2001 to 2009, the number of ships actually fell. At the same time, the Air Force budget grew by 27% in real terms but the combat airfare capacity fell by 20%. He suggests that this is the reason that drones have been used so heavily. See Chapter 22,(p 240–241) in Freedman, Lawrence. 2017. *The Future of War: A History*. New York: Public Affairs.
15. Peter Peterson provides an interesting global overview of the effect global aging will have on the world economy in Peterson, Peter G. 1999. "Gray Dawn: The Global Aging Crisis." *Foreign Affairs* 78, no. 1: 42–55.

16. Corporate pensions exist in Europe, but only one in seven people have one. Only around one percent of the European population has a 401 k plan. For a summary of this, see Peterson, Peter G. 2004. "Riding for a Fall." *Foreign Affairs* 83 (5):111–125.
17. Hall, Peter, and David Soskice. 2001. *Varieties of Capitalism: The Institutional Foundations of Comparative Advantage*. Oxford, UK: Oxford University Press.
18. "…European governments prioritize social welfare over all other considerations—including funding their armed forces." Bacevich, Andrew J. 2010. "Let Europe be Europe." *Foreign Policy* (178):1. Bacevich argues that the U.S. should withdraw from NATO since most members fail to meet minimum spending requirement and let the alliance evolve into a European organization. Coker argues that Europe's demilitarization is cultural and it approaches international relations fundamentally differently from the U.S. See Coker, Christopher. 2013. "A Farwell to Arms: Europe's Meritocracy and the Demilitarization of Europe." In *NATO's European Allies: Military Capability and Political Will*, edited by Janne Haaland Matlary and Magnus Petersson, 57–70. Hampshire: Palgrave Macmillan.
19. Sharma (2016).
20. The German economy faces several challenges that all pose a threat to Berlin; in addition to its falling labor pool and overall population decline that will lead to a smaller domestic market, there is increasing attention that the trade surpluses it maintains within the EU are problematic and pose long-term damage to European unity [Joseph Stiglitz work, in particular, has brought this issue to the public—see Stiglitz, Joseph E. 2018. *The Euro: How a Common Currency Threatens the Future of Europe*. New York: Norton.]. While Germany (together with Finland and Ireland) have enjoyed a trade surplus with China, this is likely to change to a deficit if the Chinese government is successful with its Made in China 2025 program; Beijing is specifically targeting German imports and seeking to produce Chinese-made substitutes. See Smith, Julianne, and Torrey Taussig 2019. "The Old World and the Middle Kingdom: Europe Wakes Up to China's Rise." *Foreign Affairs* (Sept/Oct 2019):112–24. This means that Germany is likely to have smaller trade surpluses when it's becoming more dependent on trade surpluses to prevent economic decline and this has two important consequences for the Western alliance: (i) Berlin would have incentives to prevent any agreement with the United States that could alleviate the tension caused by the increasingly problematic trade imbalance; (ii) it would be imperative for Germany to maintain trade imbalances within the European Union that are a source of instability and need to be eliminated to preserve unity.
21. Zachary Karabell provides an interesting review of two books that attempt to assess how the demographic changes will affect global capitalism in Karabell, Zachary. 2019. "The Population Bust: Demographic Decline and the End of Capitalism as We Know It." *Foreign Affairs* 98, no. 5: 216–221.
22. If the economic decline is great enough, states may have an easier time meeting the two percent defense spending threshold, but this will not strengthen NATO's military capacity.
23. Geoffrey McNicoll notes that there is widespread acceptance across Europe for two responses to aging: (i) raise the retirement age and (ii) increase immigration. The disagreement comes over the quality, size, mobility, and identity of the desired immigrants. See McNicoll, Geoffrey. "Reflections on Post-Transition Demography." *Population and Development Review* 38 (2013): 3–19.
24. See Karabell (20,190.
25. Demeny (2016).
26. O'Toole, Molly 2016 "Is NATO Still Relevant? Trump is not the only one asking" *Foreign Policy* April 1, 2016.
27. Davis and Jhinaoui (2024).
28. See Sheen (2013).
29. Davis and Slobodchikoff (2024).
30. It would be an interesting counter-factual study to assess whether the United States would have re-adjusted its commitment to European security earlier and started its pivot to Asia sooner if the 9/11 attacks had not taken place. The George W. Bush administration was regularly engaged in devoting more attention to confronting Beijing before the terrorist attacks took place. After

9/11, there was a new security focus that more deeply engaged American forces in the Middle East and removed the focus from East Asia. Although there is no way to empirically assess this question, there are grounds to argue that the United States would be more strategically detached from Europe had the 9/11 attacks not taken place.

31. Davis, G. Doug, Michael O. Slobodchikoff, and Brandon Stewart. "Nato's Future: The Show Must Go On." Chap. Fifteen In *Challenge to Nato: Global Security and the Atlantic Alliance*, edited by Michael O. Slobodchikoff, G. Doug Davis and Brandon Stewart, 281-88. Lincoln: Potomac Books at the University of Nebraska Press, 2021.
32. For an interesting example, see Larsen, Henrik. 2014. "The External Perceptions Literature and the Construction of Gaps in European Union Foreign Policy." *International Politics Review* 2 (1):11–8.
33. See also G. Doug Davis and Michael Slobodchikoff 2022. "Great-Power Competition and the Russian Invasion of Ukraine." *Journal of Indo-Pacific Affairs*. 5 (4): 215–226. And G. Doug Davis and Michael Slobodchikoff 2023. "Great Power Competition Following the Ukraine War." In: *The Great Power Competition Volume 5: The Russian Invasion of Ukraine and Implications for the Central Region*, edited by A. Farhadi, M. Grzegorzewski and A. J. Masys. Switzerland: Springer. Page 35–45.

References

Bacevich AJ (2010) Let Europe be Europe. Foreign Policy 178:1

Bricker DJ, Ibbitson J (2019) Empty planet: the shock of global population decline. Crown Publishers, New York

Ceccorulli M, Fassi E, Lucarelli S (2017) NATO's demographic paradox. Global Change, Peace Secur 29(3):249–271

Chesnais J-C (1998) Below-replacement fertility in the European Union (Eu-15): facts and policies, 1960–1997. Rev Popul Soc Policy 7:83–101

Coker C (2013) A farwell to arms: Europe's meritocracy and the demilitarization of Europe. In: Matlary JH, Petersson M (eds) NATO's European allies: military capability and political will, pp 57–70. Palgrave Macmillan, Hampshire

Davis GD, Jhinaoui K (2024) Africa and Russia after the Ukraine war (in press). In: Slobodchikoff MO (ed) Russia and the world following the invasion of Ukraine. Rowman & Littlefield

Davis GD, Slobodchikoff MO (2019) Cultural imperialism and the decline of the liberal order: Russian and Western soft power in Eastern Europe. Russian, Eurasian, and Eastern European Politics. Lexington Books, Lanham, Maryland

Davis GD, Slobodchikoff M (2023) Great power competition following the Ukraine war. In: Farhadi A, Grzegorzewski M, Masys AJ (eds) The great power competition Volume 5: the Russian invasion of Ukraine and implications for the central region. Springer, Switzerland, pp 35–45

Davis GD, Slobodchikoff M (2024) (forthcoming) American global leadership: ailing U.S. diplomacy and solutions for the twenty-first century. University of Tennessee Press

Davis GD, Slobodchikoff MO, Stewart B (2021) Nato's future: the show must go On Chap. Fifteen. In: Slobodchikoff MO, Davis GD, Stewart B (eds) Challenge to nato: global security and the atlantic alliance, pp 281–88. Potomac Books at the University of Nebraska Press, Lincoln

Demeny P (2016) Europe's two demographic crises: the visible and the unrecognized. Popul Dev Rev 42(1):111–120

Davis GD, Jhinaoui K (2024) Africa and Russia after the Ukraine war (in press). In: Slobodchikoff MO (ed) Russia and the world following the invasion of Ukraine, Rowman & Littlefield

Eberstadt N (2014) The human population unbound. Curr History 113(759):43–46

Freedman L (2017) The future of war: a history. Public Affairs, New York

Gray J (1992) Against the new liberalism-rawls, dworkin and the emptying of political life. TLS times Literary Suppl 4657:13–15

Hall P, Soskice D (2001) Varieties of capitalism: the institutional foundations of comparative advantage. Oxford University Press, Oxford, UK
Karabell Z (2019) The population bust: demographic decline and the end of capitalism as we know it. Foreign Aff 98(5):216–221
Kissinger H (2014) World order: reflections on the character of nations and the course of history. Allen Lane, London
Larsen H (2014) The external perceptions literature and the construction of gaps in European union foreign policy. Int Politics Rev 2(1):11–18
Lutz W, O'Neill BC, Scherbov S (2003) Europe's population at a turning point. Science 299(5615):1991–1992
McNicoll G (2013) Reflections on post-transition demography. Popul Dev Rev 38:3–19
Morland P (2019) The human tide: how population shaped the modern world. First US edition. PublicAffairs, New York
Nietzsche F (2003) Twilight of the idols and the anti-christ. Penguin Publishing Group, United Kingdom
O'Toole M (20160 Is NATO still relevant? trump is not the only one asking. Foreign Policy
Peterson PG (1999) Gray dawn: the global aging crisis. Foreign Aff 78(1):42–55
Peterson PG (2004) Riding for a fall. Foreign Aff 83(5):111–125
Sharma R (2016) The demographics of stagnation. Foreign Aff 95(2):18–24
Sheen S (2013) Demographic peace? Northeast Asia's aging population and regional security. Asian Surv 53(2):292–318
Slobodchikoff M, Davis GD, Stewart B (2021) The challenge to NATO: global security and the atlantic alliance. Potomac Books—University of Nebraska Press
Smith J, Taussig T (2019) The Old World and the Middle Kingdom: Europe wakes up to China's Rise. Foreign Aff 112–124
Stiglitz JE (2018) The Euro: how a common currency threatens the future of Europe. Norton, New York

Japan, The Gulf, and Sino-American Great Power Competition

Alice Dell'Era

Abstract Although the Indo-Pacific remains the primary area of competition between China and the United States, Sino-American rivalry has become increasingly a global phenomenon. The rapid expansion of Beijing's economic and political influence in the Arab Gulf region has raised concerns over its strategic aims in this critical area. Beyond Washington's concerns, other external actors, notably Japan, have been gradually but steadily recalibrating their relationship with Gulf partners due to apprehensions about Beijing's growing regional clout. As the scale of Sino-American competition extends to the Gulf, Tokyo has progressively reinterpreted its approach to this region. It now views the region through a lens of geostrategy, where China's influence is considered a key element in comprehending the ongoing changes. This has prompted a transition in Tokyo's Middle East policy. While the country's energy needs continue to dominate its agenda for the region, Japan has been attempting to present itself as an influential alternative partner, focusing on a distinctive niche that Beijing also seeks to occupy, that of supporting the Gulf's energy transition and diversification. This is evident when analyzing Tokyo's evolving partnerships with Saudi Arabia and the United Arab Emirates. Japan's efforts in this regard can be characterized as a form of soft competition leveraging economic statecraft initiatives. Such efforts align and share critical synergies with the Biden Administration's cooperative framework for the Gulf region. As the United States seeks to counterbalance Beijing's influence, Japan's soft initiatives offer the US an opportunity for coordinated, joint responses that better address the specific demands and needs of the Gulf countries.

Keywords Japan-Gulf relations · Sino-American rivalry · Sino-Japanese competition · Great power competition · Japan's Middle East policy

A. Dell'Era (✉)
Security Studies and International Affairs, Embry-Riddle Aeronautical University, Daytona Beach, FL, USA
e-mail: delleraa@erau.edu

A. Farhadi et al. (eds.), *The Great Power Competition Volume 6*,
https://doi.org/10.1007/978-3-031-70767-4_5

1 Introduction

China's emergence as a global political, economic and military power has prompted a shift in its relations with the United States. Under the Bush Administration, China was identified as a serious challenge to US interests (Silove 2016). However, this perception was initially seeded with the hope that Beijing would turn into a "responsible stakeholder" (Zoellick 2005). Such a view has been progressively replaced in Washington with the assumption that China is instead a revisionist power aimed at subverting the US-led international order. Following this trend, the 2017 National Security Strategy announced the return of "great power competition," stating that China, along with Russia, is attempting to compete for influence both at the regional and global levels at the expense of the United States (The White House 2017). The re-emergence of long-term strategic competition with revisionist powers has thus become "the central challenge to U.S. prosperity and security" (US Department of Defense 2018).

This perception has not waned and has actually intensified with the Biden Administration. Indeed, the 2022 NSS and NDS described China as a strategic competitor posing "the most comprehensive and serious challenge to U.S. national security" (The White House 2022c; US Department of Defense 2022). As a result of this, American strategic priorities have shifted eastward, prompting heated discussions over how to balance the distribution of US resources and assets across regions (Livermore National Laboratory, CRS report). Caught up in this transition has been the Middle East. Given that "inter-state strategic competition, not terrorism" was now Washington's declared priority (US Department of Defense 2018), efforts were initiated to reduce the American military footprint in the region in order to reinforce US force deployments in the Indo-Pacific (O'Rourke 2023).

Although the Indo-Pacific region remains the primary area of competition between the two rival powers (Chhabra et al. 2021; The White House 2022c), Sino-American rivalry is also increasingly a global phenomenon. Indeed, the 2022 NSS clearly recognizes that "around the world, the contest to write the rules of the road and shape the relationships that govern global affairs is playing out in every region and across economics, technology, diplomacy, development, security, and global governance" (The White House 2022c). In light of this, and despite claims of US withering commitments, the Middle East remains a critically strategic venue even in the context of great power competition (US House of Representatives 2022). To this end, the 2022 NSS clarified that "the United States will not allow foreign or regional powers to jeopardize freedom of navigation through the Middle East's waterways, including the Strait of Hormuz and the Bab al Mandab, nor tolerate efforts by any country to dominate another—or the region—through military buildups, incursions, or threats" (The White House 2022c). This reflected Biden's remarks at the GCC + 3 Summit Meeting, where the US President stated: "We will not walk away and leave a vacuum to be filled by China, Russia, or Iran" (The White House 2022d).

Beijing's growing influence in this region, where the US has traditionally acted as the dominant external power, has indeed raised concerns in Washington. Accordingly, the various dimensions of great power rivalry in the Middle East have been a growing topic of discussion within both academic and policy circles (Ghiselli 2023). Most of the attention has centered on Washington and Beijing's competing and/or divergent interests in the region, as well as on the perspective and agency of regional powers (Hoffman 2021; Zoubir 2023). However, the rise of China and the extension of competitive patterns into the region carry implications that go beyond these actors. Other external actors have been slowly, but steadily adjusting their approach to the Middle East as a result of concerns over Beijing's regional sway. As emphasized by Chhabra et al. (2021), such actors have the ability to shape Sino-American rivalry. Accordingly, it is critical to understand how they are repositioning their foreign policy approaches to the region and what such repositioning means for US-China global competition. Exemplary of this is the case of Japan, a country with a long history of economic and diplomatic engagement in the Middle East and whose Middle East policy has recently culminated in what some describe as a "major shift" (Suzuki 2023).

This chapter examines Japan's shifting relations with the Gulf countries and assesses the extent to which such a shift can be tied to China's growing influence as well as Sino-American great power rivalry. Great power competition is understood as the comprehensive struggle among a diverse array of global and regional powers for spheres of influence and preeminence within and across regions (DiCicco and Onea 2023). Such a struggle takes shape within distinct, and yet overlapping domains: from the political-diplomatic and ideological levels, to the military, economic and informational ones (DiCicco and Onea 2023). Although great powers compete simultaneously across these domains, the competitive mechanisms employed vary depending on the specific context in which competition manifests. As such, great powers may alternate between hard forms of military competition with softer mechanisms of competition. This is exemplified by the increasing tendency of great powers to rely on economic intervention and economic statecraft to gain influence over their competitors (Farhadi 2021). In line with this framework, the chapter focuses on competitive trends in the domain of tech investments and partnerships. The chapter argues that Japan views its ties with the Gulf from an increasingly geostrategic/geoeconomic perspective. While the country's energy needs continue to dominate its agenda for the region, Tokyo has been attempting to present itself as an influential alternative partner in a distinctive niche area eyed by Beijing as well, that of supporting the Gulf's energy transition and diversification. The chapter concludes that Japan's efforts present synergies with the US's new cooperative framework for the region. As such, these developments offer an important opportunity for the United States to coordinate joint responses that better fulfill the demands and needs of the Gulf countries.

Following the above introduction, the chapter is structured in three parts. The first section provides an overview of China's growing role in the Gulf. The second section focuses on Tokyo's shifting approach, especially towards Saudi Arabia and the UAE. The third and concluding section reflects on the opportunities that Japan's efforts in the Gulf offer to the US, as rivalry with China mounts in the region.

2 Beijing's Growing Influence in the Gulf

Over the past decade, China has intensified its engagement with the Gulf nations and has progressively forged closer relations that reflect the growing strategic importance of this region to Beijing. Such efforts are not new. To meet its rising energy needs, the Chinese Communist Party (CCP) was vigorously engaging with the Gulf region throughout the 1990 and 2000s (Leverett and Bader 2005). This engagement resulted in the launch in 2004 of the China-Arab States Cooperation Forum, a multilateral platform coordinating the direction of China-Arab economic, political, and security relations, and of the China-GCC Strategic Dialogue in 2010.

If Beijing's approach to the Middle East was originally shaped by the country's growing energy demands, the unveiling of the Belt and Road Initiative (BRI) led the CCP to view this region, including the Gulf, through the lens of the BRI. Similar to other key areas, Beijing has approached the Middle East in a manner consistent with what Pradhan (2018) describes as a "new silk road diplomacy." The BRI indeed represents one of the key strategic drivers behind China's Middle East policy. Announced in 2013, Xi's signature project envisions the development of an interconnected network of infrastructure and construction projects linking China to African and European markets. Within this framework, the Gulf region has been identified as a critical gateway for the realization of both the maritime silk road and the economic belt.

To exemplify this, Xi described Yanbu, a major Red Sea port located in western Saudi Arabia, as "the regional point of arrival of the Silk Road and the axis of the new Saudi industrialization" (Valori 2016). Furthermore, the *Vision and Actions on Jointly Building Silk Road Economic Belt and 21st-Century Maritime Silk Road*, one of the BRI key strategic documents, mentions the China-Arab States Cooperation Forum and the China-GCC Strategic Dialogue among important "cooperation mechanisms to push forward the building of the Belt and Road" (National Development and Reform Commission, 2015). Through the BRI, China has attempted to capitalize on the region's high demand for infrastructure. Accordingly, Chinese companies have actively pursued numerous investments and infrastructure projects, including in the Gulf.[1] According to Zhang (2023), the BRI has thus ushered a new chapter in Chinese infrastructure diplomacy in the Middle East.

From a maritime perspective, the Arabian Peninsula lies along critical maritime chokepoints: the Strait of Hormuz, the Strait of Bab el-Mandeb, and the Suez Canal. The Chinese export-oriented economy relies heavily on the SLOCs traversing these chokepoints for shipping goods to Africa and Europe. Beyond exports, Chinese imports, especially oil imports originating from Africa, navigate these same routes too (Bekkevold 2019). As a result, a smooth flow of commerce through these routes has become an important priority in the CCP's foreign policy agenda (Cook 2021). Beijing has thus doubled down on investing in much needed port-infrastructure along these strategic areas (Singh 2021). Such investments were complemented in 2018

[1] For details see: American Enterprise Institute (2023).

with plans to develop regional industrial parks, under the "industrial park-port interconnection" project (Fulton 2019). The plan involved establishing links between Chinese-developed industrial parks in the Arabian Peninsula to key Arab ports: included were, for instance, the Khalifa Port Free Trade Zone (KPFTZ) in Abu Dhabi, UAE; Oman's Duqm Special Economic Zone Authority (SEZAD); and the Jazan City for Primary and Downstream Industries (JCPDI) in Saudi Arabia (Fulton 2019; Chaziza 2020).

In light of BRI investments, China has ramped up its diplomatic activism. Since taking office, President Xi has visited the region three times. First, in 2016, when he visited Saudi Arabia, Iran, as well as Egypt. Then in 2018, when he visited the UAE, and once again in 2022 when he returned to Riyadh to attend a China-Arab Summit, and a China-GCC Summit. Beijing has also pursued Comprehensive Strategic Partnerships with the Gulf countries, signing such agreements with Saudi Arabia and Iran in 2016, and the UAE in 2018 (Fulton 2019). Furthermore, as of 2023, 7 Gulf countries are also members of the Asian Infrastructure Investment Bank (Bahrain, Iran, Iraq, Oman, Qatar, Saudi Arabia, and the UAE) (Asian Infrastructure Investment Bank 2023), a multilateral development bank launched by China in 2013 to facilitate, according to some, the realization of the BRI (Jordan and Deliwala 2022). On the occasion of Xi's 2016 visit, China further published its first-ever *Arab Policy Paper*, in which Beijing proposed a "1 + 2 + 3" cooperation approach, centered on deepening win–win cooperation in the energy sector (1), infrastructure development, trade and investments (2), and in the field of nuclear energy, renewables and space satellites (Ministry of Foreign Affairs of the People's Republic of China 2016). Each of these three domains of cooperation is intimately linked to the others.

As Beijing replaced the US as the largest crude oil importer in 2017, deepening energy-related projects has ranked at the top of China's Middle East Agenda. In this regard, the Gulf plays a crucial role. Indeed, half of Beijing's imports originate from the Gulf, with Saudi Arabia maintaining a 17% share (U.S. Energy Information Administration 2022). Moreover, despite importing most of its LNG (40.2% as of 2021) from Australia, its second largest supplier of this resource is located in the Gulf, Qatar (The Observatory of Economic Complexity 2021d). China is today the top trade partner of the Gulf region (Liao 2023). For example, it accounts for 19% of Saudi Arabia's exports as well as imports, 17% of the UAE imports, 13 and 14% of Qatar's exports and imports (The Observatory of Economic Complexity 2021a). Energy supply remains at the heart of China's trade as well as infrastructure development, and investments (Table 1).

Most Chinese investments in the Gulf are concentrated in Saudi Arabia and the UAE, which have respectively attracted $40 billion and $34 billion between 2013 and 2023 (American Enterprise Institute 2023). Chinese petrochemical companies and refineries have thus worked to establish partnerships with their counterparts in the Gulf. For example, in 2018, CNPC reached an agreement with the UAE for cooperation on the development of the Umm Shaif and Nasr, and the Lower Zakum oil fields (Emirates News Agency—WAM 2019). Saudi Aramco also reached a series of deals with Zhejiang Petrochemical, and Sinopec, including a joint venture project in the Chinese city of Panjin (Foley 2023; Tan and Aizhu 2023).

Table 1 Compiled by the author with data from the Observatory of Economic Complexity

Main products exported from Saudi Arabia to China		
	Trade value	Percent (%)
Crude petroleum	$38.3 B	78.6
Ethylene polymers	$2.19 B	4.49
Acyclic alcohols	$1.81 B	3.72
Main products exported from UAE to China		
	Trade value	Percent (%)
Crude petroleum	$8.36 B	36.6
Petroleum gas	$2.92 B	12.8
Broadcasting equipment	$2.24 B	9.79
Main products exported from Qatar to China		
	Trade value	Percent (%)
Petroleum gas	$5.94 B	49.5
Crude petroleum	$2.48 B	20.7
Refined petroleum	$2.23 B	18.6

Investments have, however, expanded beyond traditional energy cooperation (Young 2023). For example, in 2019, Beijing and the UAE signed various agreements and memorandums of understanding aimed at advancing cooperation in technology transfer, artificial intelligence, environmental protection, food security, culture and education (United Arab Emirates Ministry of Foreign Affairs 2019). Similarly, the Saudi government is courting China to invest in its Vision 2030, a program aimed at diversifying Riyadh's economy by transforming it into a digital, logistical, and transportation hub connecting three continents (Kingdom of Saudi Arabia 2023). Beijing and Riyadh have thus signed multiple Memorandum of Understandings worth billions of dollars involving economic and tech cooperation (Dorraj 2023). For example, Beijing has been pursuing joint projects in the space domain (Sun et al. 2023), as well as in the telecommunication sector. China is indeed seen as an important source of digital technology, as evidenced by Huawei's success in the region. The company, banned from operating in the US and many other partner nations due to concerns over the security of its 5G technology, has instead been welcomed in the Gulf, especially in Saudi Arabia where it opened its largest overseas flagship store in 2022 and where it now plans to open a cloud data center as part of its regional expansion (Satariano 2019; Arab News 2022; Reuters 2023a). In addition, China's overdependence on oil imports led it to rethink the critical role of renewable energy, which became a priority in its 13th 5 year plan. Beijing is already a leader in the manufacturing of solar equipment (Gross 2021). As a result, China is looking to diversify its investments to support the Gulf's transition towards renewable products.

China's investments fulfill both a critical economic and political role in the Gulf. Indeed, the BRI is simultaneously deepening China's connectivity with this region, while also entrenching its diplomatic and political influence without strings attached.

These efforts have been complemented by other complementary initiatives. For example, Beijing has been pressing the Gulf states to use the Chinese renminbi in place of the US dollar for Chinese oil sales (Tan 2022). It has also taken up the role of diplomatic broker, as evidenced by its role in mediating the historic agreement between Saudi Arabia and Iran in 2023. These developments are increasingly read as a challenge to US influence in the region.

From a security perspective, there have been concerns that Beijing's economic presence could lead to higher engagement at the military level. For example, Saudi Arabia held military exercises with China in 2016 and, alongside the UAE, has been purchasing drones and other weapon systems from Beijing (Dorraj 2023; Leoni 2023). The PLA is also involved in training Saudi and Emirati forces (Watanabe 2023). Still, these security ties remain minimal if compared with the level and proximity of relations the US retains with these Gulf partners. Weapons purchases should not be understood as necessarily representing a tilt towards Beijing, but rather as part of the two states' diversification efforts aimed at fulfilling needs that Washington cannot satisfy (Bekkevold 2019). Evidence of this includes reports that Saudi Arabia is pushing to join the Global Combat Air Program, the ambitious partnership among three key US allies—the U.K., Italy and Japan—aimed at developing a sixth-generation stealth fighter (Jolly 2023). Chinese infrastructure and tech investments remain, however, an increasing source of tension between the Gulf states and Washington given that they involve or are located near US assets or in countries that closely cooperate with the US military (Leoni 2023; Young 2023).

From this overview, it is clear that China's rise in the Middle East continues to retain predominantly a geoeconomic dimension. In this respect, Beijing has been engaging the Gulf countries not with the purpose of intentionally driving the US out of the region. As such, latent rather than hard power seems to characterize competition in this area (Sun and Zoubir 2021; Andersen 2023). In line with this, Young argues that in the Gulf, "China is competing in entirely different mechanisms than the United States" (Young 2023). As the next section clarifies, China is not, however, the sole external actor involved in soft forms of competition through initiatives supported by economic statecraft. The following sections examine Japan's participation in such processes and discuss the opportunities that Tokyo's efforts create for the United States with respect to its attempts to minimize China's influence in the region.

3 Japan and the Gulf

China is not the sole actor who has stepped up its diplomatic and economic engagement with the Gulf region, nor is the sole country to which regional actors have looked at in their search for investments. Beyond Beijing, Japan has been one of the protagonists of what Ehteshami and Miyagi (2015) identify as an emerging Middle East-East Asia Nexus. Tokyo has a long tradition of economic engagement and investments in the Middle East. Historically, its relations with Arab countries have been shaped by the confluence of two simultaneous, and yet often conflicting factors: Tokyo's overwhelming reliance on Gulf partners for energy resources, and alignment with Washington's Middle East policy. As a result of these two factors, Japan's approach to this region transitioned through phases in which Tokyo either prioritized one or the other (Miyagi 2008; Hinnebusch and Miyagi 2015; Kobayashi and Miyagi 2015; Choudhury 2018; Evron 2018).

When Japan emerged as a global economic power in the late 1960s, early 1970s, its rising energy demands, coupled with the negative effects of the 1973 oil crisis, prompted the Japanese government to pursue a diplomatic approach aimed at stabilizing access to oil, at the expense of pressures to conform to the US Arab policy. Alignment with Washington, Tokyo's security guarantor and ally, became instead a priority in the early years of the post-Cold War period, following intense criticism over Japan's failure to contribute adequately to the first crisis in the Gulf. As a result of such a shift, while continuing to ensure its energy needs, Tokyo's relations with the region evolved to incorporate components of conflict prevention and reconstruction (Nakamura 2023). Following the US invasion of Afghanistan and Iraq, for example, the Koizumi Administration passed a series of laws enabling the SDF to provide logistical support to its ally and to dispatch troops to Iraq in support of reconstruction efforts. Since the mid-2000s, Japan has embarked in efforts to diversify the areas of cooperation with partners in the Gulf and the Middle East (Miyagi 2008; Evron 2018). Great attention was for example paid to the promotion of educational cooperation (Yamada 2023). Central to this was the Kono Initiative, launched in 2001 to promote "multi-layered" economic, political, and cultural relations with the Arab world (Ministry of Foreign Affairs 2001).

In light of the ongoing changes in the global balance of power, and given the mounting rivalry between the US and China at the global level, Japan's Middle East policy has progressively had to contend with the extension of such competitive dynamics to this region. Indeed, as the rise of China reshapes global order, including in the Middle East, this development has prompted Japan to rethink its position and role within the newly emerging competitive order. With regards to the Middle East, the Japanese leadership has long seen a US disengagement from the region as an opportunity for Beijing to bid for greater influence. Consequently, while primarily concentrating on its surrounding region, Japan's strategy towards the Middle East, particularly the Gulf, has gradually gained a geostrategic aspect under the guidance of former Prime Minister Abe and former Prime Minister Kishida.

The region's critical role for Japan is nothing new. As a resource-poor nation, Japan imports about 90% of its crude oil from the Middle East, with Saudi Arabia accounting for 42.5%, the UAE for 29.9%, Kuwait for 8.6%, and Qatar for 8.3% of the total oil imports (Agency for Natural Resources and Energy 2022). This dependence tripled in the aftermath of the Fukushima disaster of 2011, which forced the Japanese government to idle all of its operational nuclear reactors (Ishiguro 2014). In light of this energy dependence on the Gulf, Japan's first National Security Strategy stated that "stability in the Middle East is an issue that is inseparably linked to the stable supply of energy, and therefore Japan's very survival and prosperity" (Ministry of Defense 2013). Such circumstances have raised the urgency of securing energy supply routes (Dell'Era 2022), as well as strengthening resource diplomacy with its main energy suppliers. Yet, concerns over maritime security and the stability of energy routes are accompanied by concerns over China's regional activities. Indeed, Tokyo remains wary of Beijing's ambitions in the Gulf and of its "geopolitics of ports" and infrastructure projects (Miller 2023). As early as 2013, former Minister of Defense and now governor of Tokyo Yuriko Koike pointed the finger at China claiming that Beijing was preparing to take advantage of changes in the Middle East through "Its 'string of pearls' across the Indian Ocean—a series of potential naval stations connecting China to the Middle East and Africa—[which] would support a Chinese blue-water navy able to patrol the sea-lanes of the Persian Gulf" (Koike 2013). She then urged Japan, among others, to be "proactive in creating a regional security structure that protects their national interest" *vis a vis* Beijing's efforts (Koike 2013). It is thus obvious that, as the PRC's weight in the region grows, the lens of China has become a third factor through which Tokyo understands developments in the region.

This is also evident if one compares the 2013 NSS, with the latest version issued at the end of 2022. The 2013 NSS described Beijing's increasing international presence as a cause of concern due to its assertive and non-transparent behavior (Ministry of Defense 2013). In contrast, the newly revised document reflects the increasingly widespread rhetoric of a return to great power competition by identifying Beijing as "an unprecedented and the greatest strategic challenge" to the security of Japan and the international community (Kantei 2022). Moreover, in a clear message to Beijing's BRI, the NSS denounces "opaque forms of assistance" by those who "are trying to expand their own influence by (…) providing loans to other nations in a manner that ignores their debt sustainability" (Kantei 2022). Japan has responded to Beijing's BRI by advancing its alternative framework for Quality Infrastructure Investments, and by competing with China for markets and infrastructure projects. If Sino-Chinese geo-economic competition is more marked in the Mekong and South Asian region (Wallace 2019; Insisa and Pugliese 2020), it has acquired a more global dimension, extending asymmetrically to other regions, including Africa, the Middle East (Schulze 2019), and less so Latin America (Hirono 2019; Levy and Rose 2019). In the Gulf, Tokyo has especially stepped-up cooperation with Saudi Arabia and the UAE.

3.1 Japan and Saudi Arabia

As mentioned earlier, Saudi Arabia is Japan's main supplier of crude petroleum, accounting for 42.5% of its total imports (Agency for Natural Resources and Energy 2022). Tokyo's imports from Riyadh also include refined petroleum and acyclic alcohols (The Observatory of Economic Complexity 2021b) (Table 2). Given Japan's reliance on Saudi Arabia for oil, the two countries enjoy a history of close ties (Yamada 2023). During the 1990s, as Japan's energy consumption rate was on the rise (Hosoi 2023), the Japan External Trade Organization opened an office in Riyadh and the Japanese government promoted a Japanese-Saudi Cooperation Agenda as well as

Table 2 Compiled by the author with data from the Observatory of Economic Complexity

Main products exported from Saudi Arabia to Japan[a]		
	Trade value	Percent (%)
Crude petroleum	$21.9 B	90.8
Refined petroleum	$945 M	3.9
Acyclic alcohols	$336 M	1.4

[a]Japan is the third destination of Saudi exports after China and India, accounting 9.43% of the total shares

the creation of various organizations to support investments in the Gulf, such as the Organization for the Promotion of Japanese Investments and the GCC-Japan Industrial Investment Company (Yamada 2023). In 2007, following then-Prime Minister Abe's first visit to the region,[2] the two countries established a Japan-Saudi Arabia Industrial Cooperation Taskforce with the aim of facilitating industrial investment opportunities (Ministry of Foreign Affairs 2007). An Investment Agreement was then reached in 2013, at the beginning of Abe's second term, during his second visit to the region. On this occasion, Tokyo pledged to promote "a multilayered relationship that encompasses wide-ranging economic cooperation, politics and security, and cultural and human exchanges" (Ministry of Foreign Affairs 2013). The importance attached by Japan to developing its strategic partnership with Saudi Arabia, and establishing elevated high-level exchanges led to the 2017 visit to Japan by H. M. King Salman bin Abdulaziz Al Saud, the first in 46 years.

Following Saudi Arabia's launch of its Vision 2030, Tokyo expressed interest in cooperating towards the realization of the Vision "by mobilizing Japan's expertise and wisdom to the full, as there are a number of areas for cooperation, such as investment mainly in the manufacturing sector, promotion of small and medium-sized enterprises, human resources development, and promotion of culture, entertainment and sports" (Ministry of Foreign Affairs 2016). Japan's interest in the Saudi Vision 2030 also reflected its hope to participate in the realization of infrastructure projects and special economic zones, a niche in which as pointed out above, Beijing is also interested (Ministry of Foreign Affairs 2017a). In an interview with Arab News, former Foreign Minister Kono Taro for example stated that: "there are mega-projects that His Royal Highness is taking leadership on, and we hope many Japanese companies participate in these projects" (Abbas 2019).

To this end, the Japan-Saudi Joint Group for Vision 2030, a platform guiding cooperation and development of joint projects, was launched. This framework of cooperation revolves around three main pillars: diversity, innovation and soft values, to be pursued across nine themes (competitive industry, energy, entertainment and media, healthcare and medicals, quality infrastructure, agriculture and food security, SMEs and capacity building, culture sports and education; investment and finance) (Ministry of Foreign Affairs 2017b). Following the government's push, the stock of Japanese investments in Saudi Arabia reached $5.22 billion in 2018. Even though

[2] Saudi Arabia was the first stop, followed by the UAE, Kuwait, Qatar and Egypt.

the flow of FDIs experienced a decrease between 2019 and 2020, it jumped to $420 million in 2022 (Japan External Trade Organization 2022). Japanese companies have been involved in numerous projects. The Japanese manufacturers Ebara Corporation and Unicharm Corporation respectively established a joint venture and a manufacturing plant, while Japan's Bank of Tokyo-Mitsubishi UFJ opened a branch in Riyadh (Saito 2023).

If the framework spans multiple fields, one central area of cooperation to which Tokyo attaches importance and in which it presents itself as a committed partner is renewable energy. For example, in 2018, SoftBank Vision Fund announced a solar power project capable of producing up to 200 GW by 2030 (French 2018). In 2021, Marubeni and Al Jomaih Energy and Water signed an agreement for the development of Rabigh Solar PV Park, which is co-financed through the Japan Bank for International Cooperation (JBIC), Mizuho Bank, and Al Rajhi Banking and Investment Corporation (Saudi Gulf Projects 2021). In addition to this project, the JBIC, alongside other Japanese banks including MUFG and the Bank of Yokohama, is also financing Saudi Electricity Company (SEC)'s green projects (Japan Bank for International Cooperation 2021).

Tokyo has also paid particular attention to Saudi Arabia's growing role in the hydrogen and blue ammonia markets, resources that play a central role in Japan's Green Growth Strategy and Green Transformation (GX) Basic Policy (Ministry of Economy, Trade, and Industry 2021; Kantei 2023). Japan was indeed the destination of the first-ever blue ammonia shipment completed by Saudi Aramco in partnership with Japan's Institute of Energy Economics and Saudi Basic Industries Corporation, and with support from the Japanese Ministry of Economy, Trade, and Industry (METI) (Saudi Aramco, IEEJ and SABIC 2020). This led to closer cooperation both at the corporate and governmental levels. At the corporate level, the Japan Organization for Metals and Energy Security (JOGMEC) signed a memorandum of cooperation with Saudi Aramco to sponsor projects involved in hydrogen and fuel ammonia production, whereas Japan's refiner ENEOS signed a memorandum of understanding with Saudi Aramco to develop a CO_2-free hydrogen and ammonia supply chain (Kumagai 2022). Moreover, Japan's power generation company JERA and Marubeni Corporation both agreed to partner with PIF in developing clean hydrogen production projects (Arab News 2023b). At the government level, in 2022 Japan's METI signed two new sets of cooperation agreements with Saudi Arabia in the area of hydrogen, fuel ammonia and carbon recycling (Reuters 2023b). It must be noted that Japan's efforts to expand cooperation with Saudi Arabia do not, however, encompass tech-cooperation in the realm of defense. Indicative of this is Japan's resistance (unlike the UK and to a lesser extent Italy) to Saudi's interest in participating in GCAP.

3.2 Japan and the United Arab Emirates

The United Arab Emirates represent the second main source of Japanese crude oil imports, accounting for 29.9% of its total imports (Agency for Natural Resources and

Energy 2022). In addition to crude petroleum, the two main other imports from the UAE include refined petroleum and raw aluminum (The Observatory of Economic Complexity 2021c) (Table 3). For the UAE, Japan is the second largest destination of its exports after India and before China. As in the case of Saudi Arabia, the two countries enjoy a long tradition of economic engagement, with the supply of energy resources acting as the foundation of the relationship (Horinuki 2023). A qualitative change in Japan-UAE ties was, however, initiated following the first and second Abe Cabinet's renewed activism in the region. On the occasion of Abe's second visit to the Gulf, the two issued a joint statement on *The Strengthening of the Comprehensive Partnership between Japan and the UAE toward Stability and Prosperity* in which they pledged to expand cooperation in a wide array of fields (Ministry of Foreign Affairs 2013). In 2016, for example, the two countries signed a memorandum of understanding on space exploration involving their respective space agencies (Emirates News Agency—WAM 2016). The agreements resulted in the successful launch of KhalifaSat from Japan's Tanegashima Space Center by Mitsubishi Heavy Industry in cooperation with Japan Aerospace Exploration Agency, followed by the launch in 2020 of the UAE-made Mars probe al-Amal (Mitsubishi Heavy Industries 2018; Agence France Presse 2020).

During his third visit to the UAE (2018), Abe addressed the Japan-UAE Business Forum emphasizing the importance of strengthening the partnership with the UAE beyond the energy field, including in infrastructure and cutting-edge technology. The visit also resulted in the launch of a new cooperative framework named the "Comprehensive Strategic Partnership Initiative" (Ministry of Foreign Affairs 2018). The inaugural meeting of the initiative was held only a few months after its launch. The two sides also signed an agreement for the promotion and protection of investments. Japan's investments in the UAE jumped from $1.68 billion in 2017 to $2.15 billion at the end of 2018, and $2.31 billion in 2019. Despite decreasing in 2021, by the end of 2022, investments were once again on the rise (Japan External Trade Organization 2022). Japan's JBIC is currently financing 6 projects in the UAE: Hamriyah (Natural) Gas-fired Combined Cycle Power Plant, Fujairah F3 Natural Gas-fired Combined Cycle Power Plant, Export of Facilities for Al Layyah Gas-fired, Combined Cycle Power Plant, Warsan Waste to Energy Project, High Voltage Direct Current Offshore Electricity Transmission Project (Japan Bank for International Cooperation 2023). The last two involve cooperation on renewable energy and decarbonization. For

Table 3 Compiled by the author with data from the Observatory of Economic Complexity

Main products exported from the UAE to Japan[a]		
	Trade value	Percent
Crude petroleum	$19.1 B	77%
Refined Petroleum	$2.74 B	11.1%
Raw aluminum	$972 M	3.92%

[a]Japan is the second destination of the UAE's exports after India and right before China, accounting 8.39% of the total shares

example, the Warsan Waste to Energy Project is being developed by Hitachi Zosen and Itochu Corporation in cooperation with Besix Group, to generate biopower.

As in the case of Saudi Arabia, Japan has been exploring closer industrial cooperation with the UAE on hydrogen and ammonia. For example, in 2021, METI and the Abu Dhabi National Oil Company (ADNOC) signed a Memorandum of Cooperation on fuel ammonia and carbon recycling, following which ADNOC partnered with INPEX, JERA and the Japan Oil, Gas, and Metals National Corporation for a joint study agreement on fuel ammonia (ADNOC 2021). ADNOC has also been pursuing cooperation with ENEOS and Mitsui in Abu Dhabi and initiated shipments of blue ammonia to INPEX, Idemitsu, and Itochu (Bloomberg 2021; Kumagai and Saadi 2022). It has also pursued agreements with various Japanese companies and has agreed to a collaboration scheme with METI (Arab News Japan 2023a, b).

The implementation of the Comprehensive Partnership Initiative was further revisited in 2022. In a joint statement, the two sides advanced a shared vision based on three pillars (peace and stability, prosperity and development and sustainable development) as well as a set of implementation mechanisms, including a Comprehensive Strategic Partnership Committee for Cooperation composed of five working-level sub-committees each tasked with a specific area of cooperation (Ministry of Foreign Affairs 2022) (Table 4). The following year, the two also signed an Agreement concerning the Transfer of Defense Equipment and Technology, aiming to foster greater cooperation in the defense sector, the first such agreement that Japan has reached with a Middle Eastern country (Ministry of Foreign Affairs 2023).

3.3 *Soft Competition with China in the Gulf?*

The broadening of Japan's relations with Saudi Arabia and the UAE have clearly further progressed under now former Prime Minister Kishida Fumio's leadership. At the end of 2022, the Keidanren, Japan Business Federation, issued a proposal urging the country to strengthen its strategic relations with the Gulf and to resume negotiations for a Free Trade Agreement. The proposal cited not only securing a stable energy supply but also securing a role for Tokyo in supporting the economic growth and transition of the Gulf countries amidst competition with other foreign actors (Keidanren 2022). The Japanese government's latest efforts to build relations with the Gulf reflect these sentiments. Attending the 2023 World Government Summit in Dubai, Japanese Cabinet Secretary for Public Affairs Shikata emphasized Tokyo's interest in furthering "scientific and technological cooperation and innovation with the region" (Harrison 2023).

Prime Minister Kishida's four-day visit to the Gulf in July 2023 signals that Tokyo is reevaluating its Middle East diplomatic approach (Suzuki 2023). Accompanied by about 30 Japanese company executives, Kishida set forth priorities that reflect Japan's attempt to position itself as a core enabler of the Gulf green transition through Japanese technology and investments. Tokyo is indeed offering its know-how and competitiveness in low-carbon technologies in return for building

Table 4 Source: Ministry of Foreign Affairs (2022)

Comprehensive strategic partnership committee for cooperation		
Subcommittee	Involved parties	Areas of Cooperation
Political, Diplomatic, and International Cooperation Subcommittee	Ministry of Foreign Affairs (JPN) and the Ministry of Foreign Affairs and International Cooperation (UAE)	Bilateral Engagements Multilateral Cooperation Humanitarian Aid and Development Assistance
Economic, Trade, Energy and Industry Subcommittee	e-Ministry of Economy, Trade and Industry (JPN), the Ministry of Foreign Affairs (JPN), the Ministry of Land, Infrastructure, Transport and Tourism (JPN) and the Financial Services Agency (JPN), and the Ministry of Economy (UAE), and the Ministry of Energy and Infrastructure (UAE), the Ministry of Industry and Advanced Technology (UAE), and the Ministry of Foreign Affairs and International Cooperation (UAE)	Business Environment for Trade and Investment Energy Infrastructure Healthcare Industry and Advanced Technology, SMEs and Innovation Aviation Finance
Agriculture, Environment and Climate Change Subcommittee	The Ministry of Agriculture, Forestry, Fisheries (JPN), the Ministry of the Environment (JPN), and the Ministry of Foreign Affairs (JPN) and the Ministry of Climate Change and Environment (UAE) and the Ministry of Foreign Affairs and International Cooperation (UAE)	Agriculture Environment Climate Change
Education, Science, Technology and Culture Subcommittee	The Ministry of Foreign Affairs (JPN), the Ministry of Education, Culture, Sport, Science and Technology (JPN), Ministry of Economy, Trade and Industry (JPN) and the Ministry of Foreign Affairs and International Cooperation (UAE), the Ministry of Economy (UAE), the Ministry of Education (UAE), and the Ministry of Culture and Youth (UAE)	Education Science and Technology Cultural Exchanges Sports Exchanges
Defense and Security Subcommittee	The Ministry of Defense (JPN) and the Ministry of Foreign Affairs (JPN) and the Ministry of Defense (UAE) and the Ministry of Foreign Affairs and International Cooperation (UAE)	Bilateral Cooperation Framework Defense Equipment and Technology Cooperation

"green alliances" with the Gulf countries (Calabrese 2023). Japan reached bilateral and B2B agreements with Saudi Arabia, the UAE, and Qatar in various areas, such as semiconductors, space and technology, medicine, as well as education, agriculture and tourism (Suzuki 2023). The visit also ended with the pledge to resume FTA negotiations with the GCC (Arab News 2023a). With Saudi Arabia, Japan announced two new projects, the Manar Initiative and the Lighthouse Initiative for Clean Energy Cooperation (Al Arabiya News 2023a, b). The relevance of these projects goes beyond cooperation for clean energy. In fact, the projects also entail collaboration on critical mineral resources and the resilience of their supply chain (METI 2023).

As Kishida laid out its vision for helping the region become "a next-generation energy hub" (Pollman 2023), competition over influence, especially in this niche, with other tech powers, including China, mounts. While Japan's approach to the region must still be understood in light of its reliance on the Gulf for energy resources, it cannot be divorced from the country's broader concerns over Beijing's increasing political and economic activism in the area, as well as the traditional lack of transparency surrounding such activism. Indeed, the China factor has prompted Tokyo to re-frame its relations with the Gulf through the lens of its broader foreign policy and security outlook for a Free and Open Indo-Pacific (Matsuda 2023). Even though the Middle East does not lie at the center of Japan's geographical understanding of the Indo-Pacific, the Gulf stands out as a critical point of connection linking the Indo-Pacific to Africa and Europe and as a "hub of energy resources, financial dealings, and commodity distribution" (Ministry of Foreign Affairs (MOFA) 2017c). As such, it is seen in Tokyo as a critical participant in Japan's efforts to sustain a free and open international order grounded on the rule of law. Under the FOIP framework, the prospects for deeper Japan-Gulf ties have thus opened up broader opportunities (AlBadawi 2023).

Several interpret Japan's diplomatic efforts as a "spin-off" of "intra-Asian" competition in the Gulf (Evron 2018; Matsuda 2023). Such competitive dynamics appear to be both the extension of competition in the Indo-Pacific, as well as of the Sino-American rivalry for global influence. Some have noted that Japan's involvement in strategic competition within the Gulf remains minimal, if compared to those regions where Tokyo retains more pressing interests (Evron 2018; Heng 2022). Commercial collaborations between Japanese and Chinese businesses on solar technology have been cited as evidence of this (Heng 2022). Nonetheless, as the nexus between economy and security grows deeper in shaping order, the prospects for soft competition grow higher (Fulton 2022). As in the case of other regions, such as Southeast Asia or Africa, Japan may not have the financial capacity to compete with China in terms of the scale of its investments and projects. Nonetheless, it can and is attempting to harness its industrial and technical know-how to position itself as an influential partner within a specific emerging niche, decarbonization, and energy transition, in the face of China's growing economic and technological footprint. Indeed, the trajectory of Japan's engagement with the Gulf states suggests that under former Prime Minister Abe, as well as under the leadership of Kishida, Tokyo has been working hard to "distinguish Japan from China as a trusted partner for the region" (Lamont 2020).

4 What Synergies and Opportunities for the U.S. and Japan in the Gulf?

As Tokyo rethinks its approach to the Gulf, a few considerations are warranted. As one of the closest US Allies, Japan is seen in Washington as a critical partner in addressing the global challenges posed by the rise of China. This is especially true for the Indo-Pacific region, where Tokyo and Washington's strategic interests and priorities *vis a vis* Beijing have been converging over the past decade and where this Asian ally is considered as "having an inside track on Asia issues" ('Interview with a former State Department official during the Trump Administration' 2022). While the focus of the two allies' cooperation remains centered in the Indo-Pacific, Japan's efforts to position itself as an influential tech partner in the Gulf present important synergies with the US's cooperative framework for the region.

In pledging not to leave a vacuum in the Middle East to be exploited by others, including China, the Biden Administration reaffirmed its commitment to the region and laid down the five principles guiding U.S. engagement in the Middle East: Partnerships, Deterrence, Diplomacy, Integration, Values (The White House 2022a, d). Within such a framework, the pillar of integration is critically linked to strategic investment cooperation, an issue that was high on the agenda of the 2022 US-GCC meeting. On the sidelines of the meeting in Jeddah, for example, the US signed 13 investment agreements with Saudi Arabia (Nair 2022). The agreements span multiple sectors, including renewable energy, aerospace, defense, and advanced telecommunication technologies. Similarly, the US and the UAE established a partnership in the domain of clean energy intended to spur cooperation in decarbonization as well as in safeguarding resources and supply chains necessary for clean energy production (The White House 2022b). The partnership indeed pledges to facilitate investments in the mining, production, and processing of critical minerals. Additionally, the US, alongside other regional and extra-regional partners, has recently committed to the establishment of an India-Middle East-Europe economic corridor. These initiatives bear critical synergies with those pursued by Japan and present opportunities for coordination to better fulfill the demands and needs of the Gulf states. This would not be unprecedented since Tokyo and Washington already coordinate strategic investments elsewhere as part of a variety of government-led initiatives.[3]

[3] For example, as G7 members, they have cooperated through the Build Back Better World (B3W) Initiative, then transformed into the Partnership for Global Infrastructure and Investment (PGII). In the Indo-Pacific, the two also cooperate, along with Australia, through the Blue Dot Network, and with India through the Quad Infrastructure Coordination Group.

The US approach to the Middle East continues to be guided primarily by security and defense diplomacy, as Washington still prefers to rely on military over geoeconomic power (Farhadi 2021; US House of Representatives 2022). Not surprisingly, partnerships and deterrence remain the paramount principles guiding the US engagement with the region. Such pillars reflect the historical role of Washington as the dominant security provider in the Gulf and suggest that its response to strategic competition in the Middle East will take shape predominantly within such spaces. It should be noted, however, that strategic competition is far from monolithic and can rather take multiple forms (Young 2023). To reiterate Young (2023)'s point, in the Middle East, Beijing competes through a different lens than Washington. As such, a military/security-oriented framework may not necessarily be a viable alternative, on its own, to respond to China's growing influence and footprint in the region. Within this context, Japan's soft approach to competition in the Gulf represents a well-suited opportunity for the US to cooperate with Tokyo in a domain where China seems to be gaining the upper hand. This may be necessary given that regional powers are not interested in choosing between their security and economic partner.

References

Abbas FJ (2019) The "honest broker": Japan offers to play a role in Mideast peace. Arab News. Available at: https://www.mofa.go.jp/files/000496488.pdf

ADNOC (2021) ADNOC and Three Japanese companies to explore hydrogen and blue ammonia opportunities. Available at: https://www.adnoc.ae/en/news-and-media/press-releases/2021/adnoc-and-three-japanese-companies-to-explore-hydrogen-and-blue-ammonia-opportunities

Agence France Presse (2020) UAE successfully launches Hope probe, Arab world's first mission to Mars, The Guardian. Available at: https://www.theguardian.com/science/2020/jul/20/uae-mission-mars-al-amal-hope-space

Agency for Natural Resources and Energy (2022) Energy White Paper. Available at: https://www.enecho.meti.go.jp/about/whitepaper/2022/pdf/

Al Arabiya News (2023) Saudi Arabia and Japan launch "Manar" initiative to advance clean energy. Available at: https://english.alarabiya.net/News/saudi-arabia/2023/07/17/Saudi-Arabia-and-Japan-launch-Manar-initiative-to-advance-clean-energy

AlBadawi H (2023) Japan and the Gulf States: friendship prospects under the FOIP initiative. In: Rahman MM, Al-Azm A (eds) Social change in the Gulf region: multidisciplinary perspectives. Springer Nature Singapore, Singapore, pp 563–580

American Enterprise Institute (2023) China global investment tracker. Available at: https://www.aei.org/china-global-investment-tracker/

Andersen LE (2023) China and US policy in the Persian Gulf: is a new security architecture evolving? In: Zoubir YH (ed) Routledge companion to China and the Middle East and North Africa. Taylor & Francis, pp 102–114

Arab News (2022) Huawei opens largest overseas flagship store in Riyadh. Available at: https://arab.news/4wfan

Arab News Japan (2023a) ADNOC signs agreements with Japanese partners to boost low-carbon hydrogen value chain. Available at: https://arab.news/8v8ma

Arab News Japan (2023b) Japan, ADNOC sign "Japan-UAE Collaboration Scheme for Advanced Technology". Available at: https://arab.news/83tgv

Arab News (2023a) GCC, Japan agree to resume talks on free-trade agreement. Available at: https://arab.news/r33p2
Arab News (2023b) PIF signs MoU with Japanese power major JERA to develop green hydrogen projects. Available at: https://arab.news/64yxz
Asian Infrastructure Investment Bank (2023) Members and prospective Members of the Bank, Asian infrastructure investment bank. Available at: https://www.aiib.org/en/about-aiib/governance/members-of-bank/index.html
Bekkevold JI (2019) China, Russia and the great power contest in the Middle East. In: Bekkevold JI (ed) Sino-Russian relations in the 21st century, pp 141–166
Bloomberg (2021) UAE sells another blue ammonia cargo to Japan in hydrogen push, Arabian Business. Available at: https://www.arabianbusiness.com/industries/energy/467281-uae-sells-another-blue-ammonia-cargo-to-japan-in-hydrogen-push
Calabrese J (2023) Japan looks to the Gulf as it bets big on hydrogen, The Middle East Institute. Available at: https://mei.edu/publications/japan-looks-gulf-it-bets-big-hydrogen
Chaziza M (2020) China's relationship with Egypt and Oman: a strategic framework for the implementation of China's maritime silk road initiative. In: Clarke M, Sussex M, Bisley N (eds) The belt and road initiative and the future of regional order in the Indo-Pacific. Lexington Books, London, pp 141–157
Chhabra T et al (2021) Global China: assessing China's growing role in the World. Brookings Institution Press
Choudhury SR (2018) Japan and the Middle East: an overview. Contemp Rev Middle East 5(3):181–198
Cook SA (2021) Major power rivalry in the Middle East. Council on Foreign Relations
Dell'Era A (2022) Securitizing Beijing through the maritime commons: the "China threat" and Japan's security discourse in the Abe era. Pac Rev 37(1):147–180
DiCicco JM, Onea TA (2023) Great-power competition. Oxford Research Encyclopedia of International Studies
Dorraj M (2023) The belt and road initiative and China's expanding ties with West Asia and North Africa. In: Zoubir YH (ed) Routledge companion to China and the Middle East and North Africa. Taylor & Francis, pp 64–79
Ehteshami A, Miyagi Y (2015) The emerging Middle East-East Asia Nexus. Routledge
Emirates News Agency—WAM (2019) ADNOC sales to Chinese companies in 2018 reach US$7.9 billion. Available at: https://wam.ae/en/details/1395302776058#:~:text=In%20March%202018%2C%20ADNOC%20signed,subsidiary%20PetroChina%2C%20has%20provided%20AED2
Emirates News Agency—WAM (2016) UAE Space Agency, Government of Japan sign MoU. Available at: https://wam.ae/en/details/1395295460216#:~:text=ABU%20DHABI%2C%2014th%20May%2C%202016%20%28WAM%29%20--%20The,Al%20Rumaithi%2C%20Chairman%20of%20the%20UAE%20Space%20Agency
Evron Y (2018) Implications of China's belt and road initiative for Japan's involvement in the Middle East. Contemp Rev Middle East 5(3):199–214
Farhadi A (2021) Conceptualizing the great power competition and US geoeconomic strategy for the Central and South Asia (CASA) region. In: The great power competition volume 1: regional perspectives on peace and security. Springer, pp 31–53
Foley S (2023) Saudi Arabia's relations with China. In: Zoubir YH (ed) Routledge companion to China and the Middle East and North Africa. Taylor & Francis, pp 333–359
French D (2018) SoftBank Vision Fund, Saudi Arabia to create world's biggest solar power firm, Reuters. Available at: https://www.reuters.com/article/us-saudi-softbank-group-idUSKBN1H40DN
Fulton J (2022) Systemic change and regional orders: Asian responses to a Gulf in transition. Int Spectator 57(4):1–19

Fulton J (2019) China's changing role in the Middle East. Atlantic Council. Available at: https://www.atlanticcouncil.org/wp-content/uploads/2019/06/Chinas_Changing_Role_in_the_Middle_East.pdf

Ghiselli A (2023) China and the United States in the Middle East: policy continuity amid changing competition, The Middle East Institute. Available at: https://www.mei.edu/publications/china-and-united-states-middle-east-policy-continuity-amid-changing-competition

Gross S (2021) The global energy trade's new center of gravity. In: Chhabra T et al (eds) Global China: assessing China's growing role in the world. Brookings Institution Press, pp 319–325

Harrison P (2023) Japan to expand relations in Gulf, offer help to rebuild quake-hit Turkiye: Shikata, Arab News Japan. Available at: https://www.arabnews.jp/en/business/article_89793/

Heng Y-K (2022) Japan's role in Gulf security. In: Fulton J, Sim LC (eds) Asian perceptions of Gulf security. Routledge, pp 50–69

Hinnebusch R, Miyagi Y (2015) East Asia and the Middle East: inter-regional dynamics and american hegemony. In: Ehteshami A, Miyagi Y (eds) The emerging Middle East-East Asia Nexus. Routledge, pp 43–63

Hirono M (2019) Asymmetrical rivalry between China and Japan in Africa: to what extent has Sino-Japan rivalry become a global phenomenon? Pac Rev 32(5):831–862

Hoffman J (2021) The return of great-power competition to the Middle East: a two-level game. Middle East Policy 28(1):87–104

Horinuki K (2023) Japan–UAE Relations: establishment of multifaceted interdependence based on energy. In: Nakamura S, Wright S (eds) Japan and the Middle East: foreign policies and interdependence. Springer, pp 57–86

Hosoi T (2023) Oil market and supply: from the perspective of Japan's energy policy. In: Nakamura S, Wright S (eds) Japan and the Middle East: foreign policies and interdependence. Springer, pp 207–241

Insisa A, Pugliese G (2020) The free and open Indo-Pacific versus the belt and road: spheres of influence and Sino-Japanese relations. Pac Rev 35(3):557–585

Ishiguro H (2014) The Japan maritime self-defence force and regional security in the Indian ocean and the Gulf: Looking Towards 2013, Niblock & Yang Guang [Preprint]

Interview with a former State Department official during the Trump Administration (2022)

Japan Bank for International Cooperation (2021) Loan to Saudi electricity company under GREEN operations. Available at: https://www.jbic.go.jp/en/information/press/press-2021/0708-015002.html

Japan Bank for International Cooperation (2023) Recent major overseas infrastructure projects (Loans, equity participations, and guarantees in the last five years). Available at: https://www.jbic.go.jp/en/business-areas/sectors/image/000002610.pdf

Japan External Trade Organization (2022) Japanese trade and investment statistics. Available at: https://www.jetro.go.jp/en/reports/statistics.html

Jolly J (2023) Saudis ask to join UK, Italy and Japan's joint air combat programme, The Guardian. Available at: https://www.theguardian.com/uk-news/2023/aug/11/saudi-arabia-asks-to-join-uk-italy-japan-joint-air-combat-programme-tempest-gcap

Jordan Z, Deliwala S (2022) Why Washington should care about India's support for the AIIB', Asia unbound, council on foreign relations. Available at: https://www.cfr.org/blog/why-washington-should-care-about-indias-support-aiib

Kantei (2022) National security strategy of Japan. Available at: https://www.cas.go.jp/jp/siryou/221216anzenhoshou/nss-e.pdf

Kantei (2023) GX Jitsugen ni muketa Kihon Hōshin [Basic Policy for the Realization of GX]. Available at: https://www.cas.go.jp/jp/seisaku/gx_jikkou_kaigi/pdf/kihon.pdf

Keidanren (2022) Proposal for strengthening strategic relations with the Gulf Countries in the Middle East. Available at: https://www.keidanren.or.jp/en/policy/2022/104_proposal.pdf

Kingdom of Saudi Arabia (2023) An ambitious vision for an ambitious nation: overview, Vision 2030. Available at: https://www.vision2030.gov.sa/en/vision-2030/overview/

Kobayashi Y, Miyagi Y (2015) Japan's energy policy and energy diplomacy in the Gulf. In: Miyagi Y, Ehteshami A (eds) The Emerging Middle East-East Asia Nexus. Routledge, pp 64–84
Koike Y (2013) Asia and a Post-American Middle East, Project Syndicate. Available at: https://www.project-syndicate.org/commentary/us-disengagement-and-the-middle-east-s-future-by-yuriko-koike
Kumagai T, Saadi D (2022) Japan's ENEOS, Mitsui agree to study large hydrogen project with UAE's ADNOC', S&P Global. Available at: https://www.spglobal.com/commodityinsights/en/market-insights/latest-news/energy-transition/060722-japans-eneos-mitsui-agree-to-study-200000-mtyear-hydrogen-supply-chain-with-uaes-adnoc
Kumagai T (2022) Japan bolsters ties with Saudi Arabia beyond oil to hydrogen and fuel ammonia, S&P Global. Available at: https://www.spglobal.com/commodityinsights/en/market-insights/latest-news/energy-transition/122522-japan-bolsters-ties-with-saudi-arabia-beyond-oil-to-hydrogen-and-fuel-ammonia
Lamont C (2020) Japan's evolving ties with the Middle East. Asia Society. Available at: https://asiasociety.org/asias-new-pivot/japan
Leoni Z (2023) US-China rivalry in the MENA region. In: Zoubir YH (ed) Routledge companion to China and the Middle East and North Africa. Taylor & Francis, pp 89–101
Leverett F, Bader J (2005) Managing China-US energy competition in the Middle East. Wash Q 29(1):187–201
Levy K, Rose C (2019) Are China and Japan rivals in Latin America? a rivalry perception analysis. Pac Rev 32(5):898–921
Liao JX (2023) China's energy diplomacy towards the Middle East in the BRI Era. In: Zoubir YH (ed) Routledge companion to China and the Middle East and North Africa. Taylor & Francis, pp 25–39
Matsuda K (2023) Intra-East Asian competition in the Gulf and the Indo-Pacific: the case of Japan and China. In: Ghafar AA, Baabood A (eds) Asia in the GCC: a new strategic partner? Middle East Council on Global Affairs & Waseda University, pp 49–56
METI (2023) Joint statement on KSA-Japan lighthouse initiative for clean energy cooperation. Available at: https://www.meti.go.jp/press/2023/07/20230717001/20230717001_1.pdf.
Miller R (2023) Asian powers in the Gulf maritime fomain: The Arab Gulf perspective. In: Ghafar AA, Baabood A (eds) Asia in the GCC: a new strategic partner? Middle East Council on Global Affairs & Waseda University, pp 79–90
Ministry of Defense (2013) National security strategy. Available at: https://worldjpn.net/documents/texts/JPSC/20131217.O1E.html.
Ministry of Economy, Trade, and Industry (2021) METI unveils green growth strategy to support Japan's 2050 carbon neutral goal. Available at: https://www.meti.go.jp/english/mobile/2021/20210208001en.html
Ministry of Foreign Affairs (2017c) Speech by H.E. Taro Kono, Minister for Foreign Affairs of Japan at the IISS Manama Dialogue. Available at: https://www.mofa.go.jp/mofaj/files/000315610.pdf
Ministry of Foreign Affairs (2001) Press Conference 2 February 2001. Available at: https://www.mofa.go.jp/announce/press/2001/2/202.html
Ministry of Foreign Affairs (2007) Joint Statement between the Kingdom of Saudi Arabia and Japan. Available at: https://www.mofa.go.jp/region/middle_e/pmv0704/joint_saudi.html
Ministry of Foreign Affairs (2013) Prime Minister Shinzo Abe's visit to the Middle East. Available at: https://www.mofa.go.jp/region/page6e_000028.html
Ministry of Foreign Affairs (2016) Meeting of Prime Minster Abe and Saudi Deputy Crown Prince Mohammed bin Salman. Available at: https://www.mofa.go.jp/me_a/me2/sa/page4e_000511.html
Ministry of Foreign Affairs (2017a) Japan-Saudi Arabia Summit Meeting. Available at: https://www.mofa.go.jp/me_a/me2/sa/page4e_000589.html
Ministry of Foreign Affairs (2017b) Saudi Japan Vision 2030. Available at: https://www.mofa.go.jp/files/000237093.pdf

Ministry of Foreign Affairs (2018) Joint statement on deepening and strengthening strategic partnership between Japan and the United Arab Emirates—opening a new chapter of cooperation towards prosperity and stability. Available at: https://www.mofa.go.jp/files/000359557.pdf

Ministry of Foreign Affairs (2022) Joint declaration on the implementation of Comprehensive Strategic Partnership Initiative (CSPI) between Japan and the United Arab Emirates. Available at: https://www.mofa.go.jp/files/100402681.pdf

Ministry of Foreign Affairs (2023) Signing of the Agreement between the Government of Japan and the Government of the United Arab Emirates (UAE) concerning the Transfer of Defense Equipment and Technology. Available at: https://www.mofa.go.jp/press/release/press7e_000004.html

Ministry of Foreign Affairs of the People's Republic of China (2016) China's Arab Policy Paper. People's Republic of China

Mitsubishi Heavy Industries (2018) 'MHI successfully launches UAE's KhalifaSat Satellite. Available at: https://www.mhi.com/news/1810292140.html#:~:text=Tanegashima%20Space%20center%2C%20October%2029%2C%202018%20-%20Mitsubishi,orbit%20today%20via%20the%20H-IIA%20launch%20vehicle%20F40

Miyagi Y (2008) Japan's Middle East security policy: theory and cases. Routledge

Nair D (2022) Saudi Arabia and the US sign 13 investment agreements during President Joe Biden's visit, The National News. Available at: https://www.thenationalnews.com/business/economy/2022/07/16/saudi-arabia-and-the-us-sign-13-investment-agreements-during-president-joe-bidens-visit/

Nakamura S (2023) 'Origin of Japan's Relations with Middle Eastern Countries by Practical Internationalism. In: Nakamura S, Wright S (eds) Japan and the Middle East: foreign policies and interdependence. Springer, pp 297–324

National Development and Reform Commission (2015) Vision and actions on jointly building silk road economic belt and 21st-century maritime silk road. Ministry of Foreign Affairs and Ministry of Commerce of the People's Republic of China

O'Rourke R (2023) Great Power competition: implications for defense—issues for congress. Congressional Research Service. Available at: https://crsreports.congress.gov/product/pdf/r/r43838

Pollman M (2023) Kishida's Trip to Middle East focuses on energy, defense cooperation, The Diplomat. Available at: https://thediplomat.com/2023/07/kishidas-trip-to-middle-east-focuses-on-energy-defense-cooperation/

Pradhan R (2018) The rise of China in Central Asia: The new silk road diplomacy. Fudan J Humanit Soc Sci 11:9–29

Reuters (2023a) China's Huawei opens cloud data centre in Saudi Arabia in regional push. Available at: https://www.reuters.com/technology/chinas-huawei-opens-cloud-data-centre-saudi-arabia-regional-push-2023-09-04/

Reuters (2023b) Japan's Marubeni, Saudi's PIF weighing clean hydrogen production in kingdom. Available at: https://www.reuters.com/business/sustainable-business/japans-marubeni-saudis-pif-weighing-clean-hydrogen-production-kingdom-2023-03-02/

Saito J (2023) Investment and trade promotion policies: Gulf and Japan's non-energy sector interdependence. In: Nakamura S, Wright S (eds) Japan and the Middle East: foreign policies and interdependence. Springer, pp 269–295

Satariano A (2019) U.A.E. to Use Equipment from Huawei Despite American Pressure, The New York Times. Available at: https://www.nytimes.com/2019/02/26/technology/huawei-uae-5g-network.html

Saudi Aramco, IEEJ and SABIC (2020) Joint Press Release: World's first blue ammonia shipment opens new route to a sustainable future. Available at: https://eneken.ieej.or.jp/en/press/press200927_en.pdf

Saudi Gulf Projects (2021) Marubeni led consortium signed 300MW Rabigh Solar PV Project. Available at: https://www.saudigulfprojects.com/2021/04/marubeni-led-consortium-signed-300mw-rabigh-solar-pv-project/

Schulze K (2019) Rivalry in the Middle East? Japan's CEAPAD initiative and China's rise. Pac Rev 32(5):809–830
Silove N (2016) The Pivot before the Pivot: U.S. strategy to preserve the power balance in Asia. Int. Secur. 40(4):45–88. Available at: https://doi.org/10.1162/ISEC_a_00238
Singh MK (2021) China and the United States in the Middle East: between dependency and rivalry. In: The great power competition volume 1: regional perspectives on peace and security. Springer, pp 369–395
Sun D, Zoubir YH (2021) Securing China's "latent power": the dragon's anchorage in Djibouti. J Contemp China 30(130):677–692
Sun D, Zhang Y, Lin L (2023) Building an "Outer Space Silk Road": China's Beidou navigation satellite system in the Arab World. In: Zoubir YH (ed) Routledge companion to China and the Middle East and North Africa. Taylor & Francis, pp 179–191
Suzuki K (2023) Japan Makes Major Shift in Middle East Diplomacy, The Diplomat. Available at: https://thediplomat.com/2023/07/japan-makes-major-shift-in-middle-east-diplomacy/
Tan F, Aizhu C (2023) Saudi Aramco boosts China investment with two refinery deals, Reuters. Available at: https://www.reuters.com/business/energy/saudi-aramco-open-new-china-refinery-petchem-complex-2026-2023-03-26/
Tan H (2022) China says it's moving to buy oil and gas in the yuan—a move that could threaten the dollar's global dominance in the long run, Markets Insider. Available at: https://markets.businessinsider.com/news/currencies/dollar-vs-yuan-china-buy-gulf-oil-energy-rmb-currency-2022-12
The Observatory of Economic Complexity (2021a) Qatar (QAT) exports, imports, and trade partners. Available at: https://oec.world/en/profile/country/qat?yearlyTradeFlowSelector=flow1&latestTrendsFlowSelectorNonSubnat=flow1
The Observatory of Economic Complexity (2021b) Saudi Arabia: yearly trade, historical data. Available at: https://app-quail.oec.world/en/profile/country/sau#yearly-trade
The Observatory of Economic Complexity (2021c) United Arab Emirates: yearly trade, historical data. Available at: https://oec.world/en/profile/country/are#historical-data
The Observatory of Economic Complexity (2021d) Where does China import natural gas, liquefied from? (2021). Available at: https://oec.world/en/visualize/tree_map/hs92/import/chn/show/5271111/2021/
The White House (2017) National security strategy of the United States of America. Available at: https://trumpwhitehouse.archives.gov/wp-content/uploads/2017/12/NSS-Final-12-18-2017-0905.pdf
The White House (2022a) FACT SHEET: The United States strengthens cooperation with Middle East partners to address 21st century Challenges. Available at: https://www.whitehouse.gov/briefing-room/statements-releases/2022/07/16/fact-sheet-the-united-states-strengthens-cooperation-with-middle-east-partners-to-address-21st-century-challenges/
The White House (2022b) FACT SHEET: U.S.-UAE Partnership to Accelerate Transition to Clean Energy (PACE). Available at: https://www.whitehouse.gov/briefing-room/statements-releases/2022/11/01/fact-sheet-u-s-uae-partnership-to-accelerate-transition-to-clean-energy-pace/
The White House (2022c) National security strategy. Available at: https://www.whitehouse.gov/wp-content/uploads/2022/11/8-November-Combined-PDF-for-Upload.pdf
The White House (2022d) Remarks by President Biden at the GCC + 3 summit meeting. Available at: https://www.whitehouse.gov/briefing-room/speeches-remarks/2022/07/16/remarks-by-president-biden-at-the-gcc-3-summit-meeting/
U.S. Energy Information Administration (2022) Country Analysis Executive Summary, China. Available at: https://www.eia.gov/international/analysis/country/CHN
United Arab Emirates Ministry of Foreign Affairs (2019) UAE, China issue joint statement on strengthening comprehensive strategic partnership. Available at: https://www.mofa.gov.ae/en/MediaHub/News/years/2019/7/23/23-07-2019-uae-china-statment1
US Department of Defense (2018) Summary of the 2018 national defense strategy of the United States: sharpening the American Military's competitive edge. Available at: https://dod.defense.gov/Portals/1/Documents/pubs/2018-National-Defense-Strategy-Summary.pdf

US Department of Defense (2022) 2022 National Defense Strategy of the United States. Available at: https://media.defense.gov/2022/Oct/27/2003103845/-1/-1/1/2022-NATIONAL-DEFENSE-STRATEGY-NPR-MDR.PDF

US House of Representatives (2022) Examining the U.S. interest in regional security cooperation in the Middle East and North Africa: opportunities, obstacles and objectives testimony of Linda Robinson, Former Director, Center for Middle East Public Policy, RAND Corporation, Before the House Committee on Foreign Affairs Subcommittee on the Middle East, North Africa, and Global Counterterrorism. Available at: https://docs.house.gov/meetings/FA/FA13/20220921/115154/HHRG-117-FA13-Wstate-RobinsonL-20220921.pdf

Valori GE (2016) President Xi Jinping's travel to the Middle East. Modern diplomacy. Available at: https://moderndiplomacy.eu/2016/02/12/president-xi-jinping-s-travel-to-the-middle-east/

Wallace C (2019) Japan's strategic contrast: continuing influence despite relative power decline in Southeast Asia. Pac Rev 32(5):863–897

Watanabe L (2023) China as a geo-economic and security actor in the MENA region. In: Zoubir YH (ed) Routledge companion to China and the Middle East and North Africa. Taylor & Francis, pp 80–88

Yamada M (2023) Japan's Relations with Saudi Arabia: the evolution of energy diplomacy in response to the developmental shift in the rentier state. In: Nakamura S, Wright S (eds) Japan and the Middle East: foreign policies and interdependence. Springer, pp 27–55

Young KE (2023) China and the Gulf: necessary partners. In: Zoubir YH (ed) Routledge companion to China and the Middle East and North Africa. Taylor & Francis, pp 192–204

Zhang C (2023) China's infrastructure construction in the Middle East. In: Zoubir YH (ed) Routledge companion to China and the Middle East and North Africa. Taylor & Francis, pp 40–51

Zoellick RB (2005) Whither China: from membership to responsibility?–remarks to national committee on US-China Relations. Available at: https://www.ncuscr.org/wp-content/uploads/2020/04/migration_Zoellick_remarks_notes06_winter_spring.pdf

Zoubir YH (ed) (2023) Routledge companion to China and the Middle East and North Africa. Taylor & Francis

Silk Roads and Subtle Blows: China's 'Soft Power' Influence in the Middle East and North Africa

Khaled Eid

Abstract China has historically been risk averse in developing 'soft power' influence in the Middle East and North Africa (MENA). However, an insistent need for energy, economic expansion, aspirations for increased international diplomatic support of Chinese strategic goals, and a desire to compete with eroding US influence, required China to create a robust 'soft power' strategy in MENA. Through various 'soft power' mechanisms such as trade, economic integration, diplomacy and mediation or political partnerships, China has begun to build an elaborate 'soft power' apparatus aimed at subtly replacing US influence in the region and entrenching its own. The US, largely through its own policy missteps and reflexive intervention initiatives, has seen its influence decline in the region. Given the rise of China in MENA, the US may be forced to contend with a policy rethink in order to stave off increased Chinese influence. China's 'development over democracy' approach might be appealing to authoritarian MENA countries as they seek economic modernization while trying to maintain regime stability. With the rise of China's 'soft power' influence, MENA may soon be placed in a difficult position of expanding ties with China at the expense of those with the US, leading to potential tensions. The US still maintains predominance in the region, largely through its security apparatus, and can still preserve its influence in the region by reprioritization of objectives in MENA.

Keywords Soft power · Trade · Economic integration · Mediation · Non-intervention · Stabilization · Multipolarity · Belt and road initiative · State sovereignty · Diplomatic recognition · Middle East · North Africa · China · US

K. Eid (✉)
ACERTAS, Claremont Graduate University, Claremont, CA, USA
e-mail: khaled.eid@cgu.edu

A. Farhadi et al. (eds.), *The Great Power Competition Volume 6*,
https://doi.org/10.1007/978-3-031-70767-4_6

1 Relationship Rethink

China's seemingly one-dimensional relationship with the Middle East and North Africa (MENA) is rapidly transforming. China's interest in the region is no longer purely centered around commodities and energy—MENA is now a critical component in China's plan to dominate global trade and become a global power on equal footing with the US and EU.[1] Over the last decade and increasingly so in the post-Covid era, China's strategic use of 'soft power' has seen an expansion of its geopolitical and economic influence in the region—in many instances they are the largest trade partner with and investor in many MENA nations.[2] China was initially cautious in the development of relations due to regional instability. However, the recent decline of US dominance in the region, coupled with China's insistent need for cheap natural resources and access to trade routes for the Build and Road Initiative (BRI) as well as the desire for greater diplomatic recognition and support on the world stage, has led China to accelerate into a significant stakeholder in the region—directly competing with the US. To safeguard the strategic importance of MENA to China, it aims to use its 'soft power' to gain influence, while helping stabilize the region through economic investment and development rather than military interventions—subtly eroding US influence and replacing it with their own. Their goals center around maintaining stable relations with the regional powers, diplomatic cooperation, and mediation all while deepening commercial ties through BRI and other strategic investments.

MENA's relationship with China is also not a one-way road, the region has also benefitted greatly from improved and expanded relations with China. Over the last decade, US imports of Middle East oil and liquid natural gas (LNG) have been steadily declining, preferring imports from neighboring Canada and Mexico—who now provide over half of all such exports to the US.[3] Conversely during that time China has become the largest purchaser of energy exports out of MENA, serving as an ever-willing buyer of cheap oil, despite US sanctions on countries like Iran.[4,5] Data suggests that oil imports have only increased since sanctions have been imposed.[6]

Conversely, US' primary method of engagement with MENA has taken the form of direct/indirect military intervention in the region. Be it the war in Iraq from 2003–2011, the 20 year occupation of Afghanistan, allowing the question of Palestine to remain a point of conflict for decades or the massive weapons transfers to KSA

[1] Al-Tamimi et al. (2019). China's great game in the Middle East. ECFR. https://ecfr.eu/publication/china_great_game_middle_east/#from-economic-interests-to-political-and-security-engagement [Accessed 23 Sep. 2023].

[2] IBID.

[3] Plumer (2014). China now buys more Middle East oil than the US. [online] Vox. Available at: https://www.vox.com/2014/9/3/6101885/middle-east-now-sells-more-oil-to-china-than-to-the-us [Accessed 22 Sep. 2023].

[4] IBID.

[5] Aizhu and Lawler (2022). China buys more Iranian oil now than it did before sanctions, data shows. Reuters. [online] 1 Mar. Available at: https://www.reuters.com/world/china/china-buys-more-iranian-oil-now-than-it-did-before-sanctions-data-shows-2022-03-01/.

[6] IBID.

and UAE, leading to expanded conflicts in Yemen. Really, since the 1990s, it has been nearly 30 years of endless wars, aptly summarized "Washington has discovered getting in is easy, getting out is seemingly impossible."[7,8] In an almost self-reinforcing loop, the US finds itself engaging in one military intervention after the other, to a point where it is almost a reflex of US policy in the region. The policies US has adopted over the last few decades have failed to solidify relations, stabilize the region, guarantee US influence, or protect national interests. Instead, relations with some MENA countries have frayed, paving the way China to use its 'soft power' to exploit the strained relations and gain influence. For example, after 20 years of the US fighting the Taliban in Afghanistan, the US eventually withdrew triggering the collapse of the Afghan government in 2021, allowing the Taliban to take over the country again. Two years later in 2023, China signed a 25 year multi-million dollar economic agreement with the Taliban led Afghan government, leading to very favorable resource extraction terms for Chinese firms.[9]

While not innocent, China has largely stuck to its non-intervention in sovereign states policy stemming from their 'Five Principles of Peaceful Coexistence' largely maintaining a neutral approach to conflicts or escalations—positioning itself as a would-be mediator, especially in the MENA.[10] For countries in the region this may be a welcome change, resulting in less wars, and enduring stability, while also allaying fears leaders maintain over regime stability and authoritarian control of their countries. Reaffirming mutual regime legitimacy between MENA and their Chinese counterparts. Both have seen the benefits of this arrangement in providing diplomatic cover and support in the UN and across other global initiatives, fostering a relationship "... between China and [MENA] countries [in an] attempt to hedge their bets against US predominance."[11]

[7] Wikipedia Contributors (2024). Occupation of Iraq (2003–2011). [online] Wikipedia. Available at: https://en.wikipedia.org/wiki/Occupation_of_Iraq_(2003%E2%80%932011)#:~:text=The%20Occupation%20of%20Iraq%20(2003 [Accessed 22 Sep. 2023].

[8] Riedel (2020). 30 years after our 'endless wars' in the Middle East began, still no end in sight. [online] Brookings. Available at: https://www.brookings.edu/articles/30-years-after-our-endless-wars-in-the-middle-east-began-still-no-end-in-sight/ [Accessed 23 Sep. 2023].

[9] Kumar (2024). Why has China recognised Taliban's envoy to Beijing? [online] Al Jazeera. Available at: https://www.aljazeera.com/news/2024/2/14/is-chinas-recognition-of-afghanistan-envoy-a-diplomatic-win-for-taliban [Accessed 16 Feb. 2024].

[10] Ergenc (2018). China's relations with the Middle East: a perspective from the region | IIAS. [online] www.iias.asia. Available at: https://www.iias.asia/the-newsletter/article/chinas-relations-middle-east-perspective-region [Accessed 25 Sep. 2023].

[11] Al-Tamimi et al. (2019). China's great game in the Middle East. ECFR. https://ecfr.eu/publication/china_great_game_middle_east/#from-economic-interests-to-political-and-security-engagement [Accessed 23 Sep. 2023].

2 Why is MENA Important for China?

The increased importance of MENA to China seeds itself in its geographic location, economic factors, and diplomatic incentives. China needs uninterrupted access to the strategic trade routes as well as the critical energy commodities MENA offers. The Belt-Road Initiative (BRI) places great emphasis on both maritime and land integration in the region and as China seeks to compete with US and EU diplomatically, MENA may offer welcome allies at the UN and in other inter-governmental bodies.[12]

2.1 Geographic Location—Trade and Natural Resources

Situated at the crossroads of Asia and Europe, MENA has long been a pivotal region in global trade and energy supply. With the Suez Canal (SC) and Bab El Mandeb on the west side and Strait of Hormuz on the east side of the Arabian Peninsula—maintaining good relations is vital to preserving stable trade and access to the region's resources. Approximately 12% of global trade flows through Bab El Mandeb, up the Red Sea and through the Suez Canal heading towards western markets making it the world's third most important water trade route.[13] Furthering this region's importance, is the Strait of Hormuz just 1,284 miles away, on the other side of the peninsula where 21% of global petroleum liquids flow throw, making it the world's most important oil checkpoint.[14] Given the relatively high concentration of vital trade and energy routes, developing good relations in the region will help China in maintaining and expanding access. Accordingly, disruptions in the area can pose major consequences to global market stability. Threats to trade stability in the region can be highlighted by actions the Iranian backed Houthis "[l]aunched in response to Israel's bombardment of Gaza…"[15] The Houthis, who largely gained control of the area near the Red Sea due to backing by Iran and an unsuccessful US backed Saudi military intervention in Yemen, are launching attacks at vessels attempting to cross Bab El Mandeb—adding 3000–3,500 nautical miles to ship voyages, greatly increasing costs and slowing shipping times (Fig. 1).[16]

---

[12] IBID.

[13] Salameh (2021). The Strategic Importance of the Suez Canal. [online] www.energymanagementcentre.eu. Available at: https://www.energymanagementcentre.eu/views-on-energy-news/suez-canal/ [Accessed 26 Sep. 2023].

[14] Dunn and Barden (2023). The Strait of Hormuz is the world's most important oil transit chokepoint. [online] www.eia.gov. Available at: https://www.eia.gov/todayinenergy/detail.php?id=61002#:~:text=the%20Arabian%20Sea. [Accessed 29 Nov. 2023].

[15] Partington (2024). What is the Red Sea crisis, and what does it mean for global trade? The Guardian. [online] 3 Jan. Available at: https://www.theguardian.com/world/2024/jan/03/what-is-the-red-sea-crisis-and-what-does-it-mean-for-global-trade#:~:text=Sea%20trade%20route%3F [Accessed 24 Jan. 2024].

[16] IBID.

Fig. 1 Middle East oil and gas fields—(VOX, US Energy Information Administration 2015)

Given the strategic importance of the trade routes and energy exports, the potential for continuous interruptions such as this will impact longer term strategic planning and increase uncertainty for major powers attempting to utilize the trade routes and energy resources for major projects. To mitigate the potential risks military interventions could cause, China leans more on a 'soft power' approach whenever possible, aiming to stabilize through economic investment or diplomatic partnerships such as that of the BRI and other related initiatives.[17]

2.2 *Economic Factors—Belt and Road Initiative*

To briefly summarize, the BRI, is a major Chinese infrastructure and trade project meant to utilize excess domestic labor, strengthen bilateral diplomatic relations, and enhance global trade while also attempting to create, "…an economic system

[17] Al-Tamimi e al. (2019). China's great game in the Middle East. ECFR. https://ecfr.eu/publication/china_great_game_middle_east/#from-economic-interests-to-political-and-security-engagement [Accessed 23 Sep. 2023].

outside Washington's control."[18] Essential for China to achieve this, is making sure it has strategic relations in the region, promotes stability in MENA, and incentivizes countries participation through various 'soft power' efforts.

For example, in Egypt, China plans to heavily invest primarily through deepening and widening upgrades to SC infrastructure as well as further investments into the Suez Economic and Trade Cooperation Zone (SETC-Zone) and TEDA Cooperation Zone, all central to the 'maritime silk road' aspect of the BRI. For China, this gives the country greater strategic influence over the canal and in Egypt in general. It also allows China to compete more directly with the US economically in a large market which also sees the US invest heavily, amounting to $22bn in 2022.[19] In terms of trade, these enhancements to the SC are expected to increase Chinese sea trade through the canal from approximately 100–250 million tons of container shipping.[20] For Egypt, investments in the various economic zones and the SC could generate significant income for the government, who are currently facing a deepening economic crisis, currency devaluation, and inability to service the $155.7bn external debt.[21] The potential revenue from integration to the BRI and other projects like the SETC-Zone and TEDA may serve as a way for the Egyptian government to address the economic crisis. However, Egypt may be put into an increasing tough situation, as "...[an] attempt to enhance its bilateral relations with China may be viewed by Washington as fraternizing with the enemy."[22] This is something that El-Sisi will need to grapple with as he attempts to navigate the long-standing relations with the US and deepening ties with China.

Another example is in Iran, where China constructed another project vital to the 'belt' routes of the BRI, the Tehran railway, connecting to the China-Pakistan route and further to Central Asian countries, finally continuing to Istanbul.[23] The journey for goods leaving China heading to central Asia and Iran used to take the 45-day journey from Shanghai to the port of Bandar Abbas, with the railway, the Yiwu-Tehran route, it will take less than 14 days (Fig. 2).[24]

Given the success of the initial projects, even back in 2016, China signaled increasing trade between the two countries to around $600bn over the next decade.[25]

---

[18] IBID.

[19] Al-Anani (2023). Egypt's Strategic Partnership with China: Opportunities and Implications. [online] Arab Center Washington DC. Available at: https://arabcenterdc.org/resource/egypts-strategic-partnership-with-china-opportunities-and-implications/ [Accessed 27 Sep. 2023].

[20] Rakha and El-Aasar (2024). The impact of the belt and road initiative on the Suez Canal cargo trade. Journal of Shipping and Trade, 9(1). https://doi.org/10.1186/s41072-024-00167-y.

[21] IBID.

[22] IBID.

[23] Ergenc (2018). China's relations with the Middle East: a perspective from the region | IIAS. [online] www.iias.asia. Available at: https://www.iias.asia/the-newsletter/article/chinas-relations-middle-east-perspective-region [Accessed 25 Sep. 2023].

[24] Ramachandran (2016). Iran, China and the Silk Road Train. [online] thediplomat.com. Available at: https://thediplomat.com/2016/03/iran-china-and-the-silk-road-train/ [Accessed 2 Oct. 2023].

[25] Reuters (2016). First freight train from China arrives in Iran in 'Silk Road' boost: media. [online] Reuters. Available at: https://www.reuters.com/article/idUSKCN0VP0W7/ [Accessed 4 Oct. 2023].

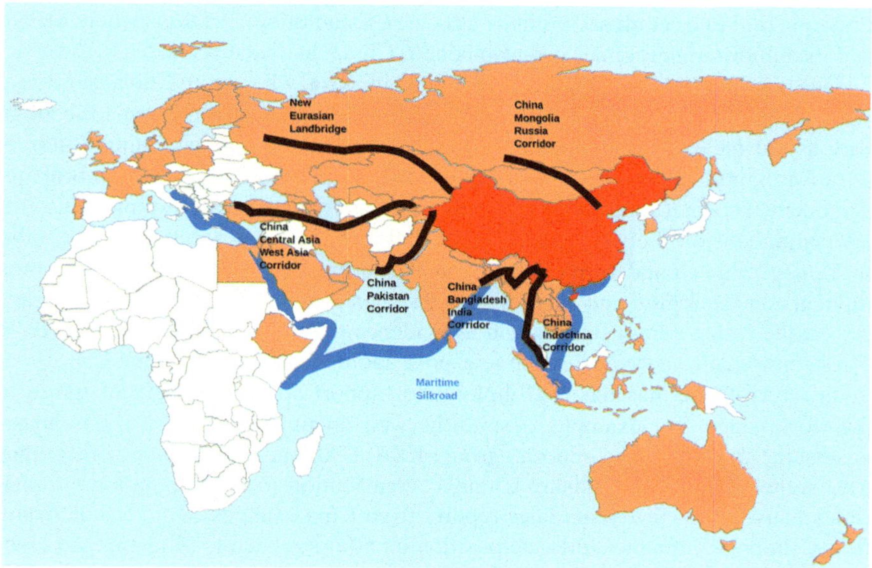

Fig. 2 Belt and road initiative sea and land routes (Wikipedia, Lommes 2017)

Iran, like Egypt, face economic instability as well as sanctions imposed by the west, could look to China to circumvent those sanctions. However also like Egypt, Iran must grapple with its ties with China and trying to get out from underneath the crushing western sanctions or getting positive outcomes from Joint Comprehensive Plan of Action (JCPOA) discussions.[26]

2.3 *Diplomatic Incentives—China's Desire for 'Multi-polarity and State Sovereignty'*

Also critical to China's desire to become a larger player on the world stage, is to strengthen its diplomatic prowess and state sovereignty/integrity claims, not just its economic might. Central to achieving this is disrupting the unipolar order and replacing it with multi-polarity in which China envisages a bloc led by them to counter the US led powers.[27] This can aid China in garnering support for their 'defense' of state sovereignty initiative for example, with regards to its One-China

[26] Pletka (2024). Whatever Happened to Biden's Iran Policy? [online] Foreign Policy. Available at: https://foreignpolicy.com/2024/03/26/bidens-iran-policy-nuclear-deal-jcpoa/ [Accessed 27 Mar. 2024].

[27] Al-Tamimi et al. (2019). China's great game in the Middle East. ECFR. https://ecfr.eu/publication/china_great_game_middle_east/#from-economic-interests-to-political-and-security-engagement [Accessed 23 Sep. 2023].

Principle (not to be confused with the US One China Policy).[28] This is where MENA and its authoritarian regimes can also benefit China, and vice versa.

One beneficiary has been the Assad regime in Syria, where both China and Russia vetoed multiple measures at the UN for example, to end fighting in Idlib or on various aid measures, helping the Syrian Regime maintain what China described as 'sovereignty' over Syria.[29] Another example is the more recent China brokered agreement between long-standing rivals KSA and Iran to restore diplomatic ties, alleviating tensions between the two especially when it comes to the proxy conflict in Yemen. This should worry the US, as they were long seen as having the most influence in the region and according to Feltman at the brooking institute, "…as a slap at the Biden administration and as evidence that China is the rising power."[30] These diplomatic successes, while few, pose additional questions to the US.

In return China has received diplomatic support across a variety of issues on the world stage. For example, despite the well documented human rights abuses against the Uyghur ethnic minority group, KSA, UAE and Egypt among other countries signed a letter "…praising China's "contribution to the international human rights cause."[31] A few years later reports from China suggested "…Saudi Arabia firmly supports China's legitimate position on affairs related to Xinjiang and Hong Kong."[32] For both China and their MENA counter parts these sort of efforts mutually enhance regime stability, legitimacy and perceived power both internationally and domestically.

3 China's Soft Power Approach to MENA

The increasing importance of the MENA region to China coupled with the region's growing distance from the US cannot be understated enough. Despite the increase of importance, China's 'soft power' approach to MENA stems as much from necessity as it does imposition—drawing from historical observations when determining its region policy. In many cases, China does not have a choice but to pursue a 'soft

[28] US Taiwan Business Council (2022). The U.S. 'One China Policy' vs. the PRC "One China Principle. [online] US-Taiwan Business Council. Available at: https://www.us-taiwan.org/resources/faq-the-united-states-one-china-policy-is-not-the-same-as-the-prc-one-china-principle/ [Accessed 23 Oct. 2023].

[29] Barber (2019). Syria: the disgraceful stain left by the UN Security Council veto. [online] Lowyinstitute.org. Available at: https://www.lowyinstitute.org/the-interpreter/syria-disgraceful-stain-left-un-security-council-veto [Accessed 3 Oct. 2023].

[30] Stewart et al. (2023). Analysis: China role in Saudi, Iran deal a tricky test for U.S. Reuters. [online] 11 Mar. Available at: https://www.reuters.com/world/china-role-saudi-iran-deal-tricky-test-us-2023-03-10/ [Accessed 3 Oct. 2023].

[31] Hoffman (2021). Analysis | Why do some Muslim-majority countries support China's crackdown on Muslims? Washington Post. [online] 4 May. Available at: https://www.washingtonpost.com/politics/2021/05/04/why-do-some-muslim-majority-countries-support-chinas-crackdown-muslims/ [Accessed 6 Oct. 2023].

[32] IBID.

power' policy approach in MENA. Despite a dissipation in power, the US are still the preeminent military force in the region, with bases in many countries and the US navy providing security for trade routes. Any significant Chinese 'hard power' military buildup would be seen as a direct challenge to US military dominance in the region. While certain circles in China want to "...shed its image as a free-rider and increase its military presence in the region..." they also understand the China largely benefits from this security arrangement as they get benefits of secure shipping lanes without incurring the financial and political costs associated with that security maintenance.[33] It is also abundantly clear that a 'hard power' dominant policy does not necessarily yield the outcomes China wants to achieve as evidenced by the US policy towards the region over the last 30 + years.

These constraints and opportunities shaped Chinese policy in the region to the point where, as Sun Degang states, "...China [preferred] developmental peace over the Western concept of democratic peace in [MENA]..." and that they felt the "...international community should focus on providing much-needed economic assistance to the region "instead of exporting ill-fitting democracy."[34] These principles, along with incentives and risks outlined above, shaped China 'soft power' policy in MENA. Centered around these principles, China is developing an ever-growing 'soft power' entrenchment policy; expansion of trade, deepening economic collaboration through long term integration (BRI), mediation and diplomacy over intervention for stabilization, and finally more robust political partnerships to foster greater diplomatic support on a world stage.

3.1 Trade

Integral to China's 'soft power' influence in MENA is its longest standing engagement with the region, trade. For the last few decades China has been positioning itself as an ever-willing buyer of MENA energy commodities, even circumventing sanctions in certain circumstances. In the last decade there has been purposeful acceleration in trade relations as data suggests, "...China-Middle East trade volume nearly doubled over the 2017 to 2022 period, from $262.5 billion to $507.2 billion," while in 2020 China overtook the EU as the GCC's largest trading partner.[35] However, the trade relationship with the region from both an exports and imports perspective is

[33] Al-Tamimi et al. (2019). China's great game in the Middle East. ECFR. https://ecfr.eu/publication/china_great_game_middle_east/#from-economic-interests-to-political-and-security-engagement [Accessed 23 Sep. 2023].

[34] Vohra (2022). Xi Jinping Has Transformed China's Middle East Policy. [online] Foreign Policy. Available at: https://foreignpolicy.com/2022/02/01/xi-jinping-has-transformed-chinas-middle-east-policy/ [Accessed 8 Oct. 2023].

[35] Zhihua (2023). Sino trade volumes soar with Middle East, Africa. [online] global.chinadaily.com.cn. Available at: https://global.chinadaily.com.cn/a/202307/17/WS64b48a76a31035260b816b72.html [Accessed 11 Oct. 2023].

relatively concentrated, with only three countries dominating over 50% of trade.[36] Further, despite the relatively similar export composition to MENA when compared to the rest of the world, imports to China from the MENA are still concentrated in hydrocarbons and are subject to price volatility.[37] In certain cases, the MENA only serves as a linkage point between China and more lucrative Western Markets.

Despite certain drawbacks, MENA and China are still benefitting from this arrangement. MENA has a large market to sell to, compensating for potential revenue lost as the US becomes the main oil exporter to the EU, while China can purchase hydrocarbons without competing as directly with other markets.[38] This also gives China added leverage over certain MENA countries, for example Iran, who is looks for alternative export avenues as increasing sanctions are imposed. In exchange for providing that export market, China can potentially leverage Iran and others into further BRI integration.

While trade may have been necessary to initially build relations between China and MENA, it is not sufficient to maintain the level of influence China seeks to have in the region. This is also true for MENA, where many countries are looking to a future beyond oil and are seeking a diversify their economies away from hydrocarbon exports.[39] China understands this and is attempting to remedy through mutually beneficial expanded economic cooperation, another facet of 'soft power.'

3.2 Economic Integration, Investment and Diversification

The next core tenant of China's regional 'soft power' policy to further entrench itself in the region is "...development is more important than democracy," a message that should resonate with the authoritarian regimes in MENA—who are incredibly concerned with maintaining complete control in their countries.[40] While efforts at economic modernization maybe met with increasing western 'democratic' guidelines from the US, China's approach requires no such commitment, making it more appealing to regimes who do not want to foster internal dissent.[41] China also sees

---

36 Shatz (2024). Middle East–China Trade Prospects Remain Robust Despite Red Sea Crisis. [online] RAND. Available at: https://www.rand.org/pubs/commentary/2024/02/middle-east-china-trade-prospects-remain-robust-despite.html [Accessed 15 Feb. 2024].

37 IBID.

38 Cooban (2023). US has replaced Russia as Europe's top crude oil supplier. CNN. [online] 28 Mar. Available at: https://www.cnn.com/2023/03/28/energy/eu-us-oil-imports-overtake-russia/index.html [Accessed 14 Oct. 2023].

39 IBID.

40 Ahmad Khan (2021). China's Increasing Influence in the Middle East. [online] E-International Relations. Available at: https://www.e-ir.info/2021/09/20/chinas-increasing-influence-in-the-middle-east [Accessed 1 Oct. 2023].

41 Al-Tamimi et al. (2019). China's great game in the Middle East. ECFR. https://ecfr.eu/publication/china_great_game_middle_east/#from-economic-interests-to-political-and-security-engagement [Accessed 23 Sep. 2023].

this approach as a way stabilize MENA. Investments over interventions should be the preferred course of action, as China argues, part of the underlying causes of instability in MENA are economic stagnation, poor infrastructure, and lack of employment opportunities.[42] Pursuing stabilization is vital for China to mitigate risks to interests in the region.

The lack of this constraint allows China to utilize economic integration and investment beyond trade, as yet another form of 'soft power' influence in the region. Chinese firms are at a distinct advantage when it comes to investing or bidding on projects in the region as they do not come with added democratic policy dictates compared to their western counterparts. For MENA, countries can now take on diversification investments with less regime exposure.[43] This is not to say that Chinese investments do not also have their own sets of stipulations.

Chinese economic 'soft power' can be seen throughout the entirety of the MENA region, be it through strategic investments or through BRI integrations as mentioned with Egypt and Iran. In 2017 China invested $33 billion in the region and are currently engaged in reconstruction projects in Iraq and Syria.[44] It has pursued strategic economic partnerships in Lebanon, Libya, Sudan, and Syria while, in 2021, China and Iran signed a 25 year strategic partnership deal.[45,46] The deal with Iran will see China invest not only in oil infrastructure, but in banking, finance, energy and even culture.[47]

In Iraq, China has been financing post-war reconstruction efforts; building 7,000 schools, health care centers, homes in Sadr city, as well a development on vital infrastructure projects such as roads, airports, and sewers.[48] In efforts to accommodate KSA desires for diversification, China has invested in Saudi renewable energy and solar sectors, while also signing an MOU on hydrogen energy development and aiding them in Saudi Vision 2030 efforts.[49] As part of further modernization efforts, China has invested heavily in the 'digital silk road,' deployed the BeiDou satellite for navigation in MENA, and developed telecommunication network in KSA, Bahrain and Egypt among others.[50]

---

[42] IBID.

[43] IBID.

[44] IBID.

[45] IBID.

[46] Rafizadeh (2023). How China became an ideal partner in the Middle East. [online] Arab News. Available at: https://www.arabnews.com/node/2342736 [Accessed 17 Oct. 2023].

[47] IBID.

[48] Vohra (2022). Xi Jinping Has Transformed China's Middle East Policy. [online] Foreign Policy. Available at: https://foreignpolicy.com/2022/02/01/xi-jinping-has-transformed-chinas-middle-east-policy/ [Accessed 8 Oct. 2023].

[49] Rafizadeh (2023). How China became an ideal partner in the Middle East. [online] Arab News. Available at: https://www.arabnews.com/node/2342736 [Accessed 17 Oct. 2023].

[50] Al-Tamimi et al. (2019). China's great game in the Middle East. ECFR. https://ecfr.eu/publication/china_great_game_middle_east/#from-economic-interests-to-political-and-security-engagement [Accessed 23 Sep. 2023].

These strategic investments across the MENA region have proved instrumental in China entrenching its influence and developing its 'soft power' in the region. Recently, Chinese firms were selected to build Lusail Stadium for the Qatar World Cup, KSA Yanbu refinery, and a high-speed rail connected Mecca and Medina.[51] Additionally, China has been active in developing ports in Arabian Sea and Red Sea as well as industrial parks, all to increase economic connectivity to the BRI and "…create an economic chain that links China to the Gulf, the Arabian Sea, the Red Sea, and the Mediterranean."[52] Given the increased investments in the region and their importance to the BRI, China finds it integral to enhance and develop regional stability and security through mediation and diplomacy, engraining longer term relations in the region.

3.3 Mediation and Diplomacy

Historically, China has been risk averse when involving itself in MENA, traditionally following a 'Zero-Enemy' and more recently a 'Four-No Foreign Policy' seeking more a non-interventionist role in the region's politics.[53] The core principles include "…non-alignment, non-interference in the internal affairs of other states, no proxies, and no filling in the power vacuum left by other countries" which can be seen as the inverse to the US approach over the decades in MENA.[54] It may be seen as a welcome sign by the region, however the world it may see it as a signal of Chinese non-committal to a strict policy but rather pure state interest optimization as they remain flexible to engage with little accountability.

Regardless, China's main concern is maintaining good relations with the GCC and MENA countries in general, seeking to use conflict resolution through diplomacy and mediation as another form of 'soft power.' China remains selective in what is chooses to get involved in, always emphasizing impartiality, and has had varying levels of success.[55]

As a result of mainly deriving their influence from 'soft power,' China's diplomatic impact is still relatively limited thus the focus is on four main engagement types: "multifaceted intervention" where they have the influence to do so, "proactive engagement" in longer running disputes, "limited intercession" where the conflict is still on going or there are many players, and lastly "indirect participation" through entities like the UN and other IGOs.[56]

---

[51] IBID.

[52] IBID.

[53] Sun (2022). China's 'Zero-Enemy Policy' in the Gulf. [online] Research Gate. Available at: https://www.researchgate.net/publication/365465988_China's_'Zero-Enemy_Policy'_in_the_Gulf [Accessed 2 Nov. 2023].

[54] IBID.

[55] IBID.

[56] IBID.

'Multifaceted intervention' can be characterized by China's diplomatic intervention Sudan and South Sudan where they have large investments and thus subsequent increased political influence in the countries. Despite relations dating back to the 1990s, Chinese diplomatic 'soft-power' influence began being examined during South Sudan's Civil War in 2013.[57] Here is where they began to trial different 'multifaceted intervention' diplomatic tactics, partially because it served as a point of national pride, as this is where was their first successful overseas oil investment was, and its geographic distance from Asia meant less scrutiny from domestic or regional entities.[58] The IGAD sponsored diplomatic efforts to resolve the civil war saw China, "...consistently engaged and supportive of the mediation process..." until a peace agreement was signed in August 2015.[59] This early success encouraged China to treat South Sudan as "...a real-world laboratory to test the boundaries of its non-interference principle..." fine tuning tactics to be used later on in other regions or conflicts. Part of these initial efforts included focusing on agenda setting or "place setting" to bring competing parties together or attempts of 'persuasion not punishment' which could be seen as a beta-version of the current MENA Policy.[60, 61] Despite the differing approaches, the lack of a full commitment to regional mediation for lasting outcomes, Chinese engagement on Sudan or South Sudan is still seen to be "...driven by self-interest although to a lesser degree [than other major powers]."[62]

Longer term or protracted issues such as solutions for Palestine has seen China utilize 'proactive engagement' in which they pursue consistent diplomatic mediation efforts in attempts to achieve lasting resolutions. China has consistently supported a two-state solution policy. In 2014 Xi Jinping launched the Four-Point Peace initiative and as recently as 2023, Xi Jinping proposed another three-point solution revolving around the establishment of a fully sovereign Palestine, economic revitalization, and addressing on going threats to religious and historical sites in occupied Palestine as well as signing a strategic partnership with Abbas.[63] [64] In the 2024 Israeli bombardment of Gaza, China has been seen as sympathetic to the Gazans and Palestinian

---

[57] International Crisis Group (2017). China's Foreign Policy Experiment in South Sudan. [online] Crisis Group. Available at: https://www.crisisgroup.org/africa/horn-africa/south-sudan/288-china-s-foreign-policy-experiment-south-sudan [Accessed 28 Sep. 2023].

[58] IBID.

[59] IBID.

[60] IBID.

[61] IBID.

[62] IBID.

[63] Jordan News (2023). Chinese President Xi Jinping proposes 3-point plan to settle Palestinian issue - Jordan News | Latest News from Jordan, MENA. [online] Jordan News | Latest News from Jordan, MENA. Available at: https://www.jordannews.jo/Section-20/Middle-East/Chinese-President-Xi-Jinping-proposes-3-point-plan-to-settle-Palestinian-issue-29241 [Accessed 14 Sep. 2023].

[64] Yahoo Finance (2023). China's Xi meets Palestinian leader Abbas, offers 3-point proposal for 'lasting solution' to conflict with Israel. [online] Yahoo Finance. Available at: https://finance.yahoo.com/news/china-pledges-support-peaceful-resolution-093000955.html [Accessed 1 Nov. 2023].

cause labeling, "...the Israeli response as 'collective punishment against civilians.'"[65] In a signal to Arab and Muslim nations of their commitment, they hosted high level meetings to discuss solutions to the Gaza war and "...challenge to America's strong support for Israel."[66] They also have used their platform at the UN to support various ceasefire legislation, while of course vetoing the US sponsored one. Despite these efforts nothing tangible has occurred. Even in the face of a UN cease fire resolution passing, at the time of this writing, the bombardment of Gaza continues. Critics argue that China's efforts are largely symbolic and are attempts exploit the war in Gaza to bolster their image in the Arab and Muslim world or to be seen as standing up for the 'global south' and to challenging the US on a world stage.[67]

'Limited intercession' tactics are generally reserved for multilateral diplomatic issues, in which China may have interests on both sides of a problem but is yet to establish substantial influence in the country or is acting with another nation.[68] During the Syrian Civil war they often exercised diplomatic 'soft power' with Russia in the United Nation Security Council on behalf of the Assad regime often against the US. They also tried to increase ties with the Syrian government having been one of the first countries involved in reconstruction efforts.[69]

Another deployment of this type of diplomatic 'soft power' was in the unexpected KSA-Iran deal. After decades of competition over regional hegemony between KSA and Iran, tensions seemed to be at an all-time high with the two engaged in a proxy war in Yemen. Iran backed the Houthi militias while the US backed KSA and UAE who intervened on the side of the Hadi government in a conflict that has raged for 9 years and destroyed large parts of Yemen.[70] Despite this, there had speculation KSA and Iran may want to restore relations. However, little to no progress had been accomplished on the mutual concessions each other demanded until China stepped in and was "...able to provide assurances to both sides."[71] The China led Joint Trilateral Statement, is the deal outlining the restoration of relations between KSA and Iran

---

[65] Jallo (2024). China Exploits the 'War on Gaza' in its Power Struggle with Washington | Wilson Center. [online] www.wilsoncenter.org. Available at: https://www.wilsoncenter.org/article/china-exploits-war-gaza-its-power-struggle-washington [Accessed 1 Feb. 2024].

[66] Brar (2023). China holds Gaza meeting in challenge to U.S. [online] Newsweek. Available at: https://www.newsweek.com/china-xi-jinping-israel-palestinian-conflict-arab-islamic-peace-talks-1845074 [Accessed 27 Nov. 2023].

[67] IBID.

[68] Al-Tamimi et al. (2019). China's great game in the Middle East. ECFR. https://ecfr.eu/publication/china_great_game_middle_east/#from-economic-interests-to-political-and-security-engagement [Accessed 23 Sep. 2023].

[69] Ergenc (2018). China's relations with the Middle East: a perspective from the region | IIAS. [online] www.iias.asia. Available at: https://www.iias.asia/the-newsletter/article/chinas-relations-middle-east-perspective-region [Accessed 25 Sep. 2023].

[70] Krause and Parker (2020). Yemen's Proxy Wars Explained | MIT Center for International Studies. [online] Mit.edu. Available at: https://cis.mit.edu/publications/analysis-opinion/2020/yemens-proxy-wars-explained [Accessed 14 Oct. 2023].

[71] Esfandiary and Jacobs (2023). How Beijing Helped Riyadh and Tehran Reach a Detente. [online] www.crisisgroup.org. Available at: https://www.crisisgroup.org/middle-east-north-africa-saudi-arabia-iran-china/how-beijing-helped-riyadh-and-tehran-reach-detente [Accessed 22 Sep. 2023].

integrating previous agreements from 1998 and 2001 between two countries.[72] In its 'non-interference' fashion, the deal also outlined a two-month communication period in which the countries would engage with each other, at times including or excluding China. While the restoration of relations is rightfully lauded as "...a significant achievement..." the deal still lacks the robust qualities of more substantial binding international agreements. Perhaps like certain other past examples, China seems to be ever willing to 'set the table' for the guests, but not ensure they 'complete the meal.' This could be attributed to a not fully mature 'soft power' apparatus to impose agreements on nations coupled with their desire to main a minimal intervention as possible.

China also seeks to utilize diplomatic 'soft power' through 'indirect participation' in the UN and UN led efforts.[73] As previously noted, China does not shy away from using their veto power in the UN to pass/block legislation favorable to powers they seek to gain influence with like with Syria, Iran and in Gaza. They are also trying to get more involved in security-oriented UN efforts to begin training their peacekeeping forces as well as increase security presence in critical regions in the area.[74] For example, helping with the maritime security efforts to remove chemical weapons in Syria or more recently pressing Iran, on behalf of UK and other western powers to help rein Houthi attacks on vessels.[75] [76]

Despite their seeming 'non-interventionist' policy and mixed results, China is engaging all over the region from conflict resolution and reconstruction to security engagements and building multilateral relations. Utilizing the ties forged through trade and economic cooperation, China may be seen as a reliable mediator in region, again subtly subverting US influence in the region. This presents a further opportunity for China to formalize these 'good will' intentions through formal political partnerships.

3.4 Political Partnerships

To further entrench relations with countries in MENA, China has proliferated its political partnerships in the region. Components of these engagements begin to mesh 'soft power' and the infancy of 'hard power' military deployments. To date China

[72] IBID.

[73] IBID.

[74] AP (2021). China touts role in UN peacekeeping, Middle East peace. [online] AP News. Available at: https://apnews.com/article/united-nations-middle-east-china-government-and-politics-6014343ada529a978436dec343c1f04e [Accessed 12 Sep. 2023].

[75] Shah (2014). Chinese ship arrives to help in removal of Syrian chemical weapons materials. [online] CNN. Available at: https://www.cnn.com/2014/01/08/world/asia/china-syria-chemical/index.html [Accessed 17 Oct. 2023].

[76] Hafezi and Hayley (2024). Exclusive: China presses Iran to rein in Houthi attacks in Red Sea, sources say. [online] Reuters. Available at: https://www.reuters.com/world/middle-east/china-presses-iran-rein-houthi-attacks-red-sea-sources-say-2024-01-26/ [Accessed 13 Feb. 2024].

has 15 + political partnerships that include anti-piracy, maritime security missions in the region, and rescue operations.[77] They now have their first overseas military base in Djibouti, which is incredibly close to Bab El Mandeb in Yemen.

Historically, China has been "...extremely careful not to become too involved, still believing that the US can take responsibility for managing security in the region..." despite this, in 2019 given tensions in the Strait of Hormuz, Chinese Ambassador to UAE conceded that "...China might participate in maritime security operations in the strait."[78] While this may signal a departure from its purely 'soft power' approach, China is still prefers a 'non-intervention' approach for the time being.

Following the 'soft power' entrenchment trajectory in the region, China initially establishes large and growing trade relations, then enhanced economic integration, followed by becoming diplomatic cover and mediation support to regimes in the region, and finally solidify relations through robust political partnerships incorporating security components.

4 Goals of Chinese 'Soft Power'

China's relatively recent acceleration of 'soft power' engagement in the MENA region serves to help accomplish their larger strategic national goals in the area. The goals relevant to China in MENA include; safeguarding trade as well as the economic expansion of the BRI, stabilization and predictable relations, promotion of multipolarity, and lastly, increase diplomatic prowess and support of its principles of state sovereignty and territorial integrity.[79]

Safeguarding access to trade routes and its extensive economic investment in MENA is vital to maintain stable growth for China. As mentioned previously, disruptions in trade or commodity flows because of geo-political instability greatly increases uncertainty in long-term strategic planning especially with high-risk projects such as the BRI. Hence 'soft power' efforts, like diplomacy or economic stabilization in MENA, are used to maintain favorable access to trade, increase influence, and secure BRI investments.[80]

China's 'soft power' deployment in MENA also seeks stabilization of the region and mending of relations between great powers while also promoting multipolarity. Through its 'non-interventionist' diplomatic engagement preference outlined above, its adoption of developmental peace, or positioning itself as a would-be impartial mediator, China is making moderate progress. As evidenced by South-Sudan ceasefire negotiations, KSA-Iran deal, or trying to leverage Iran to rein in Houthi attacks in

[77] Al-Tamimi et al. (2019). China's great game in the Middle East. ECFR. https://ecfr.eu/publication/china_great_game_middle_east/#from-economic-interests-to-political-and-security-engagement [Accessed 23 Sep. 2023].

[78] IBID.

[79] IBID.

[80] IBID.

Bab El Mandeb, Chinese 'soft power' influence is being deployed throughout region in service of this goal. The Joint Trilateral Statement between KSA-Iran referenced earlier also goes a long way in promoting stability, good relations between powers, and multipolarity in the region China seeks.[81] By stabilizing the relation between KSA and Iran, Wright states, "... it paves the way for [both] to potentially becomes a part of the BRICS group, serving as a counter weight to the G7," which further serves as a way for China to challenge the US and promote its own economic and political system.[82] This goes in line with Xi Jinping's long standing desire to, "...strengthen the bloc's voice on the world stage," and as way to counter US influence.[83] By bolstering MENA countries stability and relations while expanding multipolarity, China is hoping to receive higher levels of diplomatic support globally, especially when it comes to contentious issues Xinjiang, Hong-Kong, or South China Sea.

The fostering of a larger global diplomatic bloc in MENA will also aid China in its larger strategic national goal as it pertains to their idea of support of 'state sovereignty and territorial integrity.'[84] China has long advocated for the One China Principle, desired to make larger territorial claims in the South China Sea, or find diplomatic cover for actions in Xinjiang.[85] [86] Given the mutual authoritarianism, and a preference for a development over democracy approach, China sees the MENA countries as excellent candidates for bloc expansion and enhanced diplomatic cover on the world stage.

As mentioned above, China has already seen the benefits of this added support across various issues. There has been a concise and punctuated effort to pursue these goals throughout the region to varying degrees of success. Given the longer-term nature of Chinese strategic planning, it is only expected that China actively continues to look to the MENA region for trade, economic expansion, and diplomatic support.

---

[81] Wright (2023). China's emergence as Saudi-Iran broker signals multipolar diplomacy in Middle East - Al-Monitor: Independent, trusted coverage of the Middle East. [online] www.al-monitor.com. Available at: https://www.al-monitor.com/originals/2023/04/chinas-emergence-saudi-iran-broker-signals-multipolar-diplomacy-middle-east [Accessed 3 Oct. 2023].

[82] IBID.

[83] Hawkins (2023). Xi Jinping's wants a 'multipolar world', as China accelerates its shift away from the west. The Guardian. [online] 8 Oct. Available at: https://www.theguardian.com/world/2023/oct/09/xi-jinpings-wants-a-multipolar-world-as-china-accelerates-its-shift-away-from-the-west [Accessed 1 Dec. 2023].

[84] Al-Tamimi et al. (2019). China's great game in the Middle East. ECFR. https://ecfr.eu/publication/china_great_game_middle_east/#from-economic-interests-to-political-and-security-engagement [Accessed 23 Sep. 2023].

[85] IBID.

[86] US Taiwan Business Council (2022). The U.S. 'One China Policy' vs. the PRC "One China Principle. [online] US-Taiwan Business Council. Available at: https://www.us-taiwan.org/resources/faq-the-united-states-one-china-policy-is-not-the-same-as-the-prc-one-china-principle/ [Accessed 23 Oct. 2023].

5 Is China's 'Soft Power' Working in MENA?

The concerted effort into MENA has begun to show moderate results for China, especially in a region that has recently been firmly under US influence. This is clearly exhibited by its ever-expanding trade agreements, buy-in from MENA region countries into the BRI, acceptance of Chinese investments and MOUs to expand into different sectors, use of China as a mediator, or increased diplomatic support of China by MENA countries on a world stage.[87]

In economic and trade terms, China has seen 'soft power' policies pay dividends. As mentioned previously, trade between China and region has doubled in the last decade and Chinese firms are being selected to build many large-scale infrastructure projects.[88] In addition to this, China has also received large scale investments from MENA. Propersity7 Ventures, a KSA fund, has allocated half of its $1bn portfolio to China while also backing 20 different startups, or for example UAE's Mubadala fund co-leading a $2bn investment round for Chinese fast fashion brand Shien.[89] Further, UAE and Chinese giant Huawei have been cooperating on cyber security, despite UAE's close work with the US in intelligence and counterterrorism.[90] Additionally, China has used is 'soft power' to increase arms sales as it aims to displace Russia as the second largest arms exporter to MENA, in the last decade sales rose 80% to the region.[91] BRI projects are well underway, whether expansion of the SC, investment economic zones around the BRI or vast improvements to ports and infrastructure used to link back to China throughout the region.

Diplomatically China has been able to use 'non-interventionist' policy to increase its role as a regional mediator to stabilize certain conflicts and in return received, in certain cases, substantial diplomatic support. As previously mentioned, many MENA countries signed a letter supporting actions by China taken in Xinjiang in exchange for added investment and help in bolstering their own authoritarian regimes.[92] Another notable example was in 2016, during issues arising in the South China Sea, the Arab League issued the Doha Declaration in support of China's preferred bilateral dispute

[87] Singer (2024). How China is winning the Middle East. [online] Defense One. Available at: https://www.defenseone.com/ideas/2024/01/how-china-winning-middle-east/393483/ [Accessed 1 Feb. 2024].

[88] Xiang (2024). Tech War Crossfire: Middle Eastern Companies May Be Forced To Choose Between China And The U.S. Forbes. [online] 12 Mar. Available at: https://www.forbes.com/sites/ninaxiang/2024/03/12/tech-war-crossfire-middle-eastern-companies-may-be-forced-to-choose-between-china-and-the-us/?sh=36f018394751 [Accessed 14 Mar. 2024].

[89] IBID.

[90] Kaskanis (2021). China and the UAE: Cybersecurity Cooperation. [online] Tactics Institute for Security and Counter-Terrorism. Available at: https://tacticsinstitute.com/middle-east/uae-digital-oasis-or-cyber-oppresion-hub/ [Accessed 2 Feb. 2024].

[91] IBID.

[92] Thomas (2022). Analysis | Why Muslim countries in the Middle East support Chinese atrocities in Xinjiang. [online] The Diplomatic Pouch. Available at: https://medium.com/the-diplomatic-pouch/analysis-why-muslim-countries-in-the-middle-east-support-chinese-atrocities-in-xinjiang-f4ec7d4bea48 [Accessed 4 Jan. 2024].

resolution mechanism.[93] Having this sort of robust diplomatic support to revolve issues bilaterally, without external intervention, is invaluable given the importance of South China Sea to China's economic interests and self-perception of 'territorial integrity.'

Given the US is reducing its emphasis on expanding relations beyond the status quo with MENA to a certain extent, China may see this as a time to continue making in roads and extracting benefits. MENA countries may also continue to seek benefits of this arrangement with China as long as there is a perceived lack of US interest in making meaningful investments, or in changing is reflexive military intervention policy in the region. However, as the influence China grows and the MENA region gets further entangled economically and diplomatically, US will have to apply pressure to MENA and perhaps eventually have them chose between east or west or see their influence erode further.[94]

6 Is US Policy to Blame and Can There Be a Fix?

The US should not be surprised that China is increasing its influence through 'soft power' in MENA. Despite the US' insistent perpetuation of a global 'rules-based order' in general, MENA seems to have fallen outside of that directive. As Rad stipulates, US policy, "...can only work if the rules within this so-called 'rules-based order' are applied consistently and fairly across the board," and given past conflicts in Iraq and Afghanistan and the more recent US complete support of the Israeli bombardment of Gaza, there seems to an asymmetric application of this principle.[95] Given US policy over the last few decades and its continuation until today, China has been perceived as a potential alternative for MENA countries as they seek new avenues to growth, conflict resolution, and attempts at regional stability.

In a 2022 White House Brief, cooperation policy with MENA was outlined which on its face seemed to have begun to address some on the underlying deficiencies in the US approach to MENA and potentially begin to counter Chinese 'soft power' influence. Briefly, the core tenants of the cooperation policy can be characterized by; increasing partnerships with nations that subscribe to the 'rules-based order,' deterrence to ensure freedom of navigation of trade in the region, diplomatic engagement for conflict resolution, political and economic integration while respecting 'state sovereignty' and lastly promotion values enshrined in the UN charter.[96]

---

[93] IBID.

[94] IBID.

[95] Rad (2023). The US and its faux 'rules-based order' | Responsible Statecraft. [online] responsiblestatecraft.org. Available at: https://responsiblestatecraft.org/2023/07/12/the-us-and-its-faux-rules-based-order/ [Accessed 2 Nov. 2023].

[96] The White House (2022). FACT SHEET: The United States Strengthens Cooperation with Middle East Partners to Address 21st Century Challenges. [online] The White House. Available

In certain, very narrow circumstances, some of these policies have been implemented. GCC will invest $3bn in U.S. Partnership for Global Infrastructure and Investment, which will tackle climate change issues, develop IT infrastructure, address gender inequality, and upgrade health systems. There was also $200 m raised for Palestinian hospital networks and the US driven agreement between Iraq, Jordan, KSA and Kuwait for affordable energy infrastructure in Iraq was signed, which in turn further integrates Iraq into the GCC. The briefing also discussed additional maritime security efforts to ensure the SC, Bab El Mandeb, and the Strait of Hormuz remain free flowing.[97]

While on the face it, these policies may serve as counterweight to China's 'soft power' influence in MENA, they pale in comparison to actual US action in the region. There seemed to be a genuine change in approach to MENA, however in what is now widely deemed as a massive policy failure, unconditional support of Israel has greatly jeopardized US relations in MENA and opened the door for enhanced China-MENA influence in the region. As Heer states, "[t]he bear-hug strategy has failed in the most direct way possible: Far from being restrained, Israel is fighting one of the most ferociously murderous wars in the 21st century."[98] The policy eviscerates US moral standing in the region and allows China to exploit the war for their benefit as seen by their meetings with MENA members over the topic and support of every ceasefire resolution in the UN outside the US one—bolstering the image of China in the region.[99] In a calculated and methodical execution, the US has been backing Israel's action which many have characterized as "…a pattern of war crimes and violations of international law."[100] The 'rules-based order' the US seeks to uphold is now seen as hypocritical and is evaporating. As Callamard, remarks:

> The risk of genocide, the gravity of the violations being committed, and the flimsy justifications by elected officials in Western democracies warn of a change of eras. The rules-based order that has governed international affairs since the end of World War II is on its way out, and there may be no turning back.[101]

For MENA countries, who for years have been criticized for their own human rights violations, may be skeptical at the distorted application in the specific case. Many came out and condemned Hamas attacks, however, are perplexed at the unflinching

at: https://www.whitehouse.gov/briefing-room/statements-releases/2022/07/16/fact-sheet-the-united-states-strengthens-cooperation-with-middle-east-partners-to-address-21st-century-challenges/ [Accessed 5 Oct. 2023].

97 IBID.

98 Heer (2023). Biden's Bear Hug of Netanyahu Is a Disaster. [online] www.thenation.com. Available at: https://www.thenation.com/article/world/biden-netanyahu-bear-hug-disaster/ [Accessed 8 Dec. 2023].

99 Brar (2023). China holds Gaza meeting in challenge to U.S. [online] Newsweek. Available at: https://www.newsweek.com/china-xi-jinping-israel-palestinian-conflict-arab-islamic-peace-talks-1845074 [Accessed 27 Nov. 2023].

100 Callamard (2024). Gaza and the End of the Rules-Based Order. [online] Foreign Affairs. Available at: https://www.foreignaffairs.com/israel/gaza-and-end-rules-based-order [Accessed 21 Feb. 2024].

101 IBID.

US backing of the brutality unfolding in Gaza.[102] The fallout from this has been swift in MENA.

In the MENA region, fears the humanitarian crisis in Gaza will lead to permeant damage to that population as well as added regional instability are rising.[103] According to polling data, only about 7% of the Arab public see America's role as positive, while around 36% think China is having a positive impact.[104] In Jordan, a long time US ally, only 3% of respondents trust America compared to 25% for China.[105] This level of public disapproval is having destabilizing effects in the country, who has a large Palestinian diaspora and a population generally sympathetic to the Palestinian cause. There has been large scale, continuous demonstrations in Jordan calling for the end of their peace treaty with Israel.[106] Jordan, like other regimes in MENA, may be forced to contend with public opinion or risk wide scale challenges to regime authority—potentially at the expense of relations with the US. KSA has paused normalization efforts with Israel and the Houthis have begun disrupting trade in the red sea as a direct result of the Gaza war—costing the global economy billions.[107] Many analysts argue that the images of the war proliferating around MENA will serve as recruitment material, further increasing regional instability and threatening US interests.[108] Further as mentioned earlier, this is already causing MENA nations to adopt more pro-China positions risking the region falling deeper under Chinese influence.[109] Essentially going against every one of the points outlined in the cooperation brief mentioned above, and in certain circumstances the US themselves are accelerating the push of MENA towards China.

Despite this, there are several ways that the United States can mitigate the influence that China exercises in the MENA. Perhaps most critically, is a reexamination of its reflexive military intervention policy, and an American grand strategy that

---

[102] Waldo et al. (2023). International Reactions to the Hamas Attack on Israel. [online] The Washington Institute. Available at: https://www.washingtoninstitute.org/policy-analysis/international-reactions-hamas-attack-israel [Accessed 1 Nov. 2023].

[103] Fandy (2024). Radicalization and Regional Instability: Effects of the Gaza War. [online] The Cairo Review of Global Affairs. Available at: https://www.thecairoreview.com/essays/radicalization-and-regional-instability-effects-of-the-gaza-war/ [Accessed 14 Mar. 2024].

[104] Dagher and Kaltenthaler (2023). The United States Is Rapidly Losing Arab Hearts and Minds Through Gaza War, While Competitors Benefit | The Washington Institute. [online] www.washingtoninstitute.org. Available at: https://www.washingtoninstitute.org/policy-analysis/united-states-rapidly-losing-arab-hearts-and-minds-through-gaza-war-while [Accessed 2 Dec. 2023].

[105] IBID.

[106] Al-Khalidi (2024). Jordanians protest against peace treaty with Israel in fresh rallies. [online] Reuters. Available at: https://www.reuters.com/world/middle-east/jordanians-protest-against-peace-treaty-with-israel-fresh-rallies-2024-03-29/ [Accessed 1 Apr. 2024].

[107] Dagher and Macdonald (2023). Saudis Put Israel Normalization on Hold in Blow to US Goals. Bloomberg.com. [online] 13 Oct. Available at: https://www.bloomberg.com/news/articles/2023-10-13/saudis-put-israel-normalization-on-hold-in-blow-to-us-goals [Accessed 22 Oct. 2024].

[108] Smith (2023). Israel's bombardment of Gaza could be potent Hamas recruitment tool, experts warn. [online] NBC News. Available at: https://www.nbcnews.com/news/world/israels-bombardment-gaza-hamas-recruitment-tool-rcna123562 [Accessed 21 Nov. 2023].

[109] IBID.

reprioritizes the MENA region as a potential partner for higher levels economic and diplomatic cooperation as opposed region primarily engaged through the lens military interventions, weapons transfers, counter terrorism operations, or resource extraction.

The MENA region is desperate for economic investment as many hydrocarbon-based economies are in the process of diversification. The US should identify ways to engage on these aspects, not just to satisfy MENA desires, but to counter BRI investments.[110] The US also should lean more into cooperative engagements that seek avenues for longer term collaboration not just in security arrangements but in diplomacy and economic affairs. As Elmali argues, by providing long term, strategic diplomatic and economic alternatives to China this may, "…preempt the [MENA] nations' inclination to diversify their trade partners."[111]

Further, diplomatically, the US needs to begin recognizing MENA concerns for example on issues such as Gaza. While mostly symbolic, China has been perceived to care about the situation and hence increased its influence with MENA countries. To combat this US should consider a policy rework or risk driving MENA further to China. Lastly, US should confront China's reliance on US maritime security in the region. As mentioned previously, China largely enjoys free-rider status, only responsible for a fraction of maritime security when compared with the benefits they obtain. China benefits as much as the US from stable waterways. The US could potentially extract concession from China through their relationship with Iran. In exchange for US maritime security, China "…can hold the Iranian government accountable and compliant," through its 'soft power' and aid in Strait of Hormuz stability without necessarily having China deploy navy to region.[112] Securing trade through US 'hard power' and Chinese 'soft power' may be an area of collaboration between them, reducing competition and tension in MENA.

7 Conclusion

Historically, China was risk averse to expanding relations and its influence in MENA. However, an insistent need for cheap energy, economic expansion for the BRI, and desire for increased international diplomatic support in service of Chinese national

[110] Elmali (2023). How the U.S. Can Respond to China's Engagement in the Middle East. [online] New Lines Institute. Available at: https://newlinesinstitute.org/strategic-competition/how-the-u-s-can-respond-to-chinas-engagement-in-the-middle-east/#:~:text=Accordingly%2C%20the%20U.S.%20response%20should [Accessed 4 Dec. 2023].

[111] IBID.

[112] Sancho (2023). A US Perspective: The Effect of China's Growing Presence in the Middle East—Michigan Journal of Economics. [online] Michigan Journal of Economics. Available at: https://sites.lsa.umich.edu/mje/2023/11/08/a-us-perspective-the-effect-of-chinas-growing-presence-in-the-middle-east/ [Accessed 4 Feb. 2024].

strategic goals required China to create a robust 'soft power' strategy in MENA.[113] As China moved from a rising power to global power, MENA served as an ideal location for influence expansion. The various forms of 'soft power' influence whether, trade, economic integration, diplomacy and mediation or political partnerships have produced moderate success in building influence in MENA. Whether on infrastructure projects, fund investments, or even diplomatic cover internationally—China has begun to utilize this influence for increased benefits and entrench relations in MENA.

Despite obvious benefits, China's 'soft power' influence still has varying degrees of success. They are keen to make more in roads especially economically and diplomatically in MENA but are wary of directly challenging US, namely militarily, as they are largely benefitting from the status quo of the security arrangement. Thus Chinese 'soft power' tools and hence influence are directly incentivized and constrained by the risks and opportunities given the current MENA framework. This may change as Chinese interests becomes more entrenched in the region.

MENA also finds itself in a promising yet difficult position. While increased economic and diplomatic relations with China may aid in economic diversification, promotion of regime stability, or provide opportunities to youth in MENA, it may come at expense of increased tension with the US. The US still maintains predominance in the region and may eventually force MENA to choose sides.

The increased rise in importance of and influence in MENA by China should make the US consider pivoting its policy. A continuation of a reflexive military intervention and a lopsided application of a 'rules-based order' may see the influence the US built in the region over decades evaporate. The current US policy may be the pushing MENA into the economic and diplomatic pull of China.

The power still resides with the US, who can maintain current engagement with MENA which may lead to more regional instability. Conversely, policies can be adopted by the US in MENA preferring economic and diplomatic cooperation, equal application of human rights and 'rules-based order' to increase influence and incentivize MENA countries into developing better relations. This can potentially stave of Chinese influence in the region and maintain US interests in the MENA.

References

Ahmad Khan S (2021) China's increasing influence in the Middle East. [online] E-international relations. Available at: https://www.e-ir.info/2021/09/20/chinas-increasing-influence-in-the-middle-east. Accessed 1 Oct 2023

Aizhu C, Lawler A (2022) China buys more Iranian oil now than it did before sanctions, data shows. Reuters. [online] 1 Mar. Available at: https://www.reuters.com/world/china/china-buys-more-iranian-oil-now-than-it-did-before-sanctions-data-shows-2022-03-01/

---

[113] Al-Tamimi et al. (2019). China's great game in the Middle East. ECFR. https://ecfr.eu/publication/china_great_game_middle_east/#from-economic-interests-to-political-and-security-engagement [Accessed 23 Sep. 2023].

Al-Anani K (2023) Egypt's strategic partnership with China: opportunities and implications. [online] Arab Center Washington DC. Available at: https://arabcenterdc.org/resource/egypts-strategic-partnership-with-china-opportunities-and-implications/. Accessed 27 Sept 2023

Al-Khalidi S (2024) Jordanians protest against peace treaty with Israel in fresh rallies. [online] Reuters. Available at: https://www.reuters.com/world/middle-east/jordanians-protest-against-peace-treaty-with-israel-fresh-rallies-2024-03-29/. Accessed 1 April 2024

Al-Tamimi JF, Lons C, Sun D, Naser (2019) China's great game in the Middle East. ECFR. https://ecfr.eu/publication/china_great_game_middle_east/#from-economic-interests-to-political-and-security-engagement. Accessed 23 Sept 2023

AP (2021) China touts role in UN peacekeeping, Middle East peace. [online] AP News. Available at: https://apnews.com/article/united-nations-middle-east-china-government-and-politics-6014343ada529a978436dec343c1f04e. Accessed 12 Sept 2023

Barber R (2019) Syria: the disgraceful stain left by the UN Security Council veto. [online] Lowyinstitute.org. Available at: https://www.lowyinstitute.org/the-interpreter/syria-disgraceful-stain-left-un-security-council-veto. Accessed 3 Oct 2023

Brar A (2023) China holds Gaza meeting in challenge to U.S. [online] Newsweek. Available at: https://www.newsweek.com/china-xi-jinping-israel-palestinian-conflict-arab-islamic-peace-talks-1845074. Accessed 27 Nov 2023

Callamard A (2024) Gaza and the end of the rules-based order. [online] Foreign Affairs. Available at: https://www.foreignaffairs.com/israel/gaza-and-end-rules-based-order. Accessed 21 Feb 2024

Cooban A (2023) US has replaced Russia as Europe's top crude oil supplier. CNN. [online] 28 Mar. Available at: https://www.cnn.com/2023/03/28/energy/eu-us-oil-imports-overtake-russia/index.html. Accessed 14 Oct 2023

Dagher M, Kaltenthaler K (2023) The United States is rapidly losing Arab hearts and minds through Gaza War, While Competitors Benefit | The Washington Institute. [online] www.washingtoninstitute.org. Available at: https://www.washingtoninstitute.org/policy-analysis/united-states-rapidly-losing-arab-hearts-and-minds-through-gaza-war-while. Accessed 2 Dec 2023

Dagher S, Macdonald F (2023) Saudis Put Israel Normalization on Hold in Blow to US Goals. Bloomberg.com. [online] 13 Oct. Available at: https://www.bloomberg.com/news/articles/2023-10-13/saudis-put-israel-normalization-on-hold-in-blow-to-us-goals. Accessed 22 Oct 2024

Dunn C, Barden J (2023) The strait of Hormuz is the world's most important oil transit chokepoint. [online] www.eia.gov. Available at: https://www.eia.gov/todayinenergy/detail.php?id=61002#:~:text=the%20Arabian%20Sea. Accessed 29 Nov 2023

Elmali B (2023) How the U.S. Can Respond to China's Engagement in the Middle East. [online] New Lines Institute. Available at: https://newlinesinstitute.org/strategic-competition/how-the-u-s-can-respond-to-chinas-engagement-in-the-middle-east/#:~:text=Accordingly%2C%20the%20U.S.%20response%20should. Accessed 4 Dec 2023

Ergenc C (2018) China's relations with the Middle East: a perspective from the region | IIAS. [online] www.iias.asia. Available at: https://www.iias.asia/the-newsletter/article/chinas-relations-middle-east-perspective-region. Accessed 25 Sept 2023

Esfandiary D, Jacobs A (2023) How Beijing helped riyadh and tehran reach a detente. [online] www.crisisgroup.org. Available at: https://www.crisisgroup.org/middle-east-north-africa-saudi-arabia-iran-china/how-beijing-helped-riyadh-and-tehran-reach-detente. Accessed 22 Sept 2023

Fandy M (2024) Radicalization and regional instability: effects of the Gaza War. [online] The Cairo Review of Global Affairs. Available at: https://www.thecairoreview.com/essays/radicalization-and-regional-instability-effects-of-the-gaza-war/. Accessed 14 Mar 2024

Hafezi P, Hayley A (2024) Exclusive: China presses Iran to rein in Houthi attacks in Red Sea, sources say. [online] Reuters. Available at: https://www.reuters.com/world/middle-east/china-presses-iran-rein-houthi-attacks-red-sea-sources-say-2024-01-26/. Accessed 13 Feb 2024

Hawkins A (2023) Xi Jinping's wants a 'multipolar world', as China accelerates its shift away from the west. The Guardian. [online] 8 Oct. Available at: https://www.theguardian.com/world/2023/oct/09/xi-jinpings-wants-a-multipolar-world-as-china-accelerates-its-shift-away-from-the-west. Accessed 1 Dec 2023

Heer J (2023) Biden's Bear Hug of Netanyahu Is a Disaster. [online] www.thenation.com. Available at: https://www.thenation.com/article/world/biden-netanyahu-bear-hug-disaster/. Accessed 8 Dec 2023

Hoffman J (2021) Analysis | Why do some Muslim-majority countries support China's crackdown on Muslims? Washington Post. [online] 4 May. Available at: https://www.washingtonpost.com/politics/2021/05/04/why-do-some-muslim-majority-countries-support-chinas-crackdown-muslims/. Accessed 6 Oct 2023

International Crisis Group (2017) China's foreign policy experiment in South Sudan. [online] Crisis Group. Available at: https://www.crisisgroup.org/africa/horn-africa/south-sudan/288-china-s-foreign-policy-experiment-south-sudan. Accessed 28 Sept 2023

Jallo A (2024) China Exploits the 'War on Gaza' in its Power Struggle with Washington Wilson Center. [online] www.wilsoncenter.org. Available at: https://www.wilsoncenter.org/article/china-exploits-war-gaza-its-power-struggle-washington. Accessed 1 Feb 2024

Jordan News (2023) Chinese President Xi Jinping proposes 3-point plan to settle Palestinian issue—Jordan News | Latest News from Jordan, MENA. [online] Jordan News | Latest News from Jordan, MENA. Available at: https://www.jordannews.jo/Section-20/Middle-East/Chinese-President-Xi-Jinping-proposes-3-point-plan-to-settle-Palestinian-issue-29241. Accessed 14 Sept 2023

Kaskanis A (2021) China and the UAE: cybersecurity cooperation. [online] Tactics Institute for Security and Counter-Terrorism. Available at: https://tacticsinstitute.com/middle-east/uae-digital-oasis-or-cyber-oppresion-hub/. Accessed 2 Feb 2024

Krause P, Parker T (2020) Yemen's Proxy Wars Explained | MIT Center for International Studies. [online] Mit.edu. Available at: https://cis.mit.edu/publications/analysis-opinion/2020/yemens-proxy-wars-explained. Accessed 14 Oct 2023

Kumar R (2024) Why has China recognised Taliban's envoy to Beijing? [online] Al Jazeera. Available at: https://www.aljazeera.com/news/2024/2/14/is-chinas-recognition-of-afghanistan-envoy-a-diplomatic-win-for-taliban. Accessed 16 Feb 2024

Partington R (2024) What is the red sea crisis, and what does it mean for global trade? The Guardian. [online]. Available at: https://www.theguardian.com/world/2024/jan/03/what-is-the-red-sea-crisis-and-what-does-it-mean-for-global-trade#:~:text=Sea%20trade%20route%3F. Accessed 24 Jan 2024

Pletka D (2024) Whatever Happened to Biden's Iran Policy? [online] Foreign Policy. Available at: https://foreignpolicy.com/2024/03/26/bidens-iran-policy-nuclear-deal-jcpoa/. Accessed 27 Mar 2024

Plumer B (2014) China now buys more Middle East oil than the US. [online] Vox. Available at: https://www.vox.com/2014/9/3/6101885/middle-east-now-sells-more-oil-to-china-than-to-the-us. Accessed 22 Sept 2023

Rad A (2023) The US and its faux 'rules-based order' | Responsible Statecraft. [online] responsiblestatecraft.org. Available at: https://responsiblestatecraft.org/2023/07/12/the-us-and-its-faux-rules-based-order/. Accessed 2 Nov 2023

Rafizadeh M (2023) How China became an ideal partner in the Middle East. [online] Arab News. Available at: https://www.arabnews.com/node/2342736. Accessed 17 Oct 2023

Rakha A, El-Aasar K.(2024) The impact of the belt and road initiative on the Suez Canal cargo trade. J Shipping Trade **9**(1). https://doi.org/10.1186/s41072-024-00167-y

Ramachandran S (2016) Iran, China and the Silk road train. [online] thediplomat.com. Available at: https://thediplomat.com/2016/03/iran-china-and-the-silk-road-train/. Accessed 2 Oct 2023

Reuters (2016) First freight train from China arrives in Iran in 'Silk Road' boost: media. [online] Reuters. Available at: https://www.reuters.com/article/idUSKCN0VP0W7/. Accessed 4 Oct 2023

Riedel B (2020) 30 years after our 'endless wars' in the Middle East began, still no end in sight. [online] Brookings. Available at: https://www.brookings.edu/articles/30-years-after-our-endless-wars-in-the-middle-east-began-still-no-end-in-sight/. Accessed 23 Sep 2023

Salameh M (2021) The strategic importance of the Suez canal. [online] www.energymanagementcentre.eu. Available at: https://www.energymanagementcentre.eu/views-on-energy-news/suez-canal/. Accessed 26 Sep 2023
Sancho C (2023) A US Perspective: The effect of China's growing presence in the middle east—Michigan J Econ [online] Michigan Journal of Economics. Available at: https://sites.lsa.umich.edu/mje/2023/11/08/a-us-perspective-the-effect-of-chinas-growing-presence-in-the-middle-east/. Accessed 4 Feb 2024
Shah K (2014) Chinese ship arrives to help in removal of Syrian chemical weapons materials. [online] CNN. Available at: https://www.cnn.com/2014/01/08/world/asia/china-syria-chemical/index.html. Accessed 17 Oct 2023
Shatz H (2024). Middle East–China trade prospects remain robust despite red sea crisis. [online] RAND. Available at: https://www.rand.org/pubs/commentary/2024/02/middle-east-china-trade-prospects-remain-robust-despite.html. Accessed 15 Feb 2024
Singer P (2024) How China is winning the Middle East. [online] Defense one. Available at: https://www.defenseone.com/ideas/2024/01/how-china-winning-middle-east/393483/. Accessed 1 Feb 2024
Smith A (2023) Israel's bombardment of Gaza could be potent Hamas recruitment tool, experts warn. [online] NBC News. Available at: https://www.nbcnews.com/news/world/israels-bombardment-gaza-hamas-recruitment-tool-rcna123562. Accessed 21 Nov 2023
Stewart P, Nichols M, Stewart P (2023) Analysis: China role in Saudi, Iran deal a tricky test for U.S. Reuters. [online]. Available at: https://www.reuters.com/world/china-role-saudi-iran-deal-tricky-test-us-2023-03-10/. Accessed 3 Oct 2023
Sun D (2022) China's 'Zero-Enemy Policy' in the Gulf. [online] Research gate. Available at: https://www.researchgate.net/publication/365465988_China's_'Zero-Enemy_Policy'_in_the_Gulf. Accessed 2 Nov 2023
The White House (2022) FACT SHEET: The United States strengthens cooperation with Middle East Partners to Address 21st century challenges. [online] The White House. Available at: https://www.whitehouse.gov/briefing-room/statements-releases/2022/07/16/fact-sheet-the-united-states-strengthens-cooperation-with-middle-east-partners-to-address-21st-century-challenges/. Accessed 5 Oct 2023
Thomas H (2022) Analysis | Why Muslim countries in the Middle East support Chinese atrocities in Xinjiang. [online] The Diplomatic Pouch. Available at: https://medium.com/the-diplomatic-pouch/analysis-why-muslim-countries-in-the-middle-east-support-chinese-atrocities-in-xinjiang-f4ec7d4bea48. Accessed 4 Jan 2024
US Taiwan Business Council (2022) The U.S. 'One China Policy' versus the PRC One China principle. [online] US-Taiwan business council. Available at: https://www.us-taiwan.org/resources/faq-the-united-states-one-china-policy-is-not-the-same-as-the-prc-one-china-principle/. Accessed 23 Oct 2023
Vohra A (2022) Xi Jinping has transformed China's Middle East policy. [online] Foreign Policy. Available at: https://foreignpolicy.com/2022/02/01/xi-jinping-has-transformed-chinas-middle-east-policy/. Accessed 8 Oct 2023
Waldo C, Epstein G, Hilbush S, Zelin A (2023) International rreactions to the Hamas attack on Israel. [online] The Washington Institute. Available at: https://www.washingtoninstitute.org/policy-analysis/international-reactions-hamas-attack-israel. Accessed 1 Nov 2023
Wikipedia Contributors (2024) Occupation of Iraq (2003–2011). [online] Wikipedia. Available at: https://en.wikipedia.org/wiki/Occupation_of_Iraq_(2003%E2%80%932011)#:~:text=The%20Occupation%20of%20Iraq%20(2003. Accessed 22 Sep 2023
Wright S (2023) China's emergence as Saudi-Iran broker signals multipolar diplomacy in Middle East—Al-monitor: independent, trusted coverage of the Middle East. [online] www.al-monitor.com. Available at: https://www.al-monitor.com/originals/2023/04/chinas-emergence-saudi-iran-broker-signals-multipolar-diplomacy-middle-east. Accessed 3 Oct 2023
Xiang N (2024) Tech war crossfire: Middle Eastern companies may be forced to choose between china and the U.S. Forbes [online]. Available at: https://www.forbes.com/sites/ninaxiang/2024/

03/12/tech-war-crossfire-middle-eastern-companies-may-be-forced-to-choose-between-china-and-the-us/?sh=36f018394751. Accessed 14 Mar 2024
Yahoo Finance (2023) China's Xi meets Palestinian leader Abbas, offers 3-point proposal for 'lasting solution' to conflict with Israel [online] Yahoo Finance. Available at: https://finance.yahoo.com/news/china-pledges-support-peaceful-resolution-093000955.html. Accessed 1 Nov 2023
Zhihua L (2023) Sino trade volumes soar with Middle East, Africa [online] global.chinadaily.com.cn. Available at: https://global.chinadaily.com.cn/a/202307/17/WS64b48a76a31035260b816b72.html. Accessed 11 Oct 2023

Figures

US EIA, Fisher M (2015) EIA website (http://www.eia.gov/), https://www.vox.com/a/maps-explain-the-middle-east [online]. VOX.
Wikipedia, Lommes (2017) https://en.wikipedia.org/wiki/Belt_and_Road_Initiative#/media/File:One-belt-one-road.svg [online]. Wikipedia

China's Economic Statecraft: The Ministry of State Security, Rule by Law, and Cyberspace

Mark Grzegorzewski

Abstract The Chinese Communist Party's (CCP) stability is linked to national economic success. To meet the expectations of China's growing population, the CCP must deliver continuous economic growth. Historical precedence suggests that failure to meet these expectations can lead to social unrest, a scenario the CCP wants to avoid. This is especially true in China's current economic environment with its reliance on an export-driven economy and a relatively weak innovation sector. This backdrop makes completely achieving Chinese economic goals more aspirational than achievable. Therefore, the Chinese security sector must make up this shortfall and support national economic imperatives. The Ministry of State Security (MSS) plays a central role in achieving these national imperatives and supporting the CCP's grip on power. Notably, the MSS has been involved in several high-profile cyber espionage activities as part of China's economic statecraft aimed at fulfilling national economic objectives. This cyber espionage extends to several significant breaches, highlighted by a long-term strategic focus on the United States. The cyber tactics employed by the MSS leverage domestic legislation to support its actions in pursuit of China's economic goals. Thus, the MSS is central in executing China's cyber espionage activities and serving as a critical component of its overall national economic strategy.

1 Introduction

China's national economic development (and national security) go hand in glove with its state-led cyber espionage program. Without sufficient domestic economic growth, the Chinese Communist Party (CCP) will lose legitimacy and the public's

I would like to thank Michael Alexiev and Amara Blanton, my undergraduate research assistants, for their invaluable support in data collection and analysis. Their dedication and hard work greatly contributed to the completion of this chapter.

M. Grzegorzewski (✉)
Embry-Riddle Aeronautical University, Daytona Beach, FL, USA
e-mail: esrucdesrever@yahoo.com

A. Farhadi et al. (eds.), *The Great Power Competition Volume 6*,
https://doi.org/10.1007/978-3-031-70767-4_7

support. The Chinese economy must then continuously grow to meet the demands and expectations of its ever-growing population. As Chinese history has demonstrated,[1] when the public is deprived relative to their expectations, social upheaval can ensue, an outcome the CCP wants to avoid. With the currently frail Chinese economy,[2] and its focus on an export driven economy,[3] not to mention a weak innovation sector,[4] China's 14th 5 year plan's economic goals are more aspirational than realistic. These CCP policy documents guide the Chinese state and economy. Failure to achieve national goals could be perceived domestically as damaging to the CCP's legitimacy. Therefore, the CCP must make up losses elsewhere via cyber espionage as tool of economic statecraft[5] to meet its goals.

Originally, the MSS initially focused on separatism, terrorism, and religious extremism, commonly referred to as the three existential challenges to the Communist Party.[6] Today, the MSS is China's top civilian intelligence, security, and secret police agency responsible for counterintelligence (CI), foreign intelligence (FI), and political security, and plays a significant role in foreign intelligence operations.[7] This broad mandate includes cyber espionage, an area where the MSS has been particularly active and effective.

Signifying the MSS strategic focus on the United States, its affiliated advanced persistent threat (APT) group, APT41, has recently been indicted over exploiting vulnerabilities in internet-facing web applications to infiltrate the commercial networks of six U.S. states from 2021 to 2022.[8] There are several other notable instances of MSS cyberespionage aimed at the U.S. federal government, including

---

1 Britannica, "Taiping Rebellion | Causes, Effects, and Facts," Accessed 29 Apr. 2024, https://www.britannica.com/event/Taiping-Rebellion.

2 Kurtenbach and Soo (2023).

3 Alicia García-Herrero, "China's continued focus on the same export-driven model is bound to hit an impasse," Bruegel, 2 April 2024, https://www.bruegel.org/newsletter/chinas-continued-focus-same-export-driven-model-bound-hit-impasse.

4 International Institute for Strategic Studies, "China's Weak Cyber Defences." IISS, June 2020, https://www.iiss.org/events/2020/06/chinas-weak-cyber-defences/.

5 Andrea Limbago, "One Size Does Not Fit All: The Multifaceted Nature of Cyber Statecraft," Joint Forces Quarterly, 1 July, 2015, https://ndupress.ndu.edu/JFQ/Joint-Force-Quarterly-78/Article/607659/one-size-does-not-fit-all-the-multifaceted-nature-of-cyber-statecraft/.

6 North Atlantic Treaty Organization. "China and Cyber: Attitudes, Strategies, and Organization," Cooperative Cyber Defence Centre of Excellence, 2016, https://ccdcoe.org/library/publications/china-and-cyber-attitudes-strategies-organisation/.

7 Pukhraj Singh, "Recent Chinese cyber intrusions signal a strategic shift," The Strategist, 5 July 2023, https://www.aspistrategist.org.au/recent-chinese-cyber-intrusions-signal-a-strategic-shift/.

8 Carly Page, "China-Backed APT41 Compromised 'At Least' Six U.S. State Governments," TechCrunch, March 8, 2022; Rufus Brown et al., "Does This Look Infected? A Summary of APT41 Targeting U.S. State Governments," Mandiant, March 8, 2022; U.S. Department of Justice, Seven International Cyber Defendants, Including "Apt41" Actors, Charged in Connection with Computer Intrusion Campaigns against More than 100 Victims Globally, September 16, 2020.

the 2015 breach of the Office of Personnel Management,[9] and targeting U.S. political figures, including the Alaskan governor prior to a 2018 trade delegation visit to China.[10] These actions are persistent. In fact, in 2010 China was first implicated after intrusions into Google and RSA's networks, targeting intellectual property and user data.[11] The scale and seriousness of China's digital intrusions have recently come to a head with FBI Director Chris Wray stating, "the greatest long-term threat to our nation's information and intellectual property, and to our economic vitality, is the counterintelligence and economic espionage threat from China."[12]

While this chapter could address multiple aspects of the MSS, including its counterintelligence or political security missions, this chapter focuses specifically on the MSS' involvement in cyber espionage as a tool of economic statecraft, to include the MSS leveraging of China's domestic legislation to achieve objectives through cyberspace. Since there is overlap between the Ministry of Public Security (MPS) and the People's Liberation Army (PLA), these organizations are also addressed to demonstrate their efforts in supporting MSS cyber espionage campaigns. When partnered with Chinese domestic laws, and shared national security cyber capabilities and platforms, MSS cyber espionage provides an unrivaled advantage to China's industries and the country's long-term economic development.

In what follows, this chapter will demonstrate the competitive edge the MSS provides the PRC by drawing upon primary sources like U.S. Congressional Committee reporting and official CCP strategy and policy documents. These sources are supplemented with articles from academic journals, cybersecurity reporting, and media reports. Although no single source is completely authoritative on its own, taken together, each source presents a unified perspective on China's approach to economic development via MSS cyber operations. After the introduction, this chapter articulates the People's Republic of China's (PRC) economic goals. In the next section, I reveal how China's domestic security sector collaborates with the MSS and Ministry of Public Security (MPS) creating unclear distinctions between civilian and military sectors. In the following section, I outline China's cyber collaborations between the MSS, the People's Liberation Army (PLA), and Chinese contractors. In the proceeding section I establish how the MSS is employed to achieve China's national economic objectives. I conclude with three takeaways from this chapter.

---

[9] Ellen Nakashima, "Chinese Breach Data of 4 Million Federal Workers," Washington Post, June 4, 2015.

[10] Insikt Group, "Chinese Cyberespionage Originating from Tsinghua University Infrastructure," Recorded Future, August 16, 2018.

[11] Nicole Perlroth, "How China Transformed Into a Prime Cyber Threat to the U.S." The New York Times, 19 July 2021. https://www.nytimes.com/2021/07/19/technology/china-hacking-us.html.

[12] Christopher Wray. "Transcript: The Threat Posed by the Chinese Government and the Chinese Communist Party to the Economic and National Security of the United States," Hudson, 17 April, 2024, https://www.hudson.org/national-security-defense/transcript-the-threat-posed-by-the-chinese-government-and-the-chinese-communist-party-to-the-economic-and-national-security-of-the-united-states.

2 PRC Economic Goals

Chinese cyber espionage is part of broader economic statecraft wherein the MSS assists Chinese firms in acquiring sensitive foreign technologies, which supports China's national development goals. This has been documented in cyber activities targeting foreign firms to steal intellectual property and trade secrets that feed directly into China's strategic objectives.[13] The U.S. has charged Chinese hackers with cyber espionage aimed at acquiring technologies that are central to the "Made in China 2025" initiative.[14] However, this is just one case among many demonstrating how cyber espionage supports China's ambitions to become a leader in high-tech industries as outlined in their national plans.

The strategic plans laid out in "Made in China 2025" and the 14th Five-Year Plan are advanced through cyber espionage activities attributed to the Ministry of State Security.[15] These national plans list key technologies that China aims to dominate globally, including artificial intelligence, quantum information systems, biotechnology, semiconductors, and autonomous systems.[16] To achieve these goals, China employs a range of tactics, both legal and illegal, which include cyber espionage to acquire foreign technologies crucial for these sectors.[17]

The CCP's commitment to power is highlighted by its increased investments in both economic espionage and digital information operations. Coupling state and private sectors, initiatives like "Made in China 2025"[18] promote dual-use technologies.[19] International digital technology standard-setting by China influences global

[13] Chris Borges, "Beyond Decoupling: Managing the U.S.-China Innovation Relationship | Perspectives on Innovation," Center for Strategic and International Studies, July 6, 2023, https://www.csis.org/blogs/perspectives-innovation/beyond-decoupling-managing-us-china-innovation-relationship.

[14] Sabri Ben-Achour, "China's State-Backed Cyberattacks Are Part of a Larger Plan." Marketplace, 9 Dec. 2021, https://www.marketplace.org/2021/12/09/chinas-state-sponsored-industrial-espionage-is-part-of-a-larger-system/.

[15] John Leydon. "Behind the Great Firewall: Chinese Cyber-Espionage Adapts to Post-Covid World with Stealthier Attacks." The Daily Swig, 15 Apr. 2021, https://portswigger.net/daily-swig/behind-the-great-firewall-chinese-cyber-espionage-adapts-to-post-covid-world-with-stealthier-attacks.

[16] US Department of Defense. "Military and security developments involving the People's Republic of China" In *Annual Report to Congress*, Washington, DC: US Department of Defense, 2023, https://media.defense.gov/2023/Oct/19/2003323409/-1/-1/1/2023-MILITARY-AND-SECURITY-DEVELOPMENTS-INVOLVING-THE-PEOPLES-REPUBLIC-OF-CHINA.PDF.

[17] United States–China Economic and Security Review Commission. "Sect. 2: China's Cyber Capabilities: Warfare, Espionage, And Implications for The United States," November 14, 2022. https://www.uscc.gov/sites/default/files/2022-11/Chapter_3_Section_2-Chinas_Cyber_Capabilities.pdf.

[18] Center for Security and Emerging Technology (CSET), "Translation: Made in China 2025." CSET, 8 March, 2022, https://cset.georgetown.edu/wp-content/uploads/t0432_made_in_china_2025_EN.pdf.

[19] Nick Eftimiades, "China's Espionage Recruitment Motivations," European Intelligence Academy Research Paper, Series #5, December 2023, https://www.rieas.gr/images/editorial/EIAPaper5.pdf.

technological architecture, aligning with Chinese interests.[20] The core national interests that these digital pursuits aim to secure include China's political system (with the CCP at the top), territory, and economy.

The "Made in China 2025" initiative and the 14th Five-Year Plan are central to China's economic development strategy. While similar, "Made in China 2025" is distinctive in that: (1) it encompasses the complete manufacturing lifecycle rather than solely focusing on innovation; (2) it supports the growth of both cutting-edge and traditional industries, along with modern services; and (3) there remains substantial state involvement but the market also plays a role.[21] "Made in China 2025" focuses on transforming China into a high-tech leader, enhancing development in technology sectors like robotics, aerospace, and clean energy.[22] The China 2025 initiative is designed to push China past the "middle-income trap" by transitioning from low-wage manufacturing to a more sophisticated, high-productivity economy.

China's 14th Five-Year Plan (2021–2025) outlines broader national development strategies, emphasizing high-quality, green development. It seeks to adopt innovation-driven growth, increase urban–rural integration, and promote social inclusiveness. The 5-year plan also highlights the importance of Chinese technological self-sufficiency, particularly in semiconductor production, which is essential for China's long-term economic and strategic security.[23] Additionally, about a trillion dollars have been allocated to fund key technology initiatives under this plan, signifying the scale of investment in realizing these goals.

Released in 2021, the CCP 14th 5 year plan[24] lists 7 developmental objectives: (1) The digital infrastructure system should be more complete; (2) The digital technology innovation system should have basically taken shape; (3) Digital economy development quality and efficiency should reach world-leading levels; (4) Digital society building should advance steadily; (5) Digital government building levels should increase overall; (6) Digital welfare protection capabilities should be significantly strengthened; and (7) The digitized development environment should become more perfect every day.

The first objective aims to upgrade and expand China's digital infrastructure, moving towards a more interconnected and innovative network, with advancements

---

[20] Daniel Russel and Blake Berger, "Stacking the deck: China's influence in international technology standards setting," Asia Society Policy Institute, 2021, https://asiasociety.org/policy-institute/stacking-deck-chinas-influence-international-technology-standards-setting.

[21] Scott Kennedy, "*Made in China 2025*," Center for Strategic and International Studies, June 2015, www.csis.org, https://www.csis.org/analysis/made-china-2025.

[22] James McBride and Andrew Chatzky, "Is 'Made in China 2025' a Threat to Global Trade?" Council on Foreign Relations, 13 May, 2019, https://www.cfr.org/backgrounder/made-china-2025-threat-global-trade.

[23] Asian Development Bank. "The 14th Five-Year Plan of the People's Republic of China—Fostering High-Quality Development," Asian Development Bank, 2021. www.adb.org, https://www.adb.org/publications/14th-five-year-plan-high-quality-development-prc.

[24] Chinese Communist Party, "Translation: 14th Five-Year Plan for National Informatization—Dec. 2021." DigiChina, https://digichina.stanford.edu/work/translation-14th-five-year-plan-for-national-informatization-dec-2021/.

in 5G and the development of 6G technology.[25] This includes expanding commercial applications like the Beidou satellite network, and integrating technologies such as Internet Protocol version 6 (IPv6), the Internet of Vehicles, and industrial internet. The second objective touts a strong digital technology innovation system with significant advancements in core technology areas such as integrated circuits, software, and components. This includes nurturing a collaborative innovation ecosystem among businesses, educational institutions, and research entities. The third objective aims to lead in digital economic development by merging digital technologies with the real economy, enhancing industrial chains, and fostering secure and competitive supply chains.

The fourth objective envisions creating a service-oriented digital social governance structure, led by the CCP, that integrates resources and ensures legal guarantees.[26] The fifth objective is to develop a modernized government informatization system that supports seamless integration and coordination of services countrywide. The plan focuses on improving data sharing, enhancing government operations, and ensuring oversight through enhanced digital capabilities. The sixth objective stresses enhancing digital public services in education, healthcare, and other social services to reduce disparities and increase accessibility. The final objective is to advance a standardized and secure digital development and governance environment that supports innovation and ensures robust cybersecurity.

The strategic goals laid out in "Made in China 2025" and the 14th Five-Year Plan are advanced through cyber espionage activities attributed to the Ministry of State Security.[27] These national plans list key technologies that China aims to dominate globally, including artificial intelligence, quantum information systems, biotechnology, semiconductors, and autonomous systems.[28] To achieve these goals, China employs a range of tactics, both legal and illegal, to acquire foreign technologies crucial for these sectors.[29]

---

[25] Katja Drinhausen and John Lee, "The CCP in 2021: Smart Governance, Cyber Sovereignty and Tech Supremacy," Merics, 15 June 2021, https://merics.org/en/ccp-2021-smart-governance-cyber-sovereignty-and-tech-supremacy.

[26] Mercy Kuo, "China's Cybersecurity and Statecraft," The Diplomat, 20 February 2024 https://thediplomat.com/2024/02/chinas-cybersecurity-and-statecraft/.

[27] John Leydon, "Behind the Great Firewall: Chinese Cyber-Espionage Adapts to Post-Covid World with Stealthier Attacks, The Daily Swig, 15 April 2021, https://portswigger.net/daily-swig/behind-the-great-firewall-chinese-cyber-espionage-adapts-to-post-covid-world-with-stealthier-attacks.

[28] US Department of Defense. "Military and security developments involving the People's Republic of China" In *Annual Report to Congress*, Washington, DC: US Department of Defense, 2023, https://media.defense.gov/2023/Oct/19/2003323409/-1/-1/1/2023-MILITARY-AND-SECURITY-DEVELOPMENTS-INVOLVING-THE-PEOPLES-REPUBLIC-OF-CHINA.PDF.

[29] United States–China Economic and Security Review Commission. "Sect. 2: China's Cyber Capabilities: Warfare, Espionage, And Implications for The United States," November 14, 2022. https://www.uscc.gov/sites/default/files/2022-11/Chapter_3_Section_2-Chinas_Cyber_Capabilities.pdf.

The MSS is central in achieving China's economic goals. One objective of the MSS' cyber espionage campaign is to address the misalignment with what the domestic economy and technology sector can produce and the growth rates that the CCP can deliver. In response to this misalignment, the CCP has decided that rather than reform its market and practices, it will instead steal from its economic and technological competitors[30] and use classical authoritarian "rule by law"[31] at home to achieve its economic aims. When paired together, international economic espionage and domestic cyber laws supporting the MSS contribute to Chinese economic growth. By integrating China's cyber espionage campaigns and supporting domestic legislation, it allows the CCP to meet domestic demands and undercut the economic advantages of its competitors. These developments should serve as a signal to the U.S. that certain domestic sectors will likely need more support and government protection from the continued siege of MSS cyber economic espionage.

According to Xi Jinping, China wants to become a "cyber superpower," to include taking over the helm as global technology leader.[32] By creating and employing dual-use technologies under the guise of strategic slogans, like China 2025, the country can disguise its intentions while generating synergies between their state and "private" sectors, otherwise known as civil-military fusion.[33] To that end, China's Ministry of State Security and its domestic national security cyberspace laws are a mutually reinforcing line of effort for the CCP to achieve economic goals. Due to both the National Intelligence Law and Cyber Security Law, and other domestic laws, (Table 1), the Chinese government can compel domestic organizations and companies on-demand to work with the state in "national security investigations."[34] As a result, the MSS holds a special and powerful position within China's national security apparatus , since Chinese laws empower the government to compel all Chinese

[30] The White House. "The United States, Joined by Allies and Partners, Attributes Malicious Cyber Activity and Irresponsible State Behavior to the People's Republic of China." *The White House*, 19 July 2021, https://www.whitehouse.gov/briefing-room/statements-releases/2021/07/19/the-united-states-joined-by-allies-and-partners-attributes-malicious-cyber-activity-and-irresponsible-state-behavior-to-the-peoples-republic-of-china/.

[31] United States–China Economic and Security Review Commission. "Sect. 2: China's Cyber Capabilities: Warfare, Espionage, And Implications for The United States," November 14, 2022. https://www.uscc.gov/sites/default/files/2022-11/Chapter_3_Section_2-Chinas_Cyber_Capabilities.pdf.

[32] Elsa Kania, "Made in China 2025: Xi Jinping's Plan to Turn China into the AI World Leader," Center for a New American Security, 5 October, 2018, https://www.cnas.org/press/in-the-news/made-in-china-2025-xi-jinpings-plan-to-turn-china-into-the-ai-world-leader.

[33] United States Department of State, "The Chinese Communist Party's Military-Civil Fusion Policy," Accessed 30 April, 2024, https://2017-2021.state.gov/military-civil-fusion/.

[34] Simon Handler, "The 5 × 5—China's Cyber Operations," Atlantic Council, 30 Jan. 2023, https://www.atlanticcouncil.org/content-series/the-5x5/the-5x5-chinas-cyber-operations/.

Table 1 China cyber-relevant laws

Title	Summary	Date
Cybersecurity law	• Requires network operators to implement network security protections, backups of important data, and encryption • Requires network operators to formulate and implement emergency response plans for cybersecurity incidents • Requires operators of critical information infrastructure to meet stringent cybersecurity standards, such as annual risk reviews and mandatory testing and certification of computer equipment • Requires network operators to store sensitive data domestically • Requires network operators to cooperate with China's law enforcement and security services upon request	Effective June 2017
National intelligence law	Requires individuals, organizations, and institutions to assist the security services in carrying out intelligence work, including by lending their "communications tools, premises and buildings."	Effective June 2017
Informal prohibition on participation in foreign cybersecurity events	Media reporting indicates that the Chinese government has prohibited Chinese security researchers from sharing their knowledge at some foreign cybersecurity events, such as Pwn2Own and Capture the Flag competitions	Reported March 2018
National defense law	Asserts that the Chinese government will take necessary measures to protect its activities, assets, and other interests in cyberspace	Effective January 2021

Adapted from U.S.-China Economic and Security Review Commission Report
United States–China Economic and Security Review Commission. "Sect. 2: China's Cyber Capabilities: Warfare, Espionage, And Implications for The United States," November 14, 2022. https://www.uscc.gov/sites/default/files/2022-11/Chapter_3_Section_2-Chinas_Cyber_Capabilities.pdf

citizens, companies, and government agencies to assist with its intelligence operations upon request.[35] The impact of this law comes into perspective when understood quantitatively: China has the largest number of software developers in the world, with nearly 7 million IT professionals (whereas the U.S. has 4.4 million).[36]

[35] Murray Scot Tanner, "Beijing's New National Intelligence Law: From Defense to Offense," Lawfare, July 20, 2017.

[36] Grid dynamic Market Analyst, "How Many Developers Are in US and in the World." Grid Dynamics, 8 August 2023, https://www.griddynamics.com/blog/number-software-developers-world.

3 Cyber Collaboration: China's Domestic Security Sector

Not only has there been a historically close relationship between the MSS and the Ministry of Public Security (MPS),[37] they also frequently share the same geographic space. In fact, MSS and MPS offices are frequently co-located, while also sharing the same surveillance and censorship capabilities, including the Great Firewall.[38] Sources disagree on which governmental organization ultimately oversees the Great Firewall, but organizations like the PLA, MSS, MPS, and the Propaganda Department regularly monitor it.[39]

MSS and MPS collaboration does not stop there. The Chinse Communist Party has taken a page out of Russia's playbook and now employs arrested domestic hackers in their security services, as the case of Tan DaiLin demonstrates. DaiLin (AKA Wicked Rose) was arrested by China's MPS for hacking within China. As a consequence, DaiLin now finds himself employed by the MSS.[40] While still receiving a government salary, the MSS permits DaiLin to engage in cybercrimes targeting foreigners when outside of his official duties. DaiLin is among several domestic hackers whose fortune has improved from facing potential arrest and incarceration for domestic cybercrimes to now employing their cyber skills in support of the Chinese Communist Party.

This shift is not just a change in targeting but also a perceived change in legal standing. Publicly available sources suggests that some Chinese hackers believe their collaboration with the MSS offers them legal immunity. Demonstrating this view, a 2020 Department of Justice indictment of hackers linked to the APT41, Double Dragon, emphasized Jiang Lizhi's assertion that his ties with the MSS protect him, even from the Ministry of Public Security, unless the situation were to escalate significantly.[41]

---

[37] Peter Mattis, "Chen Wenqing: China's New Man for State Security," The National Interest, 23 October, 2015, https://nationalinterest.org/feature/chen-wenqing-china%E2%80%99s-new-man-state-security-14153.

[38] Adam Kozy, oral testimony for the U.S.-China Economic and Security Review Commission, "Hearing on China's Cyber Capabilities: Warfare, Espionage, and Implications for the United States," February 17, 2022, https://www.uscc.gov/sites/default/files/2022-11/Chapter_3_Section_2-Chinas_Cyber_Capabilities.pdf. 84; Adam Kozy, written testimony for the U.S.-China Economic and Security Review Commission, Hearing on China's Cyber Capabilities: Warfare, Espionage, and Implications for the United States, February 17, 2022, https://www.uscc.gov/sites/default/files/2022-11/Chapter_3_Section_2-Chinas_Cyber_Capabilities.pdf.

[39] Booze Allen Hamilton. "Same Cloak, More Dagger: Decoding How the People's Republic of China Uses Cyberattacks," 2022, https://www.boozallen.com/insights/cyber/chinas-cyberattack-strategy-explained.html.

[40] Simon Handler, "The 5 × 5—China's Cyber Operations," Atlantic Council, 30 Jan. 2023, https://www.atlanticcouncil.org/content-series/the-5x5/the-5x5-chinas-cyber-operations/.

[41] U.S. Department of Justice, Chengdu 404 Indictment, 2019, 2, 6.; Nalani Fraser et al., "APT41: A Dual Espionage and Cyber Crime Operation," Mandiant, August 7, 2019.

Image 1 APT41

The “dual nature” of some Chinese cyber operators conceals the intent of the operations. For instance, while DaiLin engages in profit-driven hacking, his MSS-affiliated APT41, Volt Typhoon (Image 1), targets U.S. networks not for economic theft but to strategically conditions favorable to future Chinese operations.[42] This blend of individual economic motivations and national security objectives in cyberspace complicates assigning intent and attribution to some Chinese cyberspace operations.

[42] Joe Parkinson and Drew Hinshaw. February 18, 2024. *Wall Street Journal*. “FBI Director Says China Cyberattacks on U.S. Infrastructure Now at Unprecedented Scale.” https://www.wsj.com/politics/national-security/fbi-director-says-china-cyberattacks-on-u-s-infrastructure-now-at-unprecedented-scale-c8de5983.

	Characteristics
Background	APT 41 members all have undergraduate degrees, and associations with the Chinese Technology Company Chengdu 404.
Target Selection	IT, Government, Health Industry
Objectives	Carry out espionage activities, targeting various economic sectors of interest to the CCP.
Technological Sophistication	APT 41 has mastered and integrated numerous types of malware into their attacks on targets, including gh0st RAT, KEYPLUG, Mimikatz, ROCKBOOT, and Winnti for Linux, among others.
Geopolitical Targets	France, India, Italy, Japan, Myanmar, the Netherlands, Singapore, South Korea, South Africa, Switzerland, Thailand, Turkey, United Kingdom, and United States
Collaboration Practices	Known for collaborating with other Chinese state-sponsored groups, sharing tools, techniques, and infrastructure to enhance their cyber espionage and cyberattack capabilities.

APT 41 Exploit Employment Difficulty Level

Exploit Complexity	**Low**	**Moderate**	**High**
Exploit Employed	ipconfig	Net	gh0st RAT
	pwdump	ShadowPad	ZxShell
	Enterprise	Pass-The-Hash Toolkit	PowerSploit
	KEYPLUG		
	njRAT		
	Winnti for Linux		
	ROCKBOOT		

Classification Criteria

Low	**Moderate**	**High**
Exploits in this category require beginner level skills to employ or are exploits already integrated into widely used systems that can be used with very little privilege escalation.	Exploits require intermediate to advanced skills to employ, with most exploits used to laterally compromise another user's system or network.	Exploits in this category require highly advanced and experienced hacking and software manipulation skills to effectively utilize. The exploits in this category require persistence and extensive computing knowledge to discreetly execute on multiple targets.

4 China's External Security Sector Cyber Collaboration

Historically, the MSS is one of the most secretive spy organizations in the world.[43] However despite that secrecy, with confidence we can state that China's cyber espionage program is primarily run out of its Ministry of State Security. The MSS does not report to the People's Liberation Army Strategic Support Force (PLA-SSF) but rather falls under the State Council of China, reflecting its civilian intelligence role. Despite these differences, these organizations overlap in some areas. In fact, CCP security organizations like the MSS and PLA-SSF draw upon the same personal, cyber capabilities, threat intelligence, and platforms. Perhaps unsurprisingly, there have even been reports of joint MSS-PLA cyber operations in 2011 involving an Indo-Pacific pipeline.[44] As this operation demonstrates, the MSS and PLA likely have access to the same centralized development and logistics infrastructure allowing for sharing malware across organizations.[45] While their organizational structures differ significantly due to their missions and reporting lines, the joining of their cyber capabilities and platforms demonstrates China's irregular style of cyber espionage.

Despite connections, China's cyber espionage is in most cases conducted by the Ministry of State Security. MSS target selection aligns with its mandate to gather nonmilitary foreign intelligence, conduct domestic counterintelligence, and uphold political security.[46] The MSS likely has this mission over the PLA due to the latter's cyber operations typically being "noisy." This characteristic is typical of most military organizations operating in cyberspace when compared to intelligence agencies. This qualitative difference is likely due to cyberspace activities being generally more effective for intelligence gathering and shaping activities than for conducting military operations. It is far stealthier to collect or influence information in cyberspace rather than to disrupt digital communications.

Given that cyberspace activities are already a natural fit for a spy organization, it comes as little surprise that the MSS has figured out how to seamlessly integrate human operatives into its cyberspace activities and operations. This was demonstrated in the case of Yanjun Xu who blended traditional espionage with cyberspace. Xu specifically targeted experts in the aviation sector, convincing them to travel to

---

[43] Joske (2022).

[44] Booze Allen Hamilton. "Same Cloak, More Dagger: Decoding How the People's Republic of China Uses Cyberattacks," 2022, https://www.boozallen.com/insights/cyber/chinas-cyberattack-strategy-explained.html.

[45] Lucian Constantin, "ShadowPad Has Become the RAT of Choice for Several State-Sponsored Chinese APTs," CSO Online, February 15, 2022, https://www.csoonline.com/article/572061/shadowpad-has-become-the-rat-of-choice-for-several-state-sponsored-chinese-apts.html.

[46] Kelli Vanderlee, written testimony for the U.S.-China Economic and Security Review Commission, "Hearing on China's Cyber Capabilities: Warfare, Espionage, and Implications for the United States," February 17, 2022, https://www.uscc.gov/sites/default/files/2022-11/Chapter_3_Section_2-Chinas_Cyber_Capabilities.pdf.

China under professional pretext to covertly acquire sensitive technological information for the MSS.[47] When compared to the PLA, the MSS is seen as a more professionalized agency regarding cyberspace activities, and thus they attract top-level Chinese technical talent with their generous benefits.[48]

In some cases, members of the PLA or MSS may work for the Chinese government during the day, and by night pursue a life as a cybercriminal.[49] Even more damaging, the FBI has evidence of direct contractual relations between the MSS and Chinese cybercriminals.[50] In fact, due to the closeness of this contractual relationship, it is challenging to confirm when operations are carried out by the MSS or through its front companies and contractors.[51] For example, when the U.S. Department of Justice in 2017 and 2018 indicted Chinese APT groups, APT3 and APT10 respectively, it charged them as contractors working for front companies rather than as direct MSS operatives.[52]

Due to the exquisite malware used by APT41 in espionage, it is highly unlikely the indicted were simple cybersecurity contactors. Also damaging, the U.S. and Taiwanese governments have linked the ransomware attacks to Chengdu 404 (Image 2), a company in Chengdu, China, that claims to be another cybersecurity firm.[53] Chengdu 404 is believed to engage in a combination of for-profit criminal activities and contracting for the PRC, to include surveillance and espionage, and has reportedly targeted government organizations.[54] Its cybercrime operations often overlap with its contractual work, notably through targeting video game companies to steal valuable

[47] Office of Public Affairs, "Chinese Intelligence Officer Charged with Economic Espionage Involving Theft of Trade Secrets from Leading U.S. Aviation Companies," United States Department of Justice, 10 October 2018, https://www.justice.gov/opa/pr/chinese-intelligence-officer-charged-economic-espionage-involving-theft-trade-secrets-leading.

[48] Adam Kozy, oral testimony for the U.S.-China Economic and Security Review Commission, "Hearing on China's Cyber Capabilities: Warfare, Espionage, and Implications for the United States," February 17, 2022; Adam Kozy, written testimony for the U.S.-China Economic and Security Review Commission, "Hearing on China's Cyber Capabilities: Warfare, Espionage, and Implications for the United States" February 17, 2022 https://www.uscc.gov/sites/default/files/2022-11/Chapter_3_Section_2-Chinas_Cyber_Capabilities.pdf.

[49] Booze Allen Hamilton. "Same Cloak, More Dagger: Decoding How the People's Republic of China Uses Cyberattacks," 2022, https://www.boozallen.com/insights/cyber/chinas-cyberattack-strategy-explained.html.

[50] Simon Handler, "The 5 × 5—China's Cyber Operations," Atlantic Council, 30 Jan. 2023, https://www.atlanticcouncil.org/content-series/the-5x5/the-5x5-chinas-cyber-operations/.

[51] United States–China Economic and Security Review Commission. "Sect. 2: China's Cyber Capabilities: Warfare, Espionage, And Implications for The United States," November 14, 2022. https://www.uscc.gov/sites/default/files/2022-11/Chapter_3_Section_2-Chinas_Cyber_Capabilities.pdf.

[52] Mandiant. "APT41, A Dual Espionage and Cyber Crime Operation," Report, February 2022, https://www.mandiant.com/sites/default/files/2022-02/rt-apt41-dual-operation.pdf.

[53] United States of America versus Jiang Lizhi, Qian Chuan, and Fu Qiang, "Case: 1:20-cr-00158, U.S. District Court for the District of Columbia," August 11, 2020, https://www.justice.gov/opa/press-release/file/1317206/download.

[54] Booze Allen Hamilton. "Same Cloak, More Dagger: Decoding How the People's Republic of China Uses Cyberattacks," 2022, https://www.boozallen.com/insights/cyber/chinas-cyberattack-strategy-explained.html.

Image 2 Chengdu 404

digital items for resale, while simultaneously acquiring code-signing certificates that are later repurposed for espionage and surveillance activities.[55] The economic disruptions of MSS-linked cyber espionage campaigns emphasizes the dual nature of these operations, blending state-sponsored activities with cybercrime. This duality contributes to the economic strategy of the CCP through acquiring sensitive and lucrative data and advanced technologies.

5 MSS: Supporting National Economic Objectives

"The 13th Bureau" is also part of China's security system, and part of the MSS. The Bureau operates the China Information Technology Security Evaluation Center (CNITSEC) which collects information about vulnerabilities in software, hardware, and information systems. CNITSEC evaluates software intended for government use, conducts "national security reviews" of foreign technologies before entering the Chinese market, collaborates and contracts with local cybersecurity companies,

[55] Booze Allen Hamilton. "Same Cloak, More Dagger: Decoding How the People's Republic of China Uses Cyberattacks," 2022, https://www.boozallen.com/insights/cyber/chinas-cyberattack-strategy-explained.html.

and gathers data on potential vulnerabilities in software, hardware, and information systems.[56] In this effort, it maintains the China National Vulnerability Database (CNNVD). Ideally, a national vulnerability database should provide advisories to domestic audiences so they can patch against known vulnerabilities. In China, all known vulnerabilities discovered by Chinese security researchers must be turned over within 48 h to the CNNVD, allowing the MSS, and PLA, to exploit these vulnerabilities before the wider public knows about them.[57] Exemplifying the suspect nature of the database, the CNNVD delayed the publication of high-threat vulnerabilities significantly more than low-threat vulnerabilities, with delays ranging from 21 to 156 days. By comparison, the U.S. government's National Vulnerability Database published information on vulnerabilities commonly exploited by Chinese APT groups faster than CNNVD 97 percent of the time.[58]

The Ministry of State Security also employs sophisticated cyber operations that target critical sectors worldwide to secure competitive advantages by appropriating foreign technologies and intelligence. In this pursuit, the Ministry of State Security executes cyber-enabled economic espionage targeting critical American industries and technology sectors. Tactics typically include the theft of intellectual propert to enhance China's technology sector, weaken external financial systems, and erode U.S. confidence. Ultimately, these operations economically debilitate the U.S., and strengthen China's relative economic standing, all the while not prompting open conflict. This is a common PRC tactic. In fact, when confronted with evidence of cyber-enabled espionage, China backs down to regroup and then attempts the same efforts later. This was evident in the 2016 meeting between President Obama and Xi Jinping.[59] When Obama confronted Xi about the Chinese cyber espionage campaign against U.S. industries, cyber espionage by the China's PLA briefly was redirected towards other targets, but cyber espionage against the U.S. by the MSS did not stop.[60] This incident perfectly encapsulates China's approach of persistent low-level pressure to outlast competitors while minimizing political backlash.

Extending this low-level approach, Chinese universities are integrated into the country's broader cyberespionage efforts, providing academic support and research that improve MSS and PLA capabilities. The level of Chinese educational institutional integration into national security operations is incomparable to the United

[56] Insikt Group, "China's Cybersecurity Law Gives the Ministry of State Security Unprecedented New Powers over Foreign Technology," Recorded Future, August 31, 2017.

[57] United States–China Economic and Security Review Commission. "Sect. 2: China's Cyber Capabilities: Warfare, Espionage, And Implications for The United States," November 14, 2022. https://www.uscc.gov/sites/default/files/2022-11/Chapter_3_Section_2-Chinas_Cyber_Capabilities.pdf.

[58] Priscilla Moriuchi and Bill Ladd, "China's Ministry of State Security Likely Influences National Network Vulnerability Publications," Recorded Future, 16 November, 2017, https://www.recordedfuture.com/blog/chinese-mss-vulnerability-influence.

[59] Costello and McReynolds (2018).

[60] Else Kania and John Costello, "The strategic support force and the future of Chinese information operations." The Cyber Defense Review, *3*(1), https://cyberdefensereview.army.mil/Portals/6/Documents/CDR%20Journal%20Articles/The%20Strategic%20Support%20Force_Kania_Costello.pdf.

States.[61] As an example, at Hainan University, a professor collaborated with the MSS' Hainan provincial bureau to recruit students from on-campus hacking competitions during 2013 and 2016. They also offered rewards of up to $73,000 to those who could uncover software vulnerabilities that would aid in hacking operations.[62] Meanwhile, Xidian University is reported to have a joint graduate program with the Guangdong Bureau of CNITSEC, which brings together students and graduates in tackling technical challenges that support MSS objectives.[63] In terms of research and development, the Chinese Institute of Contemporary International Relations, which operates directly under the Ministry of State Security, is also of importance to U.S. security researchers given its various government-affiliated research institutions.[64]

The emphasis on research and development has paved the way for the MSS to employ its resources in innovative ways. This includes organizing hacking competitions within the country, which not only enhance domestic cybersecurity but also helps recruit top talent for MSS cyber operations.[65] These competitions identify vulnerabilities in Chinese and foreign systems, which the MSS exploits in its own missions. The MSS also has a relationship with the Tainfu Cup, one of China's largest hacking competitions to discover commercial vulnerabilities.[66] Demonstrating the relationship between the MSS and the compeition, a vulnerability that won an award at the Tianfu Cup was later used by the MSS to target the iPhones of Uyghur Muslims.[67] This use of hacking competitions to discover commercial vulnerabilities emphasizes the MSS' role in strengthening China's global economic and technological aims.

---

[61] United States–China Economic and Security Review Commission. "Sect. 2: China's Cyber Capabilities: Warfare, Espionage, And Implications for The United States," November 14, 2022. https://www.uscc.gov/sites/default/files/2022-11/Chapter_3_Section_2-Chinas_Cyber_Capabilities.pdf.

[62] Dakota Cary, written testimony for the U.S.-China Economic and Security Review Commission, Hearing on China's Cyber Capabilities: Warfare, Espionage, and Implications for the United States, February 17, 2022, https://www.uscc.gov/sites/default/files/2022-11/Chapter_3_Section_2-Chinas_Cyber_Capabilities.pdf.; Dakota Cary, "Academics, AIs, and APTs: How Six Advanced Persistent Threat-Connected Chinese Universities Are Advancing AI Research," Georgetown University Center for Security and Emerging Technology, March 2021, https://cset.georgetown.edu/publication/academics-ai-and-apts/; Intrusion Truth, "Who Is Mr. Gu?" January 10, 2020, https://intrusiontruth.wordpress.com/2020/01/10/who-is-mr-gu/.

[63] Dakota Cary, written testimony for the U.S.-China Economic and Security Review Commission, Hearing on China's Cyber Capabilities: Warfare, Espionage, and Implications for the United States, February 17, 2022, https://www.uscc.gov/sites/default/files/2022-11/Chapter_3_Section_2-Chinas_Cyber_Capabilities.pdf.

[64] North Atlantic Treaty Organization. "China and Cyber: Attitudes, Strategies, and Organization," Cooperative Cyber Defence Centre of Excellence, 2016, https://ccdcoe.org/library/publications/china-and-cyber-attitudes-strategies-organisation/.

[65] Simon Handler, "The 5 × 5—China's Cyber Operations," Atlantic Council, 30 Jan. 2023, https://www.atlanticcouncil.org/content-series/the-5x5/the-5x5-chinas-cyber-operations/.

[66] J.D. Work, "China Flaunts Its Offensive Cyber Power," War on the Rocks, October 22, 2021. https://warontherocks.com/2021/10/china-flaunts-its-offensive-cyber-power/.

[67] Patrick Howell O'Neill, "How China Turned a Prize-Winning iPhone Hack against the Uyghurs," MIT Technology Review, May 6, 2021, https://www.technologyreview.com/2021/05/06/1024621/china-apple-spy-uyghur-hacker-tianfu/.

The MSS actively pursues commercial intellectual property that matches goals outlined in China's industrial planning. As former FBI-analyst Adam Kozy testified to Congress, Chinese leaders' view cyber espionage "as a way to bridge key technology gaps and rapidly gain parity with advanced adversaries like the U.S. in a variety of dual-use technologies... that would otherwise be unattainable without years of research and billions spent on development."[68] Kelli Vanderlee, Mandiant Senior Manager of Intelligence Analysis, went further claiming Chinese state-supported cyber espionage groups consistently target organization's commercial intellectual property, including sectors like technology, engineering, construction, transportation, and biotechnology.[69] Vanderlee added that China's cyberespionage over time, primarily by the MSS, has "steadily evolved to become stealthier and more agile."[70] The long-term, systematic, and iterative refinement of tactics, techniques, and procedures shows the criticality of the MSS in contributing to the China's national economic and technological goals.

It should also be noted that reports on MSS operations are typically more frequent the other Chinese APTs, likely reflecting the vast scale of their program. Although the CCP denies any involvement in cyber espionage, cybersecurity companies have identified multiple advanced persistent threats linked to the MSS (Table 2), primarily focusing on industrial espionage. Ultimately, China's economic and technological ambitions are closed intertwined with its economic cyberspace activities, facilitated by the Ministry of State Security. "The 13th Bureau," strategically collecting intelligence on vulnerabilities and conducting national security reviews, and the CNNVD, with delays in reporting vulnerabilities, raises commercial and national security concerns. China's approach extends beyond traditional espionage, with a focus on securing competitive advantages through the theft of foreign technologies and intel-

[68] United States–China Economic and Security Review Commission. "Sect. 2: China's Cyber Capabilities: Warfare, Espionage, And Implications for The United States," November 14, 2022. https://www.uscc.gov/sites/default/files/2022-11/Chapter_3_Section_2-Chinas_Cyber_Capabilities.pdf.

[69] Kelli Vanderlee, written testimony for the U.S.-China Economic and Security Review Commission, "Hearing on China's Cyber Capabilities: Warfare, Espionage, and Implications for the United States," February 17, 2022, https://www.uscc.gov/sites/default/files/2022-11/Chapter_3_Section_2-Chinas_Cyber_Capabilities.pdf.

[70] United States–China Economic and Security Review Commission. "Sect. 2: China's Cyber Capabilities: Warfare, Espionage, And Implications for The United States," November 14, 2022. https://www.uscc.gov/sites/default/files/2022-11/Chapter_3_Section_2-Chinas_Cyber_Capabilities.pdf.

ligence. The integration of universities into China's cyberespionage efforts emphasizes the country's unique approach to national security, encouraging innovation and talent acquisition at the local level to achieve national success. With continuous adjustments to its cyber operations, the MSS will likely remain pivotal in advancing China's economic and technological goals.

Table 2 Likely MSS-affiliated APT groups

Names and years active	Overview and targets	Typical attack vector for initial access/Associated malware
APT41 (AKA Wicked Panda, Wicked Spider, BARIUM, BRONZE ATLAS, Winnti) **2012-current**	A prolific cyber threat actor likely associated with the MSS that conducts state-sponsored espionage as well as financially motivated activity for personal gain. APT41's campaigns have targeted organizations in at least 14 countries, stealing IP from the healthcare, telecommunications, technology, and videogame sectors. The group's operations have also targeted political dissidents in Hong Kong. In March 2022, Mandiant reported that APT41 had compromised six U.S. state government networks	**Vectors**: Uses spear-phishing, SQL injection, followed by more sophisticated TTPs **Malware**: Known to use at least 46 different malware families, including backdoors, credential stealers, keyloggers, and rootkits; ransomware; cryptojacking
APT40 (AKA Kryptonite Panda, GADOLINIUM, BRONZE MOHAWK, TEMP Periscope, Leviathan) **2013-current**	A cyber threat actor associated with Hainan Xiandun Technology Development Co., Ltd, a front company for the MSS' Hainan branch, that conducts state-sponsored espionage likely facilitating China's naval modernization program. APT40 has targeted governments, companies, and universities for IP spanning a wide range of industries—including maritime research—across the United States, Canada, Europe, the Middle East, and Belt and Road Initiative countries. APT40 may be connected to or overlap with HAFNIUM	**Vectors**: Uses spear-phishing, often posing as a prominent individual of interest to the target **Malware**: Known to use at least 51 different malware families, including 37 that are nonpublic and seven of which (BADSIGN, FIELDGOAL, FINDLOCK, PHOTO, SCANBOX, SOGU, and WIDETONE) are associated with other Chinese state-sponsored groups
HAFNIUM (AKA Operation Exchange Marauder) **2020-current**	A cyber threat group associated with the MSS that exploited multiple zero-day vulnerabilities in Microsoft's Exchange Server email software to carry out a massive hack affecting thousands of organizations around the world in early 2021. The U.S. government and several allied governments jointly attributed the hack to the MSS in July 2021	**Vectors**: Exploits zero-day vulnerabilities in the internet-facing and vulnerable Microsoft Exchange servers for initial access; then uploaded web shells using these vulnerabilities and executed malicious commands **Malware**: Backdoor.Hafnium web shells
APT31 (AKA Judgment Panda ZIRCONIUM) **2014-current**	A cyber threat group associated with the MSS that has conducted cyberespionage against government, financial, and defense organizations and attempted cyberattacks against individuals involved in the 2020 U.S. presidential elections. In March 2022, Google's Threat Analysis Group warned multiple Gmail users associated with the U.S. government that they were targeted in phishing attacks conducted by APT31	**Vectors**: Exploits vulnerabilities in applications such as Java and Adobe Flash; SQL injection **Malware**: SOGU, LUCKYBIRD, SLOWGYRO, and DUCKFAT

(continued)

Table 2 (continued)

Names and years active	Overview and targets	Typical attack vector for initial access/Associated malware
APT26 (AKA Turbine Panda) **2014-current**	A cyber threat group associated with the MSS' Jiangsu branch that has conducted cyberespionage campaigns targeting the aerospace, defense, and energy sectors. In 2019, CrowdStrike revealed that the group, which it calls Turbine Panda, had stolen IP from multiple foreign companies that manufactured components for China's domestic C919 airliner between 2010 and 2015. The hackers were overseen by MSS Jiangsu intelligence officers and successfully breached the systems of suppliers like Ametek, Honeywell, Safran, Capstone Turbine, GE, and others	**Vectors**: Uses watering hole attacks (strategic web compromises) and custom backdoors once inside a victim's network **Malware**: SOGU, HTRAN, POSTSIZE, TWOCHAINS, BEACON, PlugX
APT17 (AKA Deputy Dog, Tailgatorn Team) **2013-current**	A cyber threat group associated with the MSS' Jinan bureau that targets the U.S. government, international law firms, IT companies, mining companies, and nongovernmental organizations. Among the more memorable campaigns attributed to APT17 was a 2017 spear phishing attack that used a Game of Thrones-themed lure purporting to contain spoilers for the current season to convince victims to download a remote access trojan	**Vectors**: Uses spear phishing; creates profiles and posts in forums to embed encoded command and control infrastructure for use with a variant of the malware it uses **Malware**: BLACKCOFFEE
APT10 (AKA Menupass Team, Stone Panda, POTASSIUM, Red Apollo, Cicada, CVNX) **2009-current**	A cyber threat group associated with the MSS that has historically targeted construction and engineering, aerospace, and telecom firms as well as foreign governments in support of China's national security goals. Mandiant assesses that these goals include acquiring military and intelligence information as well as confidential business data to benefit Chinese corporations. APT10 perpetrated Operation Cloud Hopper, a global cyberespionage campaign that compromised a number of managed service providers in the United States and other countries to obtain the information of their clients in the engineering, industrial manufacturing, retail, energy, pharmaceuticals, telecommunications, and government industries. APT10 has also historically targeted Japanese corporations and media organizations, though reporting by Symantec in April 2022 indicated the group is now targeting government-related institutions and nongovernmental organizations in North America, the Middle East, and Europe	**Vectors**: Uses spear phishing and access to victims' networks through managed service providers **Malware**: HAYMAKER, SNUGRIDE, BUGJUICE, QUASARRAT

(continued)

Table 2 (continued)

Names and years active	Overview and targets	Typical attack vector for initial access/Associated malware
APT3 (AKA UPS Team, Gothic Panda, TG-0110, Boyusec, Buckeye) **2010-current**	A cyber threat group associated with the Chinese cybersecurity firm Guangzhou Boyu Information Technology Company, Ltd ("Boyusec"), a known contractor for the MSS. APT3 targets organizations in the aerospace, defense, construction, engineering, high-technology, telecommunications, and transportation sectors. APT3 has carried out several high-profile cyberespionage campaigns, including Operation Clandestine Fox and Operation Double Tap. According to Symantec, since 2015 APT3 has shifted from targeting U.S.-based victims to political organizations in Hong Kong	**Vectors:** Uses phishing emails, zero-days vulnerabilities in browsers (e.g., Internet Explorer, Firefox, and Adobe Flash Player) **Malware**: SHOTPUT, COOKIECUTTER, SOGU

Adapted from U.S.-China Economic and Security Review Commission Report
United States–China Economic and Security Review Commission. "Sect. 2: China's Cyber Capabilities: Warfare, Espionage, And Implications for The United States," November 14, 2022. https://www.uscc.gov/sites/default/files/2022-11/Chapter_3_Section_2-Chinas_Cyber_Capabilities.pdf

6 Conclusion

The MSS plays an indispensable role in economic statecraft via cyber espionage and domestic cyber laws. As noted in this chapter, the MSS has essentially integrated cyber espionage into China's national economic strategy. This integration is aimed at supporting China's ambitious economic development plans, such as "Made in China 2025" and the 14th Five-Year Plan. The MSS plays a key role in acquiring foreign technologies and intellectual property essential for achieving these goals.

The MSS has been particularly proficient at leveraging China's vast pool of domestic cyber talent. By recruiting domestic top-notch hackers and cybersecurity experts, often directly from universities or through national competitions, the MSS has built an impressive network capable of executing cyber operations. This not only enhances the technical capabilities of China's intelligence services but also ensures a steady influx of skilled personnel aligned with national objectives.

The MSS also collaborates closely with universities and research institutions. This relationship is twofold: it supports the MSS' espionage activities by providing advanced research and development capabilities, and it integrates academic expertise directly into national security strategies. Chinese universities serve as a pipeline for innovation in cyber capabilities and strategies, which are then employed in MSS cyber missions. This distinctive synergy between higher education and the MSS accentuates the use of academic resources in strengthening China's state-sponsored cyber activities.

Finally, the dual nature of Chinese sectors involves the strategic use of technology developed under the guise of civilian innovation which can also be employed for intelligence purposes. This approach not only maximizes resource utilization but also blurs the lines between civilian and military. This dual-use strategy is evident in how domestic cyber capabilities are shared with the state, enabling China to leverage advancements in technologies for economic growth. Additionally, the MSS strategic use of domestic cyber talent, academic partnerships, and dual-use technologies and applications, positions China to be a rival economic competitor for the foreseeable future.

References

Costello J, McReynolds J (2018) China's strategic support force: a force for a new era. NDU Press. https://ndupress.ndu.edu/Portals/68/Documents/stratperspective/china/china-perspectives_13.pdf

Joske A (2022) Spies and lies: how China's greatest covert operations fooled the world. Hardie Grant Books

Kurtenbach E, Soo Z(2023) China's Economy is forecast to slow sharply in 2024, the world bank says, calling recovery 'Fragile'. AP News. https://apnews.com/article/china-economy-property-adb-791934f7f9b83de455e8f8aa7178b628

Dragon's Embrace: China's Growing Influence in Central Asia

Zhamilia Klycheva

Abstract Central Asia is a region of the five former Soviet Union states that appeared under the radar in international politics. With the rise of China as a global power, it is seeking to spread its influence worldwide, and Central Asia is no exception. Located roughly between Russia and China, this part of the world is rich in hydrocarbons, has a strategic location between two major powers, and borders Taliban-controlled Afghanistan. Central Asia is a crucial component of the Belt and Road Initiative, as it lies on two international corridors connecting China to Europe and the Persian Gulf. Central Asia has historically relied on Russia for economic and security support, but China's growing influence has become more prominent in the recent decade. This is particularly true given Russia's shift toward the war in Ukraine. Central Asia has adopted a multi-vector foreign policy, welcoming new foreign partners and aiming to be friends with all. This chapter focuses on China's growing influence in Central Asia and its relations with each of the five countries. It will provide a high-level overview of the region, highlight distinct features of each country, and describe China's relations with Central Asia. The following sections will focus on each country in greater detail, describing the relations through political, economic, and social lenses. Additionally, it will give a general overview of the relations between the United States and Central Asia and the growing fear and skepticism over China's rise in the region.

Keywords China and Central Asia · Former Soviet Union · Multi-vector foreign policy · C + C5 · The belt and road initiative · Debt trap policy · Kazakhstan · Kyrgyzstan · Tajikistan · Turkmenistan · Uzbekistan · Silk Road

1 Introduction

Central Asia is a small pocket of newly emerged states with a low profile in the international arena. Given its direct proximity, hydrocarbon resources, and important role in China's Belt and Road Initiative, its role in China's foreign policy has increased

Z. Klycheva (✉)
TransResearch Consortium (TRC), Claremont, CA, USA
e-mail: zhamilia.klycheva@cgu.edu

A. Farhadi et al. (eds.), *The Great Power Competition Volume 6*,
https://doi.org/10.1007/978-3-031-70767-4_8

in recent years. This chapter provides an overview of China's influence and relations with Central Asia, a land-locked region between Russia and China in the heart of Asia. It will start with a high-level overview of the region, highlight the key features of each country, and describe China's relations with Central Asia. The following sections will focus on each country in greater detail, providing political, economic, and social relations between China and each of the five countries. This chapter will also briefly discuss the relations between the United States and Central Asia and the rising fear over China's growing influence.

1.1 Central Asia Overview

Central Asia is a pocket of deeply land-locked countries located under Russia to the north, the Caspian Sea to the west, China to the east, and Afghanistan and Iran to the south. It consists of five former Soviet Union states: Kazakhstan, Kyrgyzstan, Tajikistan, Turkmenistan, and Uzbekistan, which recently gained independence in 1991. The collapse of the USSR brought significant changes to the development trajectories of the Central Asian countries, from socialism to a market economy and presidential republic systems. Each of the five states has a unique history, culture, geography, and identity, although they all share commonalities. Four out of five countries, excluding Tajikistan, share common Turkic roots, particularly reflected in the language, where Tajikistan speaks Tajik, a Persian dialect.

After the collapse of the USSR, its former member states have gone through a significant recession and have slowly been recovering and expanding their economies. All five countries are considered developing, with Kazakhstan significantly outperforming its neighbors due to its size and abundance of natural resources. One of the major issues all Central Asian countries face is corruption and the rule of law. All five Central Asian countries consider themselves democratic unitary countries with democratically elected officials. However, the recent history states otherwise. Since gaining independence, Tajikistan has been ruled by only one president, while Kazakhstan and Uzbekistan have had two, and Turkmenistan has had three. On the other hand, Kyrgyzstan, formerly known as the oasis of democracy in Central Asia, has had 6 presidents since 1991. According to the global freedom index, Kyrgyzstan is the freest in the region, although still considered 'Not Free' (27/100). Turkmenistan is notorious for its authoritarian rule, with the lowest Freedom score of 2 (Freedom House 2023).

Unsurprisingly, Russia has been actively involved in the region economically and strategically. Russian is an official language in three out of five states and is commonly spoken in all five. Russia has left a deep historical and cultural footprint on all five countries and maintains diplomatic and security relations with all. Three out of five Central Asian countries are a part of the Collective Security Treaty Organization (CSTO) with Russia. The recent war in Ukraine has shifted Russia's focus and opened a window of opportunity for China's influence in Central Asia.

Central Asian states are considered to be developing countries. Most of them have smaller sizes and smaller economies except Kazakhstan. Kazakhstan's territory is only approximately three times smaller than China's and five times larger than Turkmenistan's, which is the second largest territory in Central Asia. Kazakhstan has the world's largest uranium reserves, producing 43% of the global uranium in 2019 (World Nuclear Association 2023). Kazakhstan's GDP of $221.55B (2022) is the largest, compared to its neighbor Kyrgyzstan's GDP of only $8.24B (World Bank 2022a, b, c). As expected, less developed economies such as Tajikistan (8%) and Kyrgyzstan (7%) have higher rates of economic growth compared to more developed Kazakhstan (2%). Notably, most Central Asian states are heavily indebted, with the highest government debt-to-GDP ratio in Kyrgyzstan, at 59.6% (World Bank 2021). Central Asia remains one of the least economically developed regions (Congressional Research Service 2021).

Central Asia has been known to have historically close ties with Russia. Given its size and dependence on foreign capital, this region wants to maintain relations without being overly dependent on the country. Therefore, these countries adopted a multi-vector foreign policy, engaging in bilateral relations with Russia, China, India, and the US. The key is not to become strongly dependent on a single country; instead, diversify the risk and maintain friendly relations with all. Thus, the region keeps "all the doors open" (Gavin Helf 2023) (Fig. 1).

Indicator	China	Kazakhstan	Kyrgyzstan	Tajikistan	Turkmenistan	Uzbekistan
Territory (sq km)	9,388,210	2,724,900	199,951	144,100	488,100	447,400
Population (2022)	1,412,175,000	19,621,972	6,803,300	9,952,787	6,430,770	35,648,100
GDP (2022) (const 2015 $, bln)	18,188	221.55	8.24	13.5	44.16 (2020)	123.82
GDP pc (2022) (current US$)	12,720	11,243.7	1,606.7	1,054.2	7,297.2 (2020)	2,255.2
GDP growth (annual %) (2020)	2	3.2	7	8	3.4	5.7
FDI inflows (% of GDP) (2022)	1	2.2	2.7	1.7	3.1 (2020)	3.1
Central Gov Debt (% of GDP) (2021)	77.1(2022)	22.4	56.4	46.3	11.08	32.25
Global Freedom Index (2023)	9	23	27	7	2	12

Fig. 1 China and Central Asia's key macroeconomic figures. *Sources* Freedom House (2023), CIA (2023a, b), Statista (2022), Ministry of Economics and Finance of Uzbekistan (2022), World Bank (2022, 2023)

Despite its challenges and lower level of development, Central Asia presents a strategically important region, centrally located between the two major powers and possessing rich natural resources. From a security standpoint, Central Asia is essential for investment and monitoring as three out of five countries share a common border with a Taliban-controlled Afghanistan.

1.2 China—Central Asia Relations

Unlike Russia, a long-term geopolitical, strategic, and economic partner of the region, China's influence was more subtle and came from an economic angle. China took a strategic approach, slowly spreading its economic influence within the region.

Shortly after the collapse of the Soviet Union, China began to develop economic and commercial ties with the newly emerged Central Asian states in the early 1990s. The region presents particular importance to China from two perspectives: strategic geographic and natural resources. Given its proximity to China and its location between Russia and passage to the Persian Gulf and Europe, Central Asia plays a vital role in China's global economic integration plans reflected in the Belt and Road Initiative (BRI).

China borders three Central Asian states: Kazakhstan, Kyrgyzstan, and Tajikistan. All three countries have links to the historical Xinjian Uyghur Autonomous region. China and Russia founded the Shanghai Five, which later became the Shanghai Cooperation Organization (SCO), including Uzbekistan. Its goal was to promote regional security and economic and political integration. Later, India, Pakistan, and Iran joined the SCO, making it the world's largest regional organization by territory (Fig. 2).

Recently enhanced by the BRI, China became a major trade player in Central Asia. Figure 3 below shows the time series graph of the total trade between China and the five Central Asian countries (1995–2021). The graph shows that the overall trade level was low before 2005 and gradually increased over time, with a peak around 2013. Kazakhstan has a significantly larger trade level with China, followed by Tajikistan and Kyrgyzstan. Tajikistan has maintained a low trade level consistently over time.

Cooperation between China and Central Asia benefits China from security, natural resources, and export market perspectives. It is also a mutually beneficial process for the host states, as the developing countries see pouring investment, infrastructure development, opportunities for export, and new partnerships. However, China has drawn for its potential 'debt-trap' policy as a large creditor in the region with the potential inability to pay back its debt (Muratalieva 2023).

Figure 4 below shows the evolution of the bilateral relations between China and Central Asian states. China has established strategic, comprehensive partnerships, treaties of friendship and cooperation and cooperation under the BRI initiative with all five states. The relationships are still developing with its latest upgrade of a comprehensive strategic partnership developed with Turkmenistan in 2023.

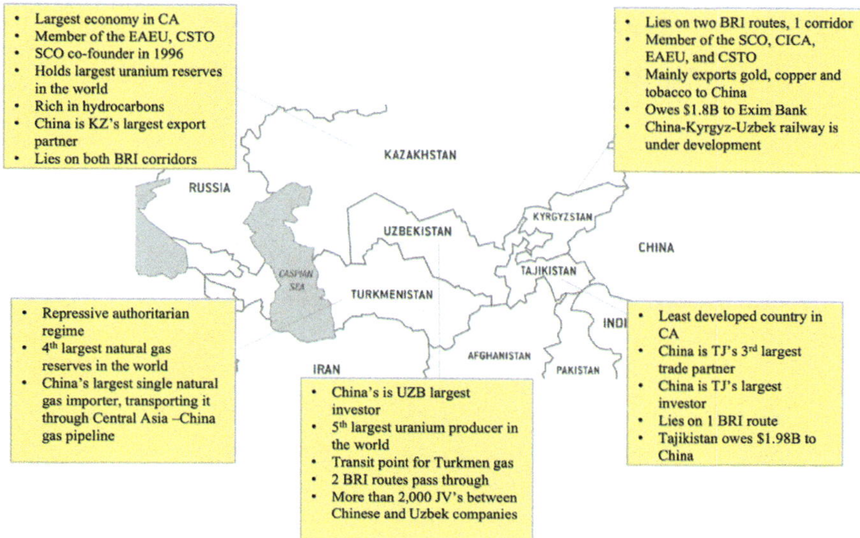

Fig. 2 Central Asia at a glance *Source* China Briefing (2021), China Daily (2003), Eurasianet (2022), Financial Times (2016), Ministry of Finance of the KR (2022), OEC (2021), Silk Road Briefing (2023), World Bank (2022, 2023), World Nuclear Association (2023)

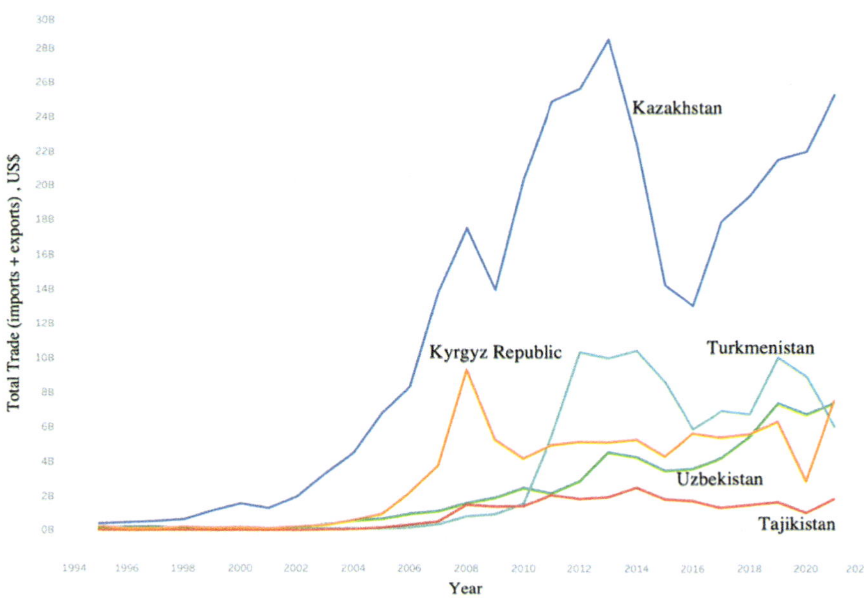

Fig. 3 Total Trade between China and Central Asia (USD), *Source* WITS (2021)

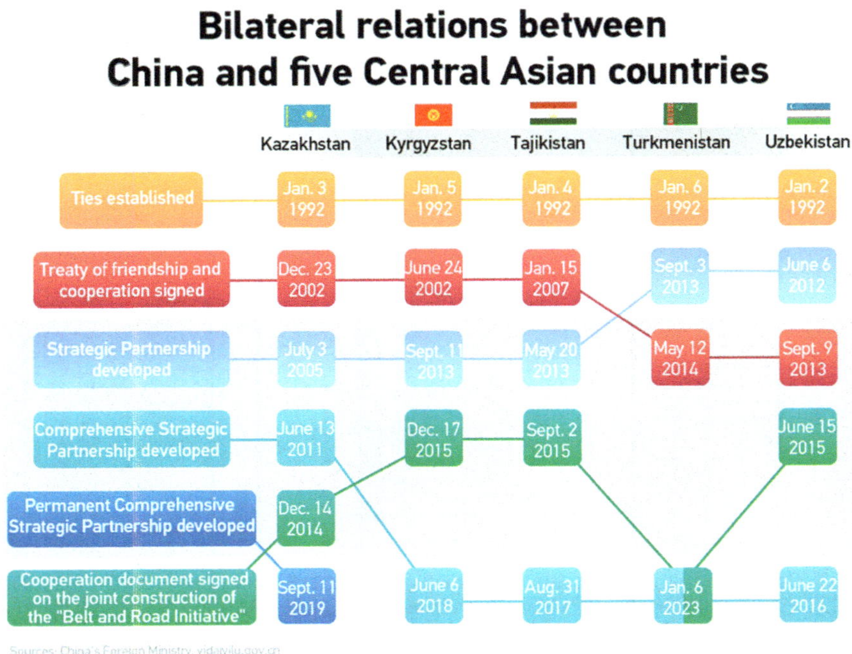

Fig. 4 Bilateral relations between China and Central Asian countries. *Source* CGTN (2023)

China has also established a new mechanism to promote cooperation with Central Asia, called China + Central Asia, or "C + C5". The first meeting between its ministers within this framework was held in 2020, followed up annually since (CGTN 2023). Moreover, China held its first China-Central Asia summit on May 18–19, 2023. It was the first time a Chinese leader met with Central Asian leaders without Russia (Hedlund 2023).

1.3 *The Belt and Road Initiative*

The BRI initiative was first announced in 2013 as a global initiative to improve transport infrastructure connectivity and promote global economic integration. Two BRI corridors pass through Central Asia: The New Eurasian Land-bridge Corridor and the China-Central Asia-West Asia Corridor. The former connects China to Europe and the latter to Iran and West Asia (World Bank 2020). The first one uses the railroad, with both routes passing through Kazakhstan. The second uses three routes and passes through Kazakhstan, Kyrgyzstan, Uzbekistan, Turkmenistan, and Tajikistan. The BRI allowed for major regional infrastructure development, efficiently connecting it via rail. Currently, most cargo between China and Europe passes via

Box 2: BRI Corridors and routes passing through CAC

The **New Eurasian Land-bridge BRI Corridor** connects **China to Europe** using two routes:

- **Route 1: China** (various cities, Urumqi, Alashankou) – **Kazakhstan** (Dostyk, Mointy, Nur-Sultan, Petropavl) – **Russia** (Yekaterinburg, Moscow) – **Belarus** (Brest) – **Poland** (Małaszewicze) – **Germany** (Duisburg) and onwards to various European cities. (Note: all are operational rail connections).
- **Route 2: China** (various cities, Urumqi, Khorgas) – **Kazakhstan** (Altynkol, Almaty, Shu, Zharyk, Zhezqazghan, Saksaulskaya, Shalkar, Beyneu, Aktau) – **Azerbaijan** (Baku/Alyat, Ganja, Beyuk Kesik) – **Georgia** (Gardabani, Tbilisi, Akhalkalaki) – **Turkey** (Kars, Istanbul) and onwards by rail/road to various European cities. (Note: all are operational rail connections except Aktau to Baku, which is the Caspian Sea ferry segment).

China-Central Asia-West Asia BRI Corridor connects **China to Iran/West Asia** using three routes:

- **Route 3: China** (various cities, Urumqi, Khorgas) - **Kazakhstan** (Altynkol, Almaty) - **Uzbekistan** (Tashkent, Samarkand, Navoi) – **Turkmenistan** (Farab, Mary, Serakhs) – **Iran** (Sarakhs, Mashad) and onwards to West Asian cities (also India through Bandar Abbas) (Note: all are operational rail connections).
- **Route 4: China** (various cities, Kashgar) – **Kyrgyz Republic** (Irkeshtam, Osh) – **Uzbekistan** (Andijan, Pap, Tashkent, Samarkand, Navoi) – **Turkmenistan** (Mary, Serakhs) – **Iran** (Sarakhs, Mashad) and to West Asia (also India through Bandar Abbas). (Note: the route is mainly rail, except Kashgar –Irkeshtam – Osh segment by road)
- **Route 5: China** (various cities, Kashgar) – **Kyrgyz Republic** (Irkeshtam, Sary Tash) – **Tajikistan** (Karamyk, Dushanbe, Vahdat, Yavan, Nizhny Panj) – **Afghanistan** (Shir Khan Bandar, Kunduz, Mazar-e-sharif, Herat, Ghurian) – **Iran** (Torbat-e Heydarieh, Tehran) to West Asia (also, India thru Bandar Abbas). (Note: route connected partly by rail with two large segments, Kashgar – Irkeshtam – Sary Tash – Karamyk – Dushanbe and Nizhny Panj - Shir Khan Bandar – Kunduz – Herat – that are connected only by road today).

Fig. 5 BRI Corridors and routes passing through Central Asia and Caucasus (CAC). *Source* World Bank (2020)

the northern route through Central Asia, Russia, and Belarus. However, the most efficient route is directly via Central Asia. Thus, the ongoing China-Kyrgyzstan-Uzbekistan project will be an important addition to the BRI initiative (Otorbaev 2023). Overall, the BRI initiative is an ambitious project with a large potential effect on the host countries' GDPs (Fig. 5).

1.4 US—Central Asia Relations

Central Asia has not been a priority in the US foreign policy. One of the main reasons for engagement in the region was due to Afghanistan following 9/11, with an increased US military presence in the region at the time being. Between 2001 and 2014, the United States had a military base in Kyrgyzstan and used Uzbekistan's K-2 airbase between 2001 and 2005 to support its mission in Afghanistan (U.S. Department of Veterans Affairs).

In recent years, the US has not been active in the region. Its overall goal in Central Asia is to promote security, long-term stability, democracy, and civil liberties. Moreover, the United States promotes economic development in the region, battles against terrorism and drug trafficking, and encourages border security.

In 2015, the US initiated a 'C5 + 1' diplomatic platform to enhance the cooperation between five Central Asian states (C5) and the United States. Since its creation, the countries have increased the inner dialogue, focusing on economic development, security, energy, and environment (US Department of State). In September 2023, leaders of all five Central Asian countries met in person for the first time with President Biden and Secretary Blinken within the C5 + 1 initiative (US Institute of Peace 2023).

Given significant Russian and Chinese influence in the region, Central Asia can view the United States as a powerful, friendly nation. Considering the region has generally favored multi-vector foreign policy, increased engagement with the United States may provide it with balancing support against Russian and Chinese dependence. Since Russia's attention shifted to Ukraine, the United States has a window of opportunity to step up its engagement in Central Asia, considering the diminishing role of the US in the region following the withdrawal from Afghanistan. This may be particularly advantageous for the US, given the growing economic dominance of China in the region. Increasing debt and dependence potentially threaten the struggling Central Asian economies.

Additionally, the United States intends to promote alternative trade routes and facilitate new connections among the US- Central Asian businesses. The US Department of State and Center for International Private Enterprise (CIPE) will work on establishing a private sector business platform to complement the C5 + 1. Moreover, the US wants to diversify the infrastructure routes and develop the Trans-Caspian Trade Corridor (The White House 2023).

2 China—Kazakhstan

2.1 Overview

Kazakhstan is the largest and the most developed country in Central Asia. It is the 9th largest country by territory in the world and the number one producer of uranium globally (Statista 2022). Kazakhstan is bordered by Russia to the north and west, Kyrgyzstan to the southeast, Uzbekistan to the south, and the Caspian Sea to the west. Kazakhstan is a unitary presidential republic. According to the Freedom House index, Kazakhstan is not free (23/100, Freedom House 2023).

Kazakhstan shares a 1,108-mile border with China (Caspian Policy Center 2022a, b, c). As the largest state in terms of territory, population, economy, and natural resources, Kazakhstan presents a strategic importance to China. The relationship between Kazakhstan and China dates back to the early 90s. Since its independence in 1991, Kazakhstan has only had two presidents in its whole history. The country's history is strongly dictated by its previous president (1991–2019), Nursultan Nazarbayev, who only resigned in 2019. Nazarbayev's foreign policy is known as 'multi-vector,' meaning the diversification of political, economic, and diplomatic ties with Russia, China, the European Union, and the United States (Russia in Global Affairs 2023). As a large enough country, it could stay relatively autonomous of other states and have a strong say in regional politics.

2.2 Political

China began diplomatic relations with Kazakhstan in 1992, a year after the collapse of the Soviet Union. The two nations share the tightest economic and security ties among all Central Asian countries. In 1996, both nations became the cofounders of the Shanghai Security Cooperation (SCO) that was fully established in 2001, including China, Russia, Kazakhstan, Kyrgyzstan, and Tajikistan, later joined by Uzbekistan, India and Pakistan (China Daily 2003). It aimed to maintain regional stability and strengthen member states' alliances (Caspian Policy Center 2022a, b, c).

Kazakhstan proposed building an oil pipeline to China to expand further the ties between the two countries and harness Kazakhstan's rich natural resources. This pipeline became the first between China and any Central Asian country (Zhao 2016). Over the years, China has expanded its roots within Kazakhstan, purchasing major Kazakhstan state-owned gas companies and investing in its natural resource extraction. By 2016, Chinese companies had invested more than US $20 billion in Kazakhstan's petroleum sector (Vakulchuk and Overland 2019). China recognizes Russia's historical geopolitical importance in the region and has kept a relatively low profile over the years.

2.3 Economic

Kazakhstan is a member of the Eurasian Economic Union (EAEU), including Russia, Belarus, Kyrgyzstan, and Armenia. The EAEU has signed a trade and economic cooperation agreement with China, enabling free trade between China and its member states (Eurasian Economic Commission 2018).

Economic ties between Kazakhstan and China have significantly changed in the past ten years. In 2021, Kazakhstan imported a staggering $13.9 billion worth of goods from China, accounting for 29.3% of its total imports and solidifying China's position as Kazakhstan's second-largest trading partner after Russia (Gaulier and Zignago 2010). The largest share of goods imported was rubber footwear (5.59%), computers (4.36%), non-knit women's coats (3.08%), toys (3.07%), motor vehicle parts and accessories (1.93%), among others (Gaulier and Zignago 2010).

China is Kazakhstan's largest export partner, resulting in $10.6 billion worth of goods traded with China in 2021, which represents 1.1% of total China's imports. The top products exported into China from Kazakhstan were mostly raw materials, such as refined copper (22.8%), crude petroleum (20.3%), copper ore (14.6%), petroleum gas (10.4%), ferroalloys (8.7%), among others (Gaulier and Zignago 2010). As shown in Fig. 6 below, bilateral trade between China and Kazakhstan in the first half of 2023 reached $16.3 billion compared to $24.1 billion for 2022, marking a significant increase.

Fig. 6 Trade and Investment between Kazakhstan and China, 2022–2023. *Source* Silk Road Briefing (2020)

According to the United Nations Conference on Trade and Development report, Kazakhstan became the top FDI destination among all post-Soviet countries (UNCTAD 2022). According to 2021, the top investors in Kazakhstan were the Netherlands, the United States, France, Japan, and China. Conversely, the top KZ FDI destination countries were the Netherlands, the Cayman Islands, and Russia (IMF 2021). As of 2021, there were 700 joint KZ-CHN enterprises, 4.7% of the total investment in the country (Caspian Policy Center 2022a, b, c). As shown in Fig. 6 above, China's gross FDI in Kazakhstan exceeded $23 billion in 2022.

2.4 *The Belt and Road Initiative*

The Belt and Road Initiative (BRI) has considerable political and economic implications for Kazakhstan. Initiated by China in 2013, it aims to connect the region from within and enhance economic cooperation. The main objective of the BRI is to improve China's land-based transportation connections to Europe and Asia via six major corridors (Silk Road Briefing 2023). Two out of six corridors pass through Central Asian states: the New Eurasian Land-bridge BRI corridor that connects China with Europe and the China-Central Asia-West Asia BRI Corridor (World Bank 2020).

Kazakhstan is well-positioned to benefit from the BRI. As the largest Central Asian state, it has a strategic role in the initiative. Both BRI corridors path through Kazakhstan, one connecting China to Europe (Poland, Germany) and another to

Iran via Kazakhstan and Turkmenistan. This can significantly benefit the country's infrastructure, efficiency with shipment time, and lower trade and agriculture growth restrictions. It can also help promote labor mobility and expand access to services within the country. Moreover, BRI has benefited Kazakhstan's connectivity through new roads, increased exports, and a potentially significant impact on the country's GDP.

Another critical strategic infrastructure development is the railroad. Currently, two China–Europe railway lines go through Kazakhstan, one via the Alashankou-Dostyk border port and another via the Khorgoz Gateway dry port (Shiphub 2023). The construction of the third railway is currently being discussed.

Despite its strategic economic and political importance, the BRI initiative faces criticism for its member states' debt burden and sustainable development. As a big enough state with good economic standing, Kazakhstan has the strongest say in the region and remains independent, although the concerns over China's influence remain.

Kazakhstan is one of the countries with the smallest government debt-to-GDP ratio in Central Asia. As of July 2023, Kazakhstan's government debt accounted for $161.8B, among which 26% is owed to the Netherlands, 8.5% to Russia, and 8% to the UK. China ranked 8th in this list, with 5.5% of the total KZ government debt owed (National Bank of Kazakhstan 2023).

2.5 *Social*

Despite the close relationship between the two nations and strong economic ties, concerns remain over China's expanding sphere of influence in Kazakhstan. This is especially prevalent among Generation Z, where young adults express concern about being overdependent on China. On the one hand, young adults criticize China's strict regime and surveillance of its citizens (Jamestown 2023). Moreover, reports of massive detention of Uyghurs and other Muslim minorities, including ethnic Kazakhs in the Xinjiang province of China, have drawn public negativity and contributed toward Sinophobia (Caspian Policy Center 2023). Concerns have been raised regarding the increased investment in developing countries and the potential for a debt trap policy, where countries may struggle to repay their debts. In fact, Kazakhstan is the sixth largest recipient of China's loans, after Angola, Brazil, and Indonesia (Malik et al. 2021).

2.6 *Kazakhstan—US Relations*

The United States was the first to acknowledge Kazakhstan's independence in 1991 and established an embassy in 1992. The two countries have strengthened their relationship, mainly through their joint efforts to reduce nuclear weapons, given

Kazakhstan's rich uranium reserves and history as the major Soviet nuclear test site. Since gaining independence from the Soviet Union, the United States and Kazakhstan have worked closely together to seal off Soviet nuclear testing sites located on Kazakhstan's territory. Kazakhstan has participated in global nuclear security summits. In 2017, the United States and Kazakhstan opened a Nuclear Security Training Center in Kazakhstan to train and inform local and regional nuclear security professionals (U.S. Department of State 2023). The United States has provided over $500 million in assistance to eliminate weapons of mass destruction under the Cooperative Threat Reduction program. The United States has also sponsored military training for Kazakhstan and conducted joint exercises such as Steppe Eagle, Viking, Eager Lion, and Shanti Prayas (U.S. Department of State 2023).

Kazakhstan is the 81st largest US trading partner, with a total of $2.5 billion in trade in 2021, roughly 9.5 times less than the trade between China and Kazakhstan (WITS 2021a, b, c, d). U. S. firms have invested tens of billions of dollars into Kazakhstan's oil and gas sector. Kazakhstan joined the World Trade Organization (WTO) in 2015 and is a member of the United Nations and OSCE (U.S. Department of State 2023).

Kazakhstan's relationship with the US primarily focuses on security matters, particularly nuclear security. In contrast to China, the US has a much lower level of economic engagement with Kazakhstan.

3 China—Kyrgyzstan

3.1 Overview

Kyrgyz Republic, known as Kyrgyzstan, is Central Asia's second-smallest country by territory and population. It borders Kazakhstan to the north, Uzbekistan to the west, Tajikistan to the south, and China to the east and southeast. Currently, Kyrgyzstan is one of the least developed countries in Central Asia (World Bank 2022a, b, c). It is a presidential unitary republic. According to the Freedom House data, Kyrgyzstan is not free, although it has the highest freedom index among all other Central Asian countries (Global Freedom Scores 2023). It is worth noting that this state has had six presidents, the highest number compared to any other Central Asian state.

China and Kyrgyzstan share a 661-mile border (CIA Factbook 2023a, b). The relationship between China and Kyrgyzstan is not nearly as strong as with Kazakhstan. This might be evident due to the lack of natural resources and the border disputes left over from the Soviet era. The two countries successfully resolved them in 1999 (Shen 2012).

The relations between Kyrgyzstan and China were passive during the 1990s; however, they began to gain momentum in the recent decade. A new phase of the relationship was the announcement of the Belt and Road Initiative (BRI) in 2013 and Kyrgyzstan's joining it as a partner. It has strategic geographic importance to the initiative since it opens up the route to Central Asia and further Europe (Hedlund 2019). Currently, Kyrgyzstan is a part of the Shanghai Cooperation Organization (China Briefing 2021a, b, c).

Overall, the relations between China and Kyrgyzstan are friendly and stable, although some regional tensions exist. This is mostly due to Kyrgyzstan's political instability in 2005, 2010, and 2020, which threatened Chinese investments. Protests against foreign investments, such as attacks on gold mines run by Chinese and Canadian companies, have concerned the Chinese government (China Briefing 2021a, b, c). Currently, the relations are at their highest since the independence of Kyrgyzstan.

3.2 Economic

Kyrgyzstan's economy mainly relies on agriculture. The country is rich in mineral resources such as coal, gold, and uranium. As of 2021, the country's largest export is gold, accounting for 38.3% of the total exports, followed by float glass and precious metals (Gaulier and Zignago 2010). Kyrgyzstan's economy heavily relies on remittances, accounting for 29.5% of its GDP (World Bank 2022a, b, c). As of September 2023, most remittances come from Kazakhstan, Russia, Belgium, and the UK (National Bank of the Kyrgyz Republic 2023). Remittances from Russia have decreased by approximately four times compared to the previous year due to the war in Ukraine (National Bank of the Kyrgyz Republic 2023). Like Kazakhstan, Kyrgyzstan is also a member of the Eurasian Economic Union, enjoying free trade within its member states.

Since joining the Belt and Road Initiative, the economic relations between China and Kyrgyzstan have significantly strengthened. China's exports to Kyrgyzstan in 2021 amounted to $6.47 billion. The primary products exported from China to Kyrgyzstan were non-knit women's coats ($720 million), non-knit men's coats ($551 million), and knit sweaters ($488 million). Over the past 26 years, China's exports to Kyrgyzstan have increased at an annualized rate of 17.2%, from $105 million in 1995 to $6.47 billion in 2021 (Gaulier and Zignago 2010). China and Russia are Kyrgyzstan's two largest import partners, with China accounting for 26.12% and Russia for 34.25% of total imports (WITS 2021).

Kyrgyzstan exported goods worth $111 M to China in 2021. The main products exported were gold ($36.8 M), copper ore ($21 M), and rolled tobacco ($16.1 M). Over the last 26 years, exports from Kyrgyzstan to China have decreased at an annualized rate of 0.088%, from $114 M in 1995 to $111 M in 2021. Kyrgyzstan's top export destinations are Russia, Kazakhstan, and the UK (Gaulier and Zignago 2010).

In 2021, Canada, Russia, and China were the top investors in Kyrgyzstan, with $1.5B, $987 M, and $774 M, respectively (IMF 2023). It is unsurprising since Canada was in charge of the largest gold extraction in Kyrgyzstan—Kumtor from 1993 till 2022.

China and Kyrgyzstan signed a Bilateral Investment Treaty (BIT) in 1992, which was enacted in 1995. Additionally, on June 24, 2002, they signed an agreement to avoid double taxation (DTA) and tax evasion. Both countries have been members of the Conference on Interaction and Confidence Building Measures in Asia (CICA) since 1992 and are also members of the Shanghai Cooperation Organization (China Briefing 2021a, b, c).

According to the National Statistic Committee of the Kyrgyz Republic, Kyrgyzstan's government debt stood at $5.5 billion in 2022, or 59.6% of its GDP, the highest among all Central Asian countries. The major creditors include the Chinese Exim Bank ($1.8 billion), Asian Development Bank ($660.97 million), World Bank ($656.96 million), and IMF ($339.3 million) (Ministry of Finance of the KR 2022). Notably, China is the largest creditor for Kyrgyzstan and the entire Central Asian region. The citizens are increasingly concerned about the state's dependence on China and its ability to repay large loans. It is worth mentioning that 84% of China's BRI investments are in countries with medium to high risk (Silk Road Briefing 2020). Thus, governments in such countries highly desire investment from the US or its allies if the funding is available.

3.3 Kyrgyzstan and the BRI

From Beijing's perspective, interaction with Kyrgyzstan with the BRI strategically connects China to South Asia, the Middle East, and Europe. China's BRI corridor in Kyrgyzstan connects China to Europe and West Asia. Routes such as China-Kyrgyzstan-Uzbekistan-Turkmenistan and the route China-Kyrgyzstan-Tajikistan-Afghanistan-Iran and the Middle East are beneficial. The construction of the BRI has become the most essential platform for practical cooperation between China and Kyrgyzstan.

In recent years, China has made significant investments in Kyrgyzstan's infrastructure. It has proposed the construction of a railway that would connect Andijan in Uzbekistan with Kashgar in China via Kyrgyzstan. The proposed China-Kyrgyzstan-Uzbekistan (CKU) railway would help to shorten the route to Europe and increase the roles of Kyrgyzstan and Uzbekistan in the BRI (Silk Road Briefing 2022). The construction was postponed indefinitely.

China's Railway 20 Bureau Group aided Kyrgyzstan's irrigation system reconstruction in 2016 (Seetao 2023). Moreover, recently, Kyrgyzstan has been seeking Chinese investment to reconstruct its major hydropower projects, although China has not commented yet (Shambetov 2023). China has invested in several projects in Kyrgyzstan, which include a household appliance manufacturing plant, the Kazakhstan-Kyrgyz highway, the repair of the Bishkek thermal power plant and wind power plant, renovation of irrigation projects, construction of the 500 kV Datka-Kemin transmission line, the Tajikistan-Kyrgyzstan project, the China-Kyrgyzstan-Uzbekistan highway, among others (Silk Road Briefing 2023).

3.4 Social

Kyrgyzstan has been considered an oasis of democracy in Central Asia. This is characterized by the number of presidents the young country has had and the various revolutions, demonstrations, and protests. Kyrgyz people have also been vocal about their growing discontent towards Chinese influence. In February 2020, around 1,000 Kyrgyz protesters gathered outside the Naryn Free Economic Zone, leased by a joint Kyrgyz-Chinese venture for 49 years. The zone was intended as a logistics center (Putz 2020). The people of Kyrgyzstan have a constant fear of China taking away their land, which is the most valuable possession of the country.

The latest government media attempted to show a positive sentiment towards China. Broadcasted on the national channel, it covered China's cultural and social influence. The report highlighted China's efforts to open medical educational facilities, conduct cultural and study exchanges, and offer Kyrgyz students full scholarships. These initiatives have helped to increase China's 'soft power' in the country.

Currently, the relations between the two countries are at their highest since the independence of Kyrgyzstan. In May 2023, Kyrgyz President Sadyr Japarov visited China for the first Central Asia-China summit, held between China and all five Central Asian states. The summit's goal was to revitalize the Belt and Road Initiative.

3.5 Kyrgyzstan—US Relations

The United States established diplomatic relations with Kyrgyzstan in 1991. The US supports the development of democracy and civil liberties, considering that Kyrgyzstan has the largest freedom score in the region.

In 2021, the US exported $69.2 million to Kyrgyzstan, primarily spacecraft, cars and gas turbines. The same year, Kyrgyzstan exported $8.64 million to the US, mainly gas turbines, motor vehicle parts, and combustion engines (Gaulier and Zignago 2010).

After 9/11, the US opened a Manas Air Base at the Manas airport in Bishkek, Kyrgyzstan, and operated it until 2014 (Kucera 2014a, b). The primary unit at the base was the 376th Air Expeditionary Wing, which supported US military operations in Afghanistan.

4 China—Tajikistan

4.1 Overview

Tajikistan is the smallest country in Central Asia by territory and the poorest in the region, with the lowest per capita income (World Bank 2022a, b, c). Tajikistan borders Afghanistan to the south, Uzbekistan to the west, Kyrgyzstan to the north, and China to the east. Tajikistan is a presidential republic. According to the Global Freedom Index, Tajikistan is Central Asia's second least free country, followed by Turkmenistan (Freedom House 2023). Tajikistan has only had one president since its independence in 1991. Additionally, it is the only non-Turkic-speaking country in Central Asia.

China and Tajikistan established formal relations in 1992 (China Briefing 2021a, b, c). Like Kyrgyzstan, Tajikistan had a rough relationship start due to disputed border delineation left by the Soviet Union. This was resolved shortly. China and Tajikistan share a relatively short 296-mile border. However, Tajikistan holds a strategic geographic location due to its Taliban-controlled border, stretching 843 miles in length (CIA Factbook 2023a, b). The Wakhan corridor is of special importance due to its access to Xinjiang from Afghanistan, located between Tajikistan, China, and Pakistan. Tajikistan's role in economic development in the region is minor. It is not rich in natural resources and does not present a strategic economic importance. Tajikistan is also a member of the Shanghai Cooperation Organization and joined the BRI in 2018.

4.2 Security

Tajikistan holds a strategic geographic location in the region due to its Taliban-controlled border with Afghanistan. This is especially relevant now with the return of the Taliban. The Wakhan corridor, largely bordered by Tajikistan, is essential for China's security (Blank 2019). In 2016, China formed a Quadrilateral Cooperation and Coordination Mechanism between Tajikistan, Pakistan, and Afghanistan to counter Islamist militancy (Radio Free Europe 2016).

Multiple reports indicate that China has heightened its military presence in recent years. In 2016, the two countries have conducted bilateral military exercises involving 10,000 troops (Jardine 2021). Although not officially, China might host a military base within the Gorno-Badakhshan province of Tajikistan overlooking the Wakhan corridor (Sukhankin 2020a, b).

4.3 Economic

Tajikistan has recently experienced accelerated economic growth, surpassing Kyrgyzstan (World Bank 2022a, b, c). Poverty was reduced from 32% of the population in 2009 to 12.4% in 2022. Similar to Kyrgyzstan, Tajikistan's main economic sector is agriculture. It is also rich in water resources and hydropower (World Bank 2023a, b, c).

Tajikistan's exports in 2021 amounted to $2.1 billion, marking a significant increase of $876 million from 2016. The country's primary export commodities include gold, raw aluminum, raw cotton, and zinc ore. The primary countries where Tajikistan exports its goods are Switzerland (25.6%), Kazakhstan (17.1%), Turkey (11.4%), and China (7.57%). Tajikistan's primary exports to China are gold (74.6%), other ores (5.77%), and raw cotton (4.63%). Meanwhile, the primary imports from China include rubber footwear (5.27%), coated flat-rolled iron (4.89%), and motor vehicle parts (3.46%) (Gaulier and Zignago 2010).

China is the third largest Tajikistan trade partner after Russia and Kazakhstan (WITS 2021a, b, c, d). Like Kyrgyzstan, Tajikistan's economy heavily relies on remittances and foreign investment and has a high government debt risk. The share of households with a migrant is 48% as of June 2023 (World Bank 2023a, b, c). As a country with a vulnerable economy, Tajikistan heavily relies on foreign investment inflows. As of 2021, the top investors in Tajikistan were China, Russia, and the UK, where China accounted for 57% or $1.7B of the total FDI in 2021 (IMF 2021). Concerns over the debt trap policy remain, especially considering Tajikistan's small size, struggling economy, and high level of corruption. In fact, as of 2022, Tajikistan owed US $1.98 billion to China's Exim Bank (Eurasianet 2022).

4.4 Tajikistan and the BRI

Tajikistan officially joined the BRI in 2018. Since that time, Chinese investors have brought $320 M in projects. Conversely, as of 2021, Tajikistan has accumulated $1.12B in debt to China's Eximbank, surpassing the largest creditor, the World Bank. It began lending to Tajikistan in 2007 to construct a 500 kW power transmission from Dushanbe to the border of Uzbekistan. China kept lending funds to Tajikistan despite the potential trouble of repayment through 2020 (Cabar Asia 2021).

Tajikistan lies only on one BRI route, particularly for the China-Central Asia-West Asia BRI corridor that connects China to Iran and West Asia. Route 5 connects China to Kyrgyzstan, Tajikistan, Afghanistan and Iran. Several railroad projects are underway in Tajikistan as part of the Belt and Road initiative. The Waya Railway is the first one, starting in Wahdat, 20 km east of Dushanbe. The railway ends in Yaway, which is 40 km south of Wahdat.

The Dushanbe-Chanak highway, built in 2010, is also part of the Belt and Road initiative. China Railway Fifth Buren is constructing the Shar-Shar Tunnel and the North–South Connection Project on the Dushanbe-Kholog highway. This is an essential link to the China-Tajikistan Highway (Seetao 2023).

4.5 Social

Given the Soviet past of the country and Russian propaganda against China's expansionism and its dislike of Muslims, Sinophobia has been present in Central Asian countries. In the past decades, China has been actively trying to pacify such hostility towards the Chinese and instead increase its regional presence. One of the ways to do it is by social means or by increasing 'soft power' (Nye 1990). In recent years, the number of Chinese workers in Tajikistan has increased. Between 2006 and 2011, the number of Chinese workers in Tajikistan increased from 3,000 to 81,000 (Eurasianet 2011). Chinese businessmen in Tajikistan have been giving preference to the Chinese workers over the Tajik due to their efficiency, promoting Chinese migration to Tajikistan. This led to increased mixed marriages between the Tajiks and the Chinese, causing concern for the Tajik authorities (Rickleton 2013).

China has been trying to improve its image by inviting young students to study in China, founding Confucius schools, and promoting Chinese traditions, food, technology, and medicine. China continues to offer scholarships for students to come and study in China (Bawa and Ashish 2023). Thus, China is the second most popular choice to study abroad after Russia (Mahmadrasul 2020).

4.6 Tajikistan—US Relations

The US and Tajikistan established diplomatic relations in 1992. The relationship between the two countries is based on security, economic development, and human rights issues, among others. In 2010, Tajikistan and the US launched an annual bilateral consultation process to enhance cooperation (US Department of State 2022). Tajikistan has actively participated in the C5 + 1 platform, which was initiated in 2015.

Tajikistan and the US are not major economic partners. In 2022, the US exported $123.8 million worth of goods to Tajikistan while importing $1.4 million back (US Census 2023).

The US and Tajikistan belong to several mutual international organizations. Tajikistan is a member of the United Nations, World Trade Organization, Organization for Security and Cooperation in Europe, Euro-Atlantic Partnership Council, International Monetary Fund, the World Bank, and the Coalition for Disaster Resilient Infrastructure. It is also a member of the North Atlantic Treaty Organization's Partnership for Peace (US Department of State 2022).

5 China—Turkmenistan

5.1 Overview

Turkmenistan is the second-largest country in Central Asia by territory and has the smallest population. It borders Kazakhstan to the northwest, Uzbekistan to the northeast, Afghanistan to the southeast, Iran to the south, and the Caspian Sea to the west. The country has a rich history and was an important stop along the Silk Road route. Turkmenistan is notorious for its repressive authoritarian rule. According to the Freedom House data, it has a Global Freedom Index of 2 (100 being the perfectly free), indicative of no individual political rights and civil liberties, tightly controlled elections, and systemic corruption (Freedom House 2023). It is a highly repressive state known for its constant surveillance, arbitrary arrests, and torture when confronting the state. From an economic standpoint, Turkmenistan has one of the world's largest reserves of natural gas. It is also a major producer of cotton.

China and Turkmenistan established diplomatic relations in 1992 (Caspian Policy Center 2022). China's relations with Turkmenistan evolved around Turkmen natural gas supplies. The era of bilateral relations began in 2006 when the Turkmen president traveled to China, leading to the construction of the Central Asia-China natural gas pipelines (Lines A, B, and C). In 2009, they started delivering 35B cubic meters of gas annually to China. In 2014, the two states began their strategic partnership (Kucera 2014a, b).

Turkmenistan has maintained a restricted foreign policy, unlike other Central Asian countries. It is not a member of China's Asian International Investment Bank (AIIB) nor an official member of the Belt and Road Initiative. However, China and Turkmenistan benefit from a comprehensive strategic partnership.

5.2 Economic

Turkmenistan publishes limited national statistics, and its data and methodology are often not credible.[1] Turkmenistan reported the 3rd largest GDP after Kazakhstan and Uzbekistan. World Bank reported a 3.4% GDP growth in 2020 (World Bank 2020). The country has the 4th largest gas reserves, accounting for approximately 10% of the world's reserves. Roughly half of the gas output is consumed domestically, and the rest is exported abroad (US Department of Commerce 2023).

Turkmenistan's primary export product is natural gas, accounting for 67.8% of its total exports. Other significant exports include refined petroleum, nitrogenous fertilizers, crude petroleum, cotton, and sulfur (World Bank 2023a, b, c). China is

[1] Given its strictly authoritarian rule, Turkmenistan data is provided by its statistical committee and does not often match other sources. The IMF and WB make their estimates of the Turkmenistan data and do not rely on the official government statistics.

Turkmenistan's primary export destination, acounting for 66.5% of total exports worth $4.79 billion in 2021. Turkey and Uzbekistan follow China as the next major export destinations, with 9.23 and 8.76%, respectively.

Turkey, China, and Russia are Turkmenistan's major import partners, with Turkey being the largest import partner at 28.2%, followed by China at 14.8%, and Russia at 13.6%. The top imported products from China to Turkmenistan are vaccines, rubber tires, coated flat-rolled iron, and iron pipes worth $513 million in 2021.

Over the past 26 years, Turkmenistan's exports to China have grown at an annual rate of 29.5%, from $5.79 million in 1995 to $4.79 billion in 2021 (Gaulier and Zignago 2010). Before 2009, Russia was the largest Turkmen export partner for natural gas. However, in the last decade, Turkmenistan shifted its exports towards China, which has now become Turkmenistan's number one natural gas destination point. Given the war in Ukraine, Russian gas is no longer wanted in Europe, which makes Turkmenistan a potential provider of inexpensive gas to Europe, given it just lost its leading supplier due to the war. However, Turkmenistan seems aligned with Russia and decided to move its gas to the east.

Central Asia—China gas pipeline is a major pipeline in Turkmenistan connecting it to China. Since its opening, it has delivered 8.4 trillion cubic feet of natural gas from Turkmenistan to China (Zhang 2021). It is China's largest single gas exporter (Reuters 2023). The pipeline is 2,277 miles long and consists of three pipelines stretching from the Turkmenistan-Uzbekistan border to Khorgos near the Kazakhstan-China border in the province of Xinjiang, connecting to China's domestic pipeline. Uzbekistan and Kazakhstan have also connected their domestic pipelines to export it to China (Hydrocarbons Technology). The ongoing construction of the pipeline's line D is experiencing technical difficulties (Fig. 7).

China's efforts to promote the Central Asia-China pipeline's development are considered a geopolitical move to invest in the Belt and Road Initiative (BRI) and move away from Russia. In recent years, China has prioritized its relations with Turkmenistan over Russia. China is Turkmenistan's largest creditor, providing $8 billion to develop natural gas infrastructure (Blackwood 2023).

5.3 *Political and Security Concerns*

Turkmenistan maintains political neutrality in the region. It rejected to join Russia's Collective Security Treaty Organization (CSTO) and China's Shanghai Cooperation Organization (SCO). Turkmenistan has maintained a policy of isolationism from the rest of the world by maintaining political neutrality. China is the major Turkmen foreign partner. Moreover, approximately 27% of military defense equipment is supplied through China (Sukhankin 2020a, b).

Most of Turkmenistan's gas exports go to China, making it its largest export partner. This leaves the economy more vulnerable as Turkmenistan can only rely on one partner and is more vulnerable to external shocks. To remain neutral and

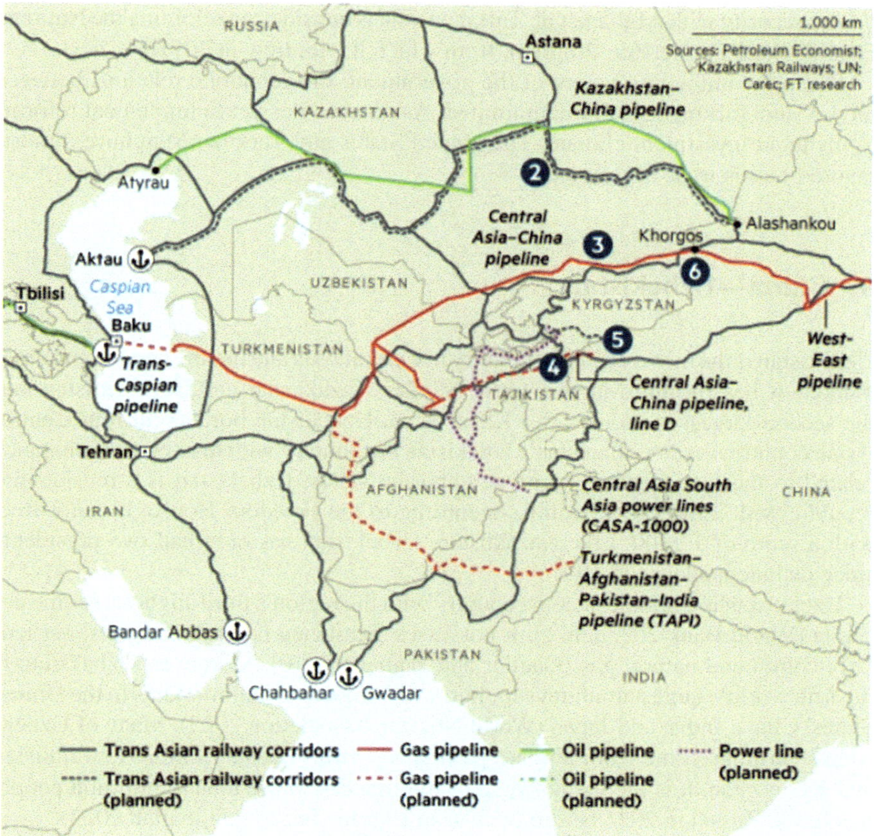

Fig. 7 3—The Central-Asia Pipeline, carrying most Turkmenistan gas to China 4—Line D of the CA Pipeline, connecting Turkmenistan to China via Tajikistan. *Source* Financial Times (2016)

independent of China, Turkmenistan can expand its relations with other countries and have multiple markets for natural gas exports.

Given a common border with Afghanistan, Turkmenistan recognizes the security concerns. It provides subsidized food, electricity, and natural gas to the northern Afghanistan provinces to promote stability (US Department of State 2022).

5.4 US—Turkmenistan Relations

The relations between the US and Turkmenistan are not deeply rooted and remain official. The main reason for US engagement is due to the security concerns of the Taliban border. Thus, main investments and support have been targeted at the border security with Afghanistan. Turkmenistan is considering rejoining the National Guard's State Partnership Program, from which it withdrew in 2011.

Given the uninviting nature of the government, the economic relations between the US and Turkmenistan remain limited. Ashkhabad has yet to implement reforms to invite an investment climate. The United States and Turkmenistan have a most-favored-nation trade agreement.

6 China—Uzbekistan

Uzbekistan is the most populated country in Central Asia and has the largest consumer market. It has the third-largest territory after Kazakhstan and Turkmenistan and the second-largest economy after Kazakhstan. Uzbekistan borders all four Central Asian countries and Afghanistan. Uzbekistan is a country with historical significance related to the Silk Road. Like its neighboring states, Uzbekistan is a presidential republic with a market economy. According to the Freedom House, it is not free, with a score of 12/100. Like Kazakhstan, Uzbekistan has only had two presidents since its independence in 1991.

Uzbekistan is a middle-income country with the region's third-highest income per capita (World Bank 2022a, b, c). It possesses significant natural resources, particularly cotton and natural gas (Gaulier and Zignago 2010). Moreover, Uzbekistan is the fifth world's largest uranium supplier, with long-term commitments to the United States, China, India, and Japan (World Nuclear Association 2023). Many of Uzbekistan's earnings come from abroad, particularly from Russia. Similar to Tajikistan and Kyrgyzstan, it is the main source of remittances. More than 4.5 million people worked in Russia in 2021 before the war in Ukraine began (Najibullah 2022).

6.1 General Relations

Uzbekistan and China established diplomatic relations in 1992, shortly after the collapse of the Soviet Union. Under President Karimov's rule, Uzbekistan developed a closed-off form of government (Pikulicka-Wilczewska 2018). Although the two countries do not share a common border, Uzbekistan is strategically important due to its proximity to the Caspian Sea and a border with a hydrocarbon-rich Turkmenistan (China Briefing 2021a, b, c). Thus, it presents an important transit point for Turkmen gas and other regional trade.

Later, when President Mirziyoyev came to power, Uzbekistan became more open to partnerships with foreign countries. Over the last 30 years, the relations between China and Uzbekistan have grown substantially. In 2001, Uzbekistan became a founding member of the Shanghai Cooperation Organization (China Briefing Team 2021a, b, c). In 2005, the two states signed a "China-Uzbek Treaty of Friendship and Cooperation". The relations were further strengthened in 2016 by turning it into a comprehensive strategic partnership. Uzbekistan has been a BRI partner since 2015 and a member of China's Asian International Investment Bank (Muratshina 2017). Over the last decade, bilateral relations between the two states have significantly grown. By the end of 2020, Uzbekistan registered about 1,800 Chinese companies operating within the country (China Briefing 2021a, b, c). Moreover, the two countries have signed over 100 bilateral agreements to enhance cooperation (Silk Road Briefing 2023). Between 1992 and 2019, the trade between China and Uzbekistan increased 100 times (The State Council of the People's Republic of China 2022).

Uzbekistan plays an essential role in the BRI initiative. In the China-Central-Asia-West Asia Corridor, two routes pass through Uzbekistan. Moreover, all four "China-Central Asia" gas pipeline corridors go through Uzbekistan. There are more than 2,000 joint ventures between Uzbekistan and Chinese companies. Large Chinese companies like Exim Bank, Huawei, and ZTE operate in Uzbekistan (Silk Road Briefing 2023). One of the most recent projects is the MoU on the China-Kyrgyzstan-Uzbekistan (CKU) railway, signed in September 2022 (Silk Road Briefing 2023).

6.2 Economic

China is one of Uzbekistan's major trade partners. Uzbekistan's primary export goods include gold (30.8%), non-retail pure cotton yarn (11%), refined copper (5.05%), and petroleum gas (4.92%) (OEC 2021). Switzerland (16.4%), China (13.2%), the UK (12.8%), and Russia (11.6%) are the top countries to which Uzbekistan exports its goods (OEC, 2021). Cotton (30.7%), petroleum gas (30%), and refined copper (11.7%) are the primary goods exported from Uzbekistan to China. Broadcasting equipment (3.87%), cars (3.04%), and coated flat-rolled iron (2.65%) are the top goods exported from China to Uzbekistan (OEC—Uzbekistan 2021). In 2021, China's exports to Uzbekistan accounted for $5.63B, while Uzbekistan's exports to China accounted for $1.94B (Gaulier and Zignago 2010).

China is Uzbekistan's largest investor, with $2.2 billion worth of FDI in 2021, followed by Russia ($2.1B), Turkey ($1.18B), and Germany ($800.7 M) (Uzbekistan Strategic Reforms Agency 2021).

Uzbekistan's government debt is at its highest level of $20.4B, equaling 36.4% of its total GDP (Central Bank of the Republic of Uzbekistan 2023). Its main creditors are the Asian Development Bank, World Bank, China Development Bank, and China Exim Bank (Ministry of Economy and Finance of the Republic of Uzbekistan 2022).

There is a potential challenge for bilateral relations due to the negative social sentiment surrounding China, especially among religious groups like Hizb ut-Tahrir and the Islamic Movement of Uzbekistan. On the other hand, the Belt and Road Initiative (BRI) and economic cooperation with China offer a great opportunity for trade expansion and improved infrastructure projects, leading to shortened access to the Persian Gulf.

6.3 Social

Like in other Central Asian countries, China has been trying to promote social and cultural presence in Uzbekistan. This includes the promotion of the language, education, and culture. A delegation led by the head of the Uzbek Ministry of Information recently visited China. They engaged with Chinese research institutions to promote collaborative research and strengthen relations between Uzbek and Chinese think tanks (Eurasianet 2023a, b).

China might have been a little late spreading its language influence in Central Asia. Therefore, it instead focused on establishing educational institutions in the region. The first Confucius Institute was established in Tashkent as early as 2004, with the second being established ten years later (Saida 2022). Tashkent State University of Oriental Studies became the first in Central Asia to launch a Chinese Studies program in 2014, occasionally receiving materials and technical support from China. The Chinese embassy has also funded the establishment of the Chinese Center for Cultural Language and Culture at Tashkent State University. Furthermore, China has held talks to introduce Chinese language courses at Uzbek kindergartens to promote the Chinese language among young learners (Saida 2022).

6.4 Uzbekistan—US Relations

The diplomatic relations between Uzbekistan and the US were officially established in 1992. The broad relationship encompasses security and border concerns, economic development, civil society, political issues, and cultural-educational activity.

In 2021, Uzbekistan exported $107 million worth of goods to the US, mainly silver, copper wire, aircraft parts, and pepper. In 2022, the trade between the US and Uzbekistan amounted to $436.8 million, making the US Uzbekistan's 16 largest trading partner (International Trade Administration 2023).

The United States and Uzbekistan cooperate with the United Nations, the International Monetary Fund, and the Organization for Security and Cooperation in Europe. Uzbekistan is a member of the North Atlantic Treaty Organization's Partnership for Peace and an observing country to the World Trade Organization (US Department of State 2023).

7 Conclusion

Often appearing under the radar, Central Asia is a region of high value for China. From an economic standpoint, the region is rich in hydrocarbons, minerals, and hydropower. Turkmenistan is China's biggest natural gas supplier, while Kazakhstan and Uzbekistan supply crude oil and gas. Kyrgyzstan and Tajikistan may be less wealthy in natural resources, but they are suppliers of minerals like gold. As the largest investor in Central Asia, China has been focusing on infrastructure development through the Belt and Road Initiative. Two international BRI corridors pass through Central Asia, connecting China to the Middle East, South Asia and Europe.

Aside from economic importance, Central Asia plays an important role from a security standpoint. Three out of five states share a common border with the Taliban-controlled Afghanistan, where Tajikistan controls a large portion of Afghanistan's Wakhan corridor border. With the rise of the Taliban, China became keener on the border security question.

Historically, Russia has been a major dominant power in the region, appearing as Central Asia's primary security and economic partner. However, due to the recent events, its shift to the Ukraine crisis has opened a window of opportunity for China to spread its influence further in the region. This works well for Central Asia, known for its multi-vector foreign policy. Indeed, Central Asia has been trying to adopt an approach of being friendly with multiple global powers, including Russia, China, Europe, and the US. China is one of Central Asia's largest trading and investment partners. While this may be beneficial for not being too dependent on Russia, fears over China's ever-increasing influence in the region remain. In fact, China is the largest Central Asian creditor, raising concern over the debt trap policy. This is problematic for vulnerable local economies, coupled with corruption and strict rule of law. Thus, it remains a major concern for the local economies when dealing with China.

Central Asia has not been a priority in US foreign policy, especially in light of recent events in the Middle East and the ongoing conflict in Ukraine. However, due to China's increasing global influence, it is crucial to monitor the region, given its proximity to China and Russia. China's presence in Central Asia is currently at its peak and is expected to grow in terms of political, economic, and social influence. With growing concerns over China's influence and skepticism towards communist China, the US has a unique opportunity to increase its involvement in Central Asian politics.

Central Asia may never appear as the most important region in the world arena. However, its unique history and the current position between major powers make it an enticing region for spreading influence. For its own sake, it would continue being friends with all and keep its doors open, hoping one day to become a region with its own political and economic will.

References

Anon (n.d.) https://invest.gov.uz/mediacenter/news/which-countries-and-how-much-invested-in-uzbekistan-in-2021/. [online] Available at: https://invest.gov.uz/mediacenter/news/which-countries-and-how-much-invested-in-uzbekistan-in-2021/

Bawa J, Ashish (2023) China's grand strategy for Tajikistan: challenges and threats. Asian J Political Sci 31:2, 125–139. https://doi.org/10.1080/02185377.2023.2226880

Blackwood M (2023) Turkmenistan. [online] Congressional Research Service. Available at: https://crsreports.congress.gov/product/pdf/IF/IF10303

Blank S (2019) China's Military Base in Tajikistan: what does it mean? The Central Asia-Caucasus Analyst. Available at: https://www.cacianalyst.org/publications/analytical-articles/item/13569-chinas-military-base-in-tajikistan-what-does-it-mean?.html

Briefing C (2021a) China and Kyrgyzstan: bilateral trade and future outlook. [online] China Briefing News. Available at: https://www.china-briefing.com/news/china-and-kyrgyzstan-bilateral-trade-and-future-outlook/

Briefing C (2021b) China-Tajikistan: bilateral investment and trade ties. [online] China Briefing News. Available at: https://www.china-briefing.com/news/china-tajikistan-bilateral-investment-and-trade-ties/

Briefing C (2021c) China-Uzbekistan: bilateral trade and future outlook. [online] China Briefing News. Available at: https://www.china-briefing.com/news/china-uzbekistan-bilateral-trade-and-future-outlook/

Briefing SR (2023) Kazakhstan Marks Ten Years of China's Belt and Road Initiative. [online] Silk Road Briefing. Available at: https://www.silkroadbriefing.com/news/2023/10/05/kazakhstan-marks-ten-years-of-chinas-belt-and-road-initiative/

CABAR.asia (2021) Living in Debt: to whom and how much does Tajikistan Owe? [online] Available at: https://cabar.asia/en/living-in-debt-to-whom-and-how-much-does-tajikistan-owe

Caspian Policy Center (2022a) CPC | China-Kyrgyzstan Relations. [online] Available at: https://www.caspianpolicy.org/research/security-and-politics-program-spp/china-kyrgyzstan-relations#_edn6 [Accessed 11 Nov 2023]

Caspian Policy Center (2022b) CPC | China-Turkmenistan Bilateral Relations. [online] Available at: https://www.caspianpolicy.org/research/security-and-politics-program-spp/china-turkmenistan-bilateral-relations

Caspian Policy Center (2022c) CPC | China-Kazakhstan Bilateral Relations. [online] Available at: https://www.caspianpolicy.org/research/security-and-politics-program-spp/china-kazakhstan-bilateral-relations

Central Bank of the Republic of Uzbekistan (2023) External Debt. [online] Available at: https://cbu.uz/en/statistics/e-gdds/data/111575/

Central Intelligence Agency (2023a) Kyrgyzstan. The World Factbook. https://www.cia.gov/the-world-factbook/countries/kyrgyzstan/summaries

Central Intelligence Agency (2023b) Tajikistan. The World Factbook. https://www.cia.gov/the-world-factbook/countries/tajikistan/

CGTN (2023) Chart of the Day: China-Central Asia relations enter a new era of mutual trust and benefit. [online] news.cgtn.com. Available at: https://news.cgtn.com/news/2023-05-12/Chart-of-the-Day-China-Central-Asia-relations-enter-a-new-era-1jKaw8iUiT6/index.html [Accessed 11 Dec. 2023]

China Railway Group builds high-quality projects in Central Asia--Seetao. [online] Available at: https://www.seetao.com/details/214693.html [Accessed 14 Oct. 2023]

Congressional Research Service (2021) Central Asia: Background and U.S. Relations. [online] Available at: https://sgp.fas.org/crs/row/R46924.pdf

China Daily (2003) Brief introduction to relations between China and Kazakhstan. China Daily. Archived from the original on April 2, 2015. Accessed 10 Oct 2008, from http://www.chinadaily.com.cn/en/doc/2003-05/27/content_166588.htm

CEIC Data (2023) Government Debt (% of Nominal GDP). CEIC. https://www.ceicdata.com/en/indicator/kazakhstan/government-debt--of-nominal-gdp

English.mofcom.gov.cn (2015) China and Uzbekistan Sign Agreement on Jointly Building the Silk Road Economic Belt. [online] Available at: http://english.mofcom.gov.cn/article/newsrelease/significantnews/201506/20150601016814.shtml [Accessed 11 Dec. 2023]

Eurasian Economic Commission (2018) AGREEMENT ON ECONOMIC AND TRADE COOPERATION BETWEEN THE EURASIAN ECONOMIC UNION AND ITS MEMBER STATES, OF THE ONE PART, AND THE PEOPLE'S REPUBLIC OF CHINA, OF THE OTHER

PART. [online] Available at: https://eec.eaeunion.org/upload/medialibrary/5b9/Tekst-angiyskiy-_EAEU-alternate_-final.pdf
Eurasianet (2022) China Debt: A Problem for Tajikistan. [online] CPECWire. Available at: https://cpecwire.com/connectivity/belt-road-bri/china-debt-tajikistan/
Eurasianet (2011) Tajikistan: China's advance causing increasing unease among Tajiks. https://eurasianet.org/tajikistan-chinas-advance-causing-increasing-unease-among-tajiks
Eurasianet (2023) China-Kyrgyzstan-Uzbekistan railway: who is going to pay? https://eurasianet.org/china-kyrgyzstan-uzbekistan-railway-who-is-going-to-pay, https://eurasianet.org/china-kyrgyzstan-uzbekistan-railway-who-is-going-to-pay
Eurasianet (2023) China looks to expand soft-power presence in Central Asia. [online] Available at: https://eurasianet.org/china-looks-to-expand-soft-power-presence-in-central-asia
Gaulier G, Zignago S (2010) BACI: international trade database at the product-level. The 1994–2007 Version. CEPII Working Paper, No. 2010–2023
Gavin Helf (2023) United States Institute of Peace New Central Asian Leaders Look to Balance Relations with Major Powers. [online] Available at: https://www.usip.org/publications/2023/09/new-central-asian-leaders-look-balance-relations-major-powers
Hedlund S (2019) China finds investment in Kyrgyzstan a risky necessity. [online] GIS reports. Available at: https://www.gisreportsonline.com/r/kyrgyzstan-china
Hedlund S (2023) Tajikistan: a litmus test for China in Central Asia. [online] GIS reports. Available at: https://www.gisreportsonline.com/r/china-central-asia-tajikistan/
International Monetary Fund (IMF) (2021) Coordinated Direct Investment Survey (CDIS). Accessed 10 Dec 2023, from https://data.imf.org/?sk=40313609-F037-48C1-84B1-E1F1CE54D6D5
International Trade Administration (2023) Uzbekistan—Country Commercial Guide. [online] Available at: https://www.trade.gov/country-commercial-guides/uzbekistan-market-overview
In Kazakhstan, Generation Z on Alert Over China (2023) Jamestown. [online] Available at: https://jamestown.org/program/in-kazakhstan-generation-z-on-alert-over-china/
Jardine B (2021) Emerging forms of Pax Sinica in Tajikistan and Cambodia. Foreign Policy Research Institute. Accessed 11 Dec 2023, from https://www.fpri.org/article/2021/03/emerging-forms-of-pax-sinica-in-tajikistan-and-cambodia/
Kucera J (2014a) U.S. Formally Closes Its Kyrgyzstan Air Base. [online] Eurasianet. Available at: https://eurasianet.org/us-formally-closes-its-kyrgyzstan-air-base
Kucera J (2014b) With Turkmenistan, China Now Has 'Strategic Partnerships' With All Five Central Asian States. Eurasianet, 13 May 2014, eurasianet.org., from https://eurasianet.org/with-turkmenistan-china-now-has-strategic-partnerships-with-all-five-central-asian-states
Mahmadrasul M (2020) Chinese Ambassador Liu Bin: China ranks second after Russia in the number of Tajik students. Accessed 22 Aug 2022, from Vecherka: https://vecherka.tj/archives/42040
Malik A, Parks B, Russell B, Lin J, Walsh K, Solomon K, Zhang S, Elston T-B, Goodman S, Mary W (2021) Banking on the Belt and Road: Insights from a new global dataset of 13,427 Chinese development projects. [online] Available at: https://docs.aiddata.org/ad4/pdfs/Banking_on_the_Belt_and_Road__Insights_from_a_new_global_dataset_of_13427_Chinese_development_projects.pdf
Ministry of Economy and Finance of the Republic of Uzbekistan (2022) State Debt. [online] Available at: https://anhor.uz/economy/national-debt/
Muratalieva N (2023) Kyrgyzstan's path to peak repayments to China: context and dilemmas. [online] The Diplomat. Available at: https://thediplomat.com/2023/08/kyrgyzstans-path-to-peak-repayments-to-china-context-and-dilemmas/
Muratshina K (2017) Asian infrastructure investment bank and post-Soviet Central Asia: New Multilateral Bank. Central Euro J Int Secur Stud 11(3). https://www.proquest.com/docview/2137488938?pq-origsite=gscholar&fromopenview=true
Najibullah F (2022). Central Asia Takes Economic Hit from Russian War in Ukraine Sooner Than Expected. RFE/RL. https://www.rferl.org/a/central-asia-migrants-ruble-impact/31730968.html

National Bank of the Kyrgyz Republic (2010) Remittances of individuals made through the money transfer system. https://www.nbkr.kg/index1.jsp?item=1785&lang=ENG

National Bank of the Republic of Kazakhstan (2023) External Debt. National Bank of Kazakhstan: https://nationalbank.kz/en/news/vneshniy-dolg

National Statistic Committee of the Kyrgyz Republic—State debt—Open Data—Statistics of the Kyrgyz Republic. [online] Available at: https://www.stat.kg/en/opendata/category/34/ [Accessed 7 Oct. 2023]

Nye JS (1990) Soft power. Foreign Policy (80):153–171. Vancouver

Otorbaev D (2023) Railway can drive Central Asia's development faster. [online] Chiina Daily. Available at: https://www.chinadaily.com.cn/a/202310/09/WS65233ac3a310d2dce4bb9717.html

Pikulicka-Wilczewska A (2018) On the reform path: Uzbekistan opens up after years of isolation. [online] www.aljazeera.com. Available at: https://www.aljazeera.com/features/2018/10/14/on-the-reform-path-uzbekistan-opens-up-after-years-of-isolation

Public Debt and State Guarantees of the Kyrgyz Republic for 2022. (2022). [online] Ministry of Finance of the Kyrgyz Republic. Available at: https://www.minfin.kg/storage/files/shares/%D0%94%D0%BE%D0%BA%D1%83%D0%BC%D0%B5%D0%BD%D1%82%D1%8B/50fb45f0b8bc7f71c17bb3c9332565fe.pdf

Putz C (2020) Kyrgyz-Chinese Joint venture scrapped after protests. [online] thediplomat.com. Available at: https://thediplomat.com/2020/02/kyrgyz-chinese-joint-venture-scrapped-after-protests/ [Accessed 11 Dec. 2023]

Reuters (2023) China's Xi calls for greater cooperation with Turkmenistan on natural gas. https://www.reuters.com/world/asia-pacific/chinas-xi-calls-greater-cooperation-with-turkmenistan-natural-gas-2023-01-06/

Rickleton C (2013) Tajik MP is against mixed marriages "Especially with Chinese". https://globalvoices.org/2013/07/04/tajik-mp-is-against-mixed-marriages-especially-with-chinese/#

Russia in Global Affairs (2023) Kazakhstan's multi-vector foreign policy. [online] Available at: https://eng.globalaffairs.ru/articles/kazakhstan-multi-vector/ [Accessed 11 Dec. 2023]

Saida N (2022. The language of the China-Russia soft power rivalry in Uzbekistan. [online] The Diplomat. Available at: https://thediplomat.com/2022/05/the-language-of-the-china-russia-soft-power-rivalry-in-uzbekistan/

Shambetov T (2023) Kyrgyzstan turns to Chinese companies for major hydropower project. radio free europe/radio liberty. [online] 1 Aug. Available at: https://www.rferl.org/a/kyrgyzstan-hydropower-megaproject-china/32529949.html [Accessed 14 Nov. 2023]

Shen W V (2012) China and its Neighbours: troubled relations. [online] Available at: http://drtktopecollege.in/pol/sites/default/files/university%20question%20papers/China%20and%20its%20Neighbours_%20troubled%20relations.pdf [Accessed 11 Dec. 2023]

ShipHub (2023) New China-Kazakhstan border railway port. https://www.shiphub.co/new-china-kazakhstan-border-railway-port/, https://www.shiphub.co/new-china-kazakhstan-border-railway-port/

Silk Road Briefing (2020) 84% of China's BRI investments are in medium-high risk countries. https://www.silkroadbriefing.com/news/2020/07/02/84-chinas-bri-investments-medium-high-risk-countries/

Silk Road Briefing (2022) Kyrgyzstan to Uzbekistan exports up as growing trends justify the China-Kyrgyz-Uzbek Railway. [online] Available at: https://www.silkroadbriefing.com/news/2022/06/13/kyrgyzstan-to-uzbekistan-exports-up-as-growing-trends-justify-the-china-kyrgyz-uzbek-railway/ [Accessed 11 Dec. 2023]

Silk Road Briefing (2023) Uzbekistan's opportunities and challenges from China's belt and road initiative. [online] Available at: https://www.silkroadbriefing.com/news/2023/03/20/uzbekistans-opportunities-and-challenges-from-chinas-belt-road-initiative/

Simoes AJG, Hidalgo CA (2011) The economic complexity observatory: an analytical tool for understanding the dynamics of economic development. In: Workshops at the twenty-fifth AAAI conference on artificial intelligence

South Caucasus and Central Asia: The Belt and Road Initiative Kazakhstan Country Case Study. (2020). [online] World Bank. Available at: https://documents1.worldbank.org/curated/en/471731593499938164/pdf/South-Caucasus-and-Central-Asia-The-Belt-and-Road-Initiative-Kazakhstan-Country-Case-Study.pdf

Sukhankin S (2020) The security component of the BRI in Central Asia, part two: China's (Para) Military Efforts to Promote Security in Tajikistan and Kyrgyzstan. Jamestown Foundation. Accessed 1 Nov 2023, from https://jamestown.org/program/the-security-component-of-the-bri-in-central-asia-part-two-chinas-paramilitary-efforts-to-promote-security-in-tajikistan-and-kyrgyzstan/

Sukhankin S (2020) The security component of the BRI in Central Asia, part three: China's (Para) military efforts to promote security in Kazakhstan, Uzbekistan and Turkmenistan. The Jamestown Foundation. https://jamestown.org/program/the-security-component-of-the-bri-in-central-asia-part-three-chinas-paramilitary-efforts-to-promote-security-in-kazakhstan-uzbekistan-and-turkmenistan/

China Forms Antiterror Alliance With Pakistan, Tajikistan, Afghanistan (2016) RadioFreeEurope/RadioLiberty. [online] Available at: https://www.rferl.org/a/china-forms-anti-terrorism-alliance-pakistan-tajikistan-afghanistan/27901474.html [Accessed 14 Oct. 2023]

The Observatory of Economic Complexity (2022) Uzbekistan (UZB) Exports, Imports, and Trade Partners. [online] Available at: https://oec.world/en/profile/country/uzb#:~:text=Yearly%20Trade&text=During%20the%20last%20five%20reported [Accessed 24 Dec. 2023]

The State Council The People's Republic of China. (2022) China-Central Asia trade grew by 100 times over 30 years. [online] Available at: http://english.www.gov.cn/statecouncil/ministries/202201/18/content_WS61e60de2c6d09c94e48a3cf9.html

The White House (2023) C5+1 Leaders' joint statement. [online] Available at: https://www.whitehouse.gov/briefing-room/statements-releases/2023/09/21/c51-leaders-joint-statement/

U.S. Department of Commerce, International Trade Administration. (2023, December). Turkmenistan—market overview. International Trade Administration. https://www.export.gov/article?series=a0pt0000000PAv2AAG&type=Country_Commercial__kav

U.S. Department of State (2023) U.S.-Kazakhstan relations. [online] Available at: https://www.state.gov/u-s-kazakhstan-relations/#:~:text=Kazakhstan%27s%20military%20participates%20in%20U.S.,Eager%20Lion%2C%20and%20Shanti%20Prayas.&text=Kazakhstan%20is%20the%2081st%20largest,two%2Dway%20trade%20in%202021

U.S. Department of Veterans Affairs (n.d.) Karshi Khanabad (K-2) air base. [online] Available at: https://www.publichealth.va.gov/exposures/karshi-khanabad.asp

UNCTAD (2022) Investment statistics and trends. [online] Available at: https://unctad.org/topic/investment/investment-statistics-and-trends.

United States Census Bureau (2023) Trade in goods with Tajikistan. [online] Available at: https://www.census.gov/foreign-trade/balance/c4642.html

United States Department of State (2022) U.S. Relations with Turkmenistan. [online] Available at: https://www.state.gov/u-s-relations-with-turkmenistan/

United States Institute of Peace (2023) U.S. Institute of peace announces pétronille vaweka the 2023 women building peace award recipient. [online] Available at: https://www.usip.org/publications/2023/09/new-central-asian-leaders-look-balance-relations-major-powers

US Department of Stat. (n.d.. Key topics—Bureau of South and Central Asian Affairs. [online] Available at: https://www.state.gov/key-topics-bureau-of-south-and-central-asian-affairs/

US Department of State (2022) U.S. relations with Tajikistan. [online] Available at: https://www.state.gov/u-s-relations-with-tajikistan/

US Department of State (2023) U.S.-Uzbekistan Relations. [online] Available at: https://www.state.gov/u-s-uzbekistan-relations/#:~:text=Since%20establishing%20diplomatic%20relations%20in,issues%2C%20and%20English%20language%20training

Uzbekistan Strategic Reforms Agency (2021) Investment attractiveness of Uzbekistan through the prism of the Agency for Strategic Development. [online] Available at: https://asr.gov.uz/ru/news/3531

Vakulchuk R, Overland I (2019) China's belt and road initiative through the lens of Central Asia. In: Cheung FM, Hong Y-Y (eds) Regional connection under the belt and road initiative: the prospects for economic and financial cooperation. Routledge, pp 115–133
WITS (2021) China exports by country in US$ Thousand 1995–2021. [online] Available at: https://wits.worldbank.org/CountryProfile/en/Country/CHN/StartYear/1995/EndYear/2021/TradeFlow/Export/Partner/BY-COUNTRY/Indicator/XPRT-TRD-VL
WITS (2021) China Imports by country in US$ Thousand 1995–2021. [online] Available at: https://wits.worldbank.org/CountryProfile/en/Country/CHN/StartYear/1995/EndYear/2021/TradeFlow/Import/Partner/BY-COUNTRY/Indicator/MPRT-TRD-VL
WITS (2021) Kazakhstan Exports by country in US$ Thousand 1995–2021. [online] Available at: https://wits.worldbank.org/CountryProfile/en/Country/KAZ/StartYear/1995/EndYear/2021/TradeFlow/Export/Partner/BY-COUNTRY/Indicator/XPRT-TRD-VL
WITS (2021) Kyrgyzstan Imports by country in US$ Thousand 1995–2021. [online] Available at: https://wits.worldbank.org/CountryProfile/en/Country/KGZ/StartYear/1995/EndYear/2022/TradeFlow/Import/Partner/WLD/Indicator/MPRT-PRTNR-SHR
WITS (2021) Tajikistan imports by country in US$ Thousand 2000–2021. [online] Available at: https://wits.worldbank.org/CountryProfile/en/Country/TJK/StartYear/2000/EndYear/2021/TradeFlow/Import/Partner/BY-COUNTRY/Indicator/MPRT-TRD-VL
World Bank (2020) South Caucasus and Central Asia: the belt and road initiative Kazakhstan country case study. [online] Available at: https://documents1.worldbank.org/curated/en/471731593499938164/pdf/South-Caucasus-and-Central-Asia-The-Belt-and-Road-Initiative-Kazakhstan-Country-Case-Study.pdf
World Bank (2023a) Tajikistan Overview: Development news, research, data. https://www.worldbank.org/en/country/tajikistan
World Bank (2023b) Turkmenistan overview: development news, research, data. https://www.worldbank.org/en/country/turkmenistan/overview
World Bank (2023) Personal remittances, received (% of GDP)—Kyrgyz Republic. [online] Available at: https://data.worldbank.org/indicator/BX.TRF.PWKR.DT.GD.ZS?locations=KG
World Integrated Trade Solution (WITS) (n.d.) China imports by country 1995–2021 | WITS Data. [online] Available at: https://wits.worldbank.org/CountryProfile/en/Country/CHN/StartYear/1995/EndYear/2021/TradeFlow/Import/Partner/BY-COUNTRY/Indicator/MPRT-TRD-VL [Accessed 11 Oct. 2023]
World Nuclear Association (2023) Uzbekistan uranium. [online] Available at: https://world-nuclear.org/information-library/country-profiles/countries-t-z/uzbekistan.aspx
Zhang R (2021) China looks to Turkmenistan for more gas as it cuts Australian supplies. South China morning post. https://www.scmp.com/news/china/diplomacy/article/3133084/china-looks-turkmenistan-more-gas-it-cuts-australian-supplies
Zhao H (2016) Central Asia in Chinese strategic thinking. In: Fingar T (ed) The new great game: China and South and Central Asia in the era of reform. Stanford University Press, pp. 173. (ISBN: 978-0-8047-9764-1, OCLC: 939553543)

Figure Sources

CIA Factbook (2023a) Kazakhstan. [online] Available at: https://www.cia.gov/the-world-factbook/countries/kazakhstan/#geography
CIA Factbook (2023b) Kyrgyzstan. [online] Available at: https://www.cia.gov/the-world-factbook/countries/kyrgyzstan/
CIA World Factbook (2023a) Tajikistan. [online] Available at: https://www.cia.gov/the-world-factbook/countries/tajikistan/
CIA World Factbook (2023b) Turkmenistan. [online] Available at: https://www.cia.gov/the-world-factbook/countries/turkmenistan/

CIA World Factbook (2023c) Uzbekistan. [online] Available at: https://www.cia.gov/the-world-factbook/countries/uzbekistan/

Freedom House (2023) Countries and Territories. [online] Available at: https://freedomhouse.org/countries/freedom-world/scores

Map: Connecting central Asia (2016) Financial Times. [online] 9 May. Available at: https://www.ft.com/content/ee5cf40a-15e5-11e6-9d98-00386a18e39d

Ministry of Economics and Finance of Uzbekistan: Uzbekistan's Government Debt (2022) Overview of the state and dynamics of the state debt of the Republic of Uzbekistan. [online] Available at: https://api.mf.uz/media/filestore/DB_2022.pdf

Sputnik (2022) Billion dollar amount: Tajikistan increased external debt. [online] Available at: https://tj.sputniknews.ru/20220718/tajikistan-vneshniy-dolg-pervoe-polugodie-2022-1050066844.html

Statista (2022) Turkmenistan—gross domestic product (GDP) 1997–2027. [online] Available at: https://www.statista.com/statistics/1034327/gross-domestic-product-gdp-in-turkmenistan/

The World Bank (2021) Central government debt, total (% of GDP) | Data. [online] Available at: https://data.worldbank.org/indicator/GC.DOD.TOTL.GD.ZS

The World Bank (2022a) Foreign direct investment, net inflows (% of GDP) | Data. [online] Worldbank.org. Available at: https://data.worldbank.org/indicator/BX.KLT.DINV.WD.GD.ZS

The World Bank (2022b) GDP (constant 2015 US$) | Data. [online] Worldbank.org. Available at: https://data.worldbank.org/indicator/ny.gdp.mktp.kd

The World Bank (2022c) GDP Growth (annual %) | Data. [online] Worldbank.org. Available at: https://data.worldbank.org/indicator/ny.gdp.mktp.kd.zg

Geopolitical Game-Changer: China's Expanding Role in Latin America's Defense, Aerospace, and Telecommunications

Alvaro Mendez and Gaspard Estrada

Abstract This chapter delves into the expanding influence of the People's Republic of China (PRC) in Latin America and the Caribbean (LAC), with an emphasis on defense, aerospace, and telecommunications. It explores Beijing's pursuit of natural resources in the region and its bid to extend its economic sway there, and how this is reshaping regional power dynamics, notably in relation to the United States. It highlights the PRC's recent strides in the global arms trade, in space exploration initiatives, and in advances in telecommunications technology. It is revealed that Beijing's engagement in these sectors is redefining international partnerships, often placing China in the dominant role. This development is acutely observable in South America: Venezuela emerges as a key collaborator with the PRC in military cooperation and arms procurement, while Brazil's involvement in space and telecommunications reveals a combination of overambition and naivety. Moving beyond merely documenting the PRC's growing presence, the chapter also offers policy recommendations for Latin America and the U.S., recognizing the significant shifts in regional relationships. These insights are vital for grasping the evolving dynamics of U.S.-Latin American relations; they signal major changes in the geopolitical landscape of the Western Hemisphere.

Keywords China · Latin America · United States · Defense · Aerospace · Telecommunications · Geopolitics

A. Mendez (✉)
London School of Economics and Political Science (LSE), LSE Global South Unit, London, UK
e-mail: alvaro.mendez@ubi.edu

G. Estrada
Advisory Board Member, Global South Unit, London School of Economics, London, UK

A. Farhadi et al. (eds.), *The Great Power Competition Volume 6*,
https://doi.org/10.1007/978-3-031-70767-4_9

1 Introduction

The People's Republic of China (PRC) has significantly expanded its presence in Latin America and the Caribbean (LAC) in recent years, concentrating on specific strategic sectors with economic and geopolitical importance. This growing penetration of LAC is motivated by China's quest for natural resources, a desire to extend its economic influence, and a wish to counterbalance the United States' sway in the region and in the world (Mendez and Alden 2021).The chapter relies on a wide-ranging analysis of the academic literature, governmental documents, and media reports to understanding developments in the most strategic sectors, in the following order: the defense industrial and technological base (DITB), space affairs, and telecommunications. In these sectors, China's involvement often resembles a compromise more than a partnership, with the People's Republic of China (PRC) setting the terms and driving the narrative.

Tracking Beijing's activity in these sectors is intricate and challenging due to the shortage of available data and the rapid pace of its proliferation throughout the region. The information presented here is intended as a broad overview rather than an exhaustive account (which may not be possible) of China's presence. Hence, all developments in other strategic sectors, such as infrastructure, agriculture, and energy, where Beijing's influence is deeply entrenched, cannot be covered.

In the DTIB sector, for instance, official data from 2019 shows that the PRC exported $4.9 billion in military goods and services. This figure compares with Russia's exports of $10.4 billion and the United States' $170 billion. Thus, China's exports in 2019, while modest, were substantial enough to establish it as a major supplier of military equipment; viz., 622 ships, 359 aircraft, 271 armored vehicles, and 159 missiles (U.S Department of State 2021). In terms of arms export volumes, China still lags behind the United States and Russia; however, the rapid expansion of its military industry, especially into regions like LAC, underscores the country's potential to become a more significant player in the future. Other reports also highlight the sheer growth of China's defense sector.

A similar trend is seen with regards to aersospace. China figures ever more prominently in the development of space architecture and space cooperation. While these ambitions are not new, they are being strengthened by President Xi Jinping's determination to transform the PRC into a "space power" (Xinhua 2019). A 2022 white paper concluded that the aerospace industry will be part of China's "national strategy" (The State Council PRC 2022). The official version of this strategy emphasizes enhancing China's capabilities in space in order to support its national security, to foster self-reliance in science and technology, to stimulate high-quality economic and social development, and to advocate for responsible governance in outer space. U.S. analysts argue that the strategy is part of a broader effort to boost Beijing's military capability to become on par with some regional space powers like Russia and India, and with farther-reaching powers like the U.S.

Following the discussion on China's defense sector and space exploration initiatives, our final variable of analysis is the country's telecommunications sector, specifically its place in the world of Information and Communications Technology (ICT). As noted by Xia (2022), China has made significant strides in reducing its dependency on foreign technology, a trend predating the 2018 US-China trade war. Companies like Huawei have been on the leading edge of developing 5G,—crucial for technologies like Smart Cities and the Internet of Things. Bartholomew (2020) raises China's intent to influence global technology standards, which poses substantial economic and national security challenges to the US and its allies. This has led the Five Eyes alliance—the UK, the US, Canada, Australia, and New Zealand—to exclude Huawei from their 5G networks, demonstrating the international concern over China's strategy in the telecommunications sector.

Considering the significant developments outlined above, analysis can now shift to the specific context of the strategically important region of LAC. Here, this chapter will explore how China's prowess in the defense, aerospace, and telecommunications industries is reshaping the *geopolitical* landscape. The discussion aims to explore the nuances of Beijing's engagement, to reveal both the opportunities and challenges of Beijing's presence in these key economies in Latin America. This evolution brings profound geopolitical consequences for Washington. Beijing's deepening involvement challenges long-standing U.S. influence in what it traditionally considers its backyard. The implications are far-reaching, potentially altering the regional, even the global balance of power as well as the nature of U.S.-Latin American relations in the coming decades.

2 Defense Technological and Industrial Base (DTIB)

The Center for Strategic and International Studies (CSIS), analyzing data from the Stockholm International Peace Research Institute (SIPRI) from 2010 to 2020, ranked China fifth in military exports during that decade (CSIS 2020). But since then, China has climbed to second, just behind the United States (Forrest 2020). Of the value of the PRC's military exports over that ten-year period 77.3% were sold to Asian countries, followed by 19.1% to African countries (RAND Corporation 2022).

Around the turn of the millennium military cooperation with China became the preferred choice of certain Latin American governments due to China's willingness to offer more affordable deals with fewer conditions attached (Paz 2006). Also, since 2000, PRC military personnel have made more than 200 visits to LAC to meet with their counterparts. China has set up a high-level forum on defense affairs with the Community of Latin American and Caribbean States (CELAC), and provides professional education for military personnel from LAC. The PLA sends its own military personnel to LAC for jungle warfare training (Lazarus and Berg 2022). China has sold weapons, airplanes, tanks, and other military equipment to many countries in LAC,—Venezuela, Argentina, and Bolivia, among others. The People's

Liberation Army Navy (PLAN) has visited many ports in the region, some as close to the United States as Cuba (Lazarus and Berg 2022).

Unlike the Cold War era, when countries were clearly aligned with either the West or the East, today's geopolitical dynamics are more complex. It is no longer a simple dichotomy as between nations in military partnership with China versus those allied with the West. Beijing has notably refrained from formal military alliances or taken a stance directly opposed to the United States. Neither have they tried to establish permanent military bases in the region. Of course, LAC countries ruled by more left-leaning politicians have shown a propensity to deal with the PRC on military matters. It is worth noting, too, that Venezuela, Bolivia, and Ecuador account for most of LAC's arms imports from China, indicating their significant presence in the market (Gurrola 2018). The next section gives an in-depth, country-specific account of China's military presence in LAC.

2.1 South America: Andean Region

Venezuela under Hugo Chavez (1999–2013) and now Nicolas Maduro (2013-present) developed deep military ties with Beijing. Caracas was among the first in LAC to purchase Chinese military hardware; viz., K-8W Karakorum jet trainers, Y-8 transport aircraft, VN-4 armored personnel carriers, and military radars (Ellis 2011). Venezuelan personnel have been trained in China and participated in Chinese military exercises like the 2017 airborne troops contest (Ellis 2022b). And the PLA has paid visits to Venezuela to participate in military parades and to provide training and support for Chinese-made military systems. These engagements have caused concern in the United States, which has placed Venezuela under sanctions and accused China of propping up the Maduro regime (Ellis 2022b).

Building on this foundation, Venezuela's military acquisitions from China have broadened in recent years; Beijing is currently the most important supplier of weapons to Caracas (Control Ciudadano 2021). The expanded arsenal encompasses mortars, rocket launchers, sundry combat and amphibious vehicles, and heavy machine guns. Specialized weaponry such as anti-tank and anti-submarine arms have also been added to Venezuela's arsenal, courtesy of China (Dialogo Americas 2022). The chronic political instability in Venezuela intersecting with this amplified military might poses a potential threat to its neighbor Colombia, should tensions boil over to the point of armed conflict (Ellis 2017).

Ecuador during Rafael Correa's administration (2007–2017) developed significant military ties to China, like Venezuela but with more challenges (Alden and Mendez 2023). In 2009 Quito purchased MA-60 transport aircraft and contracted with the China Electronics Technology Group Corporation (CETC) for air defense radars. Issues with equipment suitability and integration, however, has led to many disputes, and in 2013 CETC sued the Ecuadorian government for $280 million (Redaccion El

Universo 2016). Despite this, Ecuador continued procuring Chinese military equipment. In 2015 they bought 709 military vehicles, mainly from Sinotruk, the Chinese state-owned truck manufacturer, in a deal worth $81 million (Ellis 2018).

Ecuador's increasingly close military ties with China have raised concerns in the United States, which accused Ecuador of undermining regional security and democracy (Gurrola 2018). In response, the U.S. has tried to counteract China's influence in Ecuador by cultivating relationships with other regional partners. Meanwhile, Ecuador continued its training exchanges with China, although less extensively than Venezuela, while still predominantly relying on U.S. and European military doctrines. This shift in Ecuador's strategy has leveraged China to pursue an alternative path in politics and development (Ellis 2018).

Turning to another Andean country, Bolivia witnessed a significant deepening of military ties with the People's Republic of China during President Evo Morales's leadership from 2006 to 2019. The Morales government actively pursued military procurement from China, acquiring six Harbin H-425 helicopters and a diverse array of military vehicles. Notably, the acquisitions included 31 armored cars, among other equipment (Ellis 2016). But the bilateral relation extended beyond acquisitions and donations; Bolivia and China cooperated in other areas, too, with La Paz participating in institutional visits, training exercises, and professional military education (PME) in China (Ellis 2018). In a notable show of support, the Chinese government made a significant donation to Bolivia in mid-July 2016: the Bolivian Armed Forces received 31 assault and anti-riot vehicles with a total value of $8 million (Prensa Latina 2016). The exact extent of Bolivia's current military cooperation with China remains unknown even after Morales's resignation in 2019 (Ellis 2022b).

Shifting the focus to Peru, another Andean nation, a more strategic approach to military cooperation and procurement becomes evident. Peru has maintained relationships with both the United States and China. Lima has sent military personnel on official visits to China as part of training and PME (Professional Military Education) initiatives, and in 2011 conducted a joint military exercise with the PLA (Ramírez Bullón 2021). In 2014 Peru made a significant move when it contracted with China for 40 Type-90B Multiple Rocket Launch Vehicles (Guevara 2015). Although only 27 were delivered, this transaction marked the first step toward military cooperation. It also led to a corruption investigation by the Peruvian authorities, indicating the murkiness of bilateral relations (Watson 2017). More recently, military cooperation between Lima and Beijing has been diminishing. Nevertheless, Peru seems to be showing more interest in granting the PRC control over critical infrastructure, such as electricity supply and ports, which has raised concerns in Washington as well.

2.2 *South America: Southern Cone*

Countries of the Southern Cone have substantially grown closer to Beijing in recent years. Argentina, for instance, under the leftist Peronist governments of Nestor Kirchner (2003–2007) and Cristina Fernandez (2007–2015) purchased four Chinese

WMZ-551 armored vehicles and negotiated for 20 FC-1 fighters and 110 VN-1 armored vehicles. A deal for five P-18 offshore patrol vessels was not completed (Valiente 2018). Argentina also sent its military personnel to China for PME and on institutional exchanges, while hosting two PLAN missile frigates on a port call in 2013. Argentina's military cooperation with China waned under the government of Mauricio Macri (2015–2019), who limited purchases of military equipment and other ties with the PRC to avoid undermining its relationship with Washington (Ellis 2022b). But reports have surfaced that Beijing was lobbying Buenos Aires to buy newly made Chengdu JF-17 Thunder Block Three fighters from China built in cooperation with Pakistan. This competition also involved Washington, which lobbied hard for Buenos Aires to acquire F-16s instead (from Denmark) (Johnson 2023). In the end, Argentina opted for the F-16s, with Washington authorizing the sale of 33 of them in mid-October 2023 (Sanders 2023).

Beijing has also lobbied for permission to build a naval base in Ushuaia, in Tierra del Fuego Province. Such a base would essentially cede control of the passage between the Atlantic and the Pacific oceans to Beijing, while also serving as China's door to Antarctica (Saavedra 2023).

Other countries of the Southern Cone have had limited dealings with China; Chile, for instance, has engaged in exchanges and exercises with Beijing but has purchased no Chinese military goods. In 2022 Uruguay announced its intention to replace its A-37 planes with fast-flying aircraft with a capability to intercept. It was reported that China offered the Hongdu L-15 LIFT (Lead-In Fighter Trainer) aircraft (Rivas 2022). Around the same time, Washington offered to sell aircraft, helicopters, weaponry, and other military goods to Montevideo. According to reports, the US wanted to dissuade Uruguay from procuring such goods from China (AeroNaves 2023). This decision underscores Washington's growing concern over China's increasing influence in South America. Given that the U.S. has historically viewed Latin America as its sphere of influence, China's expanding presence in the region is viewed with apprehension.

2.3 Caribbean: Cuba and Beyond

Under the successive governments of Fidel Castro (1959–2008), Raul Castro (2008–2018), and Miguel Diaz-Canel (2018-present), Cuba has maintained continuous military relations with China. High-level talks between Chinese and Cuban defense ministers are routine, as are pledges to strengthen defense cooperation. Cuba has hosted PLAN warships and the hospital ship Peace Ark on port visits (Franks 2011). While specific purchases are not widely publicized, Cuba is known to have purchased Chinese-made tanks, anti-aircraft systems, and other military hardware (Horta 2008); however, it is not known with certainty whether these purchases were made recently or deeper in the past, nor the exact extent of Cuba's current military cooperation with the PRC (Ellis 2022b).

The first LAC country to establish diplomatic relations with the PRC, Cuba's relationship with the PRC is a close one, as was evident during the recent state visit of Diaz-Canel to Beijing in November 2022. It was not reported whether military affairs were part of their discussion. Following the visit, an 18-point joint statement was released by Havana and Beijing, committing themselves to strengthening their cooperation. In that joint statement it was noted that "China firmly supports the just struggle of the Cuban people to defend national sovereignty and oppose foreign interference and blockade", a significant point given Cuba's geographical proximity to and history with the United States (Socialist China 2022).

There have been other Caribbean nations such as Barbados, Guyana, Trinidad and Tobago, and Jamaica which have benefited from Chinese military aid in the form of equipment donations or training assistance for their military and police forces. There have also been direct transactions, like the sale of an offshore patrol vessel (OPV) to Trinidad and Tobago in 2014. Additionally, there have been some sizable donations, such as $1.1 million worth of non-lethal gear to the Jamaica Defense Force in 2011 (Roy 2022).

3 Aerospace

Beijing has moved to involve LAC in space cooperation, cultivating close ties with several countries in the region. A number of Latin American countries have been operating space programs for decades. Argentina and Brazil were the pioneers, having established space programs as early as 1960, with Mexico following suit in 1962 (Guzman 2021). These programs were set up during the Cold War, when Washington was still the leading external actor promoting them (Mendez 2017). Interestingly, the scope of Beijing's space relations with LAC is ampler than is usually understood. It is widely recognized that contemporary China-LAC relations in general began in the 2000s, yet the evidence is that specifically outer space relations date back at least two decades earlier (Li 2007). In this context, it is no surprise to learn that, as at 2021, China has become the second-largest investor in LAC's aerospace sector (Guzman 2021). In 2021 a constitutive agreement at a CELAC summit in Mexico established the Latin American and Caribbean Space Agency (ALCE) to coordinate space exploration from the LAC region, with headquarters in Mexico (SRE 2021). Considering the need for significant funding and the limited engagement of the US, coupled with China's frequent interaction with CELAC, it is possible that Beijing may play a role in this initiative someday.

Beijing's engagement with LAC utilizes entities associated with the Chinese government, primarily state-owned enterprises (SOEs) and private Chinese multinationals, in order to develop, launch, and operate satellites. China has established ground-control centers and facilities for personnel training in the region. China's presence in LAC's space sector is diverse, with significant involvement in the central-eastern part of South America (Brazil); in the Andes region (Bolivia, Ecuador, and

Venezuela); and in the Southern Cone (Argentina and Chile). The following discussion gives more specific details about these activities, which vary in depth across countries.

3.1 Brazil

The PRC and Brazil have engaged in long-standing collaborations dating back to 1984's joint agreement to undertake a China-Brazil Earth Resources Satellite program (CBERS), with the objective to reduce reliance on the United States and attain independence by acquiring their own satellite data on their respective territories (Klinger 2018). Since 1999, this program has launched six high-resolution earth observation satellites, all at the Taiyuan Satellite Launch Center in the northern Province of Shanxi, China. The satellites now provide valuable data for agriculture, urban planning, and environmental monitoring (USCC 2018). The China Great Wall Industry Corporation (CGWIC), a Chinese SOE, plays a key role in Sino-Brazilian cooperation, helping launch CBERS-01 (October 1999), CBERS-02 (October 2003), and CBERS-2B (September 2007) atop Long March-4 rockets. The CBERS-3 satellite launch failed in December 2013 due to a launch vehicle malfunction (Reuters 2013).

The program recovered with the successful launch of CBERS-4 in December 2014 and with the most recent satellite, CBERS-4A, in December 2019 atop an LM-4B rocket at Shanxi (Chin 2023). CBERS-4A is an optical imaging satellite, collaboratively run by Brazil's *Instituto Nacional del Pesquisas Espaciais* (INPE) and China's Center for Resources Satellite Data and Application (CRESDA), which "monitors water resources, agriculture, urban growth, land use, and contributes to the preservation of the Amazon rainforest" (Chin 2023). Brasilia announced in April 2023 that the two countries had prepared new agreements to continue their cooperation in the CBERS Program framework (Oliveira 2023).

Other satellite activities include a joint venture announced in 2020 between Brazil's Alya Nanosatellites and Beijing's Tianlian Space Technology to establish a ground station in Tucano, a municipality in the north-eastern Brazilian State of Bahia (Forum China PLP 2020). Alya, a firm specializing in advanced hyperspectral imaging via nanosatellites, enables optimization of satellite orbits covering Brazilian territory, and offers other services from data processing to support for LEO and GEO rockets. With partner Tianlian the venture focuses on collaboration and knowledge-sharing in space technology (Raquel 2020). Tianlian is an affiliate of China Academy of Space Technology (CAST) and China Aerospace Science and Technology (CASC) (NASA 2019). Recently, Washington has blacklisted CAST and CASC, along with other Chinese companies, on suspicion of aiding the People's Liberation Army (PLA) with intelligence and reconnaissance (BIS 2023). It is believed that the Tucano Ground Station may already be operational, but its exact location is not publicly known (Funaiole et al. 2022).

Other aerospace collaborations include those classified as research and development. This is led by the Chinese Academy of Sciences (CAS), which is directly under the supervision of the State Council (CSET 2022). In 2013 the China-Brazil Joint Laboratory for Space Weather (CBJLSW) was set up as a research joint venture between the Brazilian National Institute for Space Research (INPE) and the National Space Science Center (NSSC), an organ of the CAS (Nature 2022). According to the Chinese, this facility has two primary remits. First, it conducts ground-based observations and collects data along the meridian chain in the Western hemisphere. Secondly, it undertakes space environment exploration and research, focusing on low-latitude areas in both the Eastern and Western hemispheres (NSSC 2014).

China is keen to expand this sort of collaboration by partnering with academic institutions in Brazil. One example is the astronomical observatory at the State University of Paraiba (UEPB), which has been under construction since January 2022 in collaboration with the Astronomical Observatory of Shanghai, which has agreed to pay for equipment and setup costs (UEPB 2022b). A delegation from the NSSC visited the UEPB in October 2022 to announce China's desire to expand the project by installing new instruments to observe the upper atmosphere. The partnership is to monitor space debris, track asteroids, and conduct research on atmospheric phenomena (UEPB 2022a).

China and Brazil's collaboration in satellite technology research and development has raised security concerns in some quarters of Washington, who perceive the partnership as a "threat to U.S. national security" (Stokes et al. 2020). Some US analysts fear the sharing of rocket launch expertise for high-resolution imaging capabilities will enable China to target US satellites using cutting-edge developments. The fear is amplified by the access China may have to Brazil's space tracking facilities (Johnson 2005). It remains uncertain whether these apprehensions will materialize.

3.2 Andes Region: Venezuela, Bolivia, and Ecuador

China has played a crucial role in enabling Venezuela's entry into space exploration, with bilateral ties tracing back to 2005, when Caracas enlisted CGWIC to create its inaugural satellite (Hermoso and Fermín 2019). The agreement encompassed not only satellite development but also the construction of the *El Sombrero* satellite ground station, located in Guárico in central Venezuela (Funaiole et al. 2022). The PRC was next involved in making and launching in 2008 a communications relay satellite named VENESAT-1, which has been inoperative since 2020 (Space Watch Global 2020). The PRC also built two remote-sensing satellites for Caracas, VRSS-1, launched 2012, and VRSS-2, launched in 2017 (World Metereological Organization 2023). Both satellites serve multiple purposes from data collection and disaster prevention to services in remote areas. They also support environmental, agricultural, and health applications as well as planning, emergency management, and defense (Guzman 2021).

CGWIC again played a crucial role in building ground control facilities for these satellites as well as training Venezuelan personnel in outer space operations (Ellis 2022a). In Bolivia the government of Evo Morales set up the Bolivian Space Agency in 2010 with the objectives of launching Bolivia's first satellite; developing novel space projects; and training Bolivian personnel in space technology (Ellis 2016). La Paz then negotiated an agreement with Beijing for the latter to help it construct the Tupac Katari satellite (UMSA 2013). It was developed by CGWIC and launched from Xichang, China on December 20, 2013 (Ellis 2016). This state-of-the-art communications satellite cost $302 million in total and is currently operational (Guzman 2021, see also: USCC 2018). Beijing also provided extensive training to Bolivians in satellite technology and operations, and helped them set up ground control facilities (Ellis 2022a).

In Ecuador the PRC supported the government of Rafael Correa in launching a micro-satellite called Pegasus in 2013. Ecuadorian officials claim that it was 100% manufactured by the Ecuadorian Space Agency (EXA), but it is unclear that Quito had the knowhow to do this (EoPortal 2013). It is likely that the Ecuadorians received help from Beijing, as the micro-satellite was actually launched from the Jiuquan Satellite Launch Center in China on 26 April 2013 (Global Security 2020). In November 2013 Ecuador launched a second satellite from Russia named Kryasor (a twin of Pegasus). The satellite is expected to re-enter the Earth's atmosphere in 2025 (EXA 2023). In 2014 Ecuador announced that it wanted to develop and launch its own satellite with support from China, for which the Ecuadorian defense minister travelled to China to ink a strategic alliance for this purpose (Infoespacial 2014).

It is notable that China's engagement with these three Andean nations had a common ideological basis: each was governed by a left-leaning regime that experienced tensions with Washington. Against this backdrop, Beijing saw these countries as more amenable sites for its projects. In turn, the three identified Beijing as a provider of space technology, a resource the US seemed reluctant to offer. The intensity of collaboration differed across all three, reflecting their varying economic and political landscapes. Ecuador in particular, given its political and economic standing during this period, established a deeper relationship with China than the others. This was especially evident during the Rafael Correa years (2008–2017), when ties between China and Ecuador reached unparalleled heights (Alden and Mendez 2023).

3.3 Southern Cone: Argentina and Chile

China's aerospace activities in the Southern Cone, including key projects in Argentina and some in Chile, are particularly important. Argentina's national Space Agency (CONAE) joined hands in 2012 with China Satellite Launch & Tracking Control General (CLTC), an organization under the Commission of Science, Technology and Industry for National Defense (COSTIND) for collaboration on space matters (Funaiole et al. 2022). This laid the groundwork for a cutting-edge $50 million satellite and mission control center in Neuquén, a western Argentine province on the northern edge of Patagonia (USCC 2018). Aerial photography in December 2013 confirmed that construction of the ground station was underway (AidData 2023). Called *Espacio Lejano*, it was contracted under a 50-year, equity-free agreement. The geopolitically strategic location is optimal for space exploration due to its open skies, free of radio-electric interference, remote from urban centers, and favorably positioned geologically, climatically, and topographically (Ramirez Aburto 2023). The Argentine government under President Cristina Kirchner negotiated the deal in 2014 to cement a strategic partnership with China (Goñi 2015). However, the deal reportedly limits Argentina's sovereignty; concedes extensive tax exemptions; and permits migrant Chinese laborers to work under Chinese labor laws (Watson-Lynn 2020).

Inaugurated in March 2018, the facility is managed by CLTC. Its primary mission is to support civilian space activities and deep space exploration. Critics warn, however, that it is equally fitted-out for intelligence gathering and even, potentially, for interfering with communications, electronic networks, and electromagnetic systems throughout the Western Hemisphere (USCC 2018; Ellis 2022a). US officials like Admiral Craig Faller have voiced concerns about the Chinese military's use of this deep-space ground station to track and target American and allied satellites (Selligman 2019). And its secrecy—surrounded by an 8-foot-high barbed wire fence and accessible only by appointment—has alarmed locals (Ramirez Aburto 2023). Critics worry that the Argentine government's oversight is marginalized, and question the station's true purpose, if it exceeds the claimed remit of peacefully using outer space (Garrison 2019).

China's steps in Argentina are not only taken by government agencies and SOEs, but also by private companies close to the PRC. One example is Emposat, a company associated with the Chinese Academy of Sciences. Emposat is collaborating with the Argentine firm Ascentio in a joint venture to construct a ground station in Río Gallegos, Argentina, located in the southern part of Patagonia. Upon completion, the facility is expected to feature four to six variably sized antennas to enhance the tracking of satellites orbiting Earth due to its proximity to the South Pole. US satellite imaging has yet to determine its exact location (so far as the author has been able to learn), and other information about the joint venture is still ambiguous (Funaiole et al. 2022).

China operates a satellite ground station in Chile which has been instrumental in the launch of six satellites to support its satellite navigation system Beidou, and it has built advanced satellite tracking stations in Chile. This demonstrates once again its commitment to the development of space infrastructure in LAC (USCC 2018). Chile's aerospace sector features the notable Santiago Satellite Station, managed by the Swedish Space Corporation (SSC), which collaborates with the likes of NASA and the European Space Agency. It has a longstanding relationship with Beijing via CLTC, dating back to when it leased equipment to the University of Chile's Space Research Center. Since SSC's acquisition of the Station in 2008, CLTC has been operating two antennas in the station's northwest sector (Funaiole et al. 2022); however, SSC's relations with Beijing eroded in 2019 when the Swedish Defence Research Agency (FOI) warned that Sweden was facing a growing security challenge from China, saying one of its satellite stations in northern Sweden could be serving the Chinese military (Elmer 2019). These findings stoked fears of intelligence collection and surveillance, leading to SSC's 2020 decision not to renew any contracts with China nor accept new Chinese business at any of their stations around the world, including the one in Santiago. SSC stated that the decision was made because of geopolitical shifts (Barrett and Ahlander 2020). Some analysts believe that China's increasing influence in Chile is driven by geopolitical motives (Arostica Fernandez 2022). Due to these concerns, some local politicians and a few media outlets have voiced apprehensions. However, the pro-China lobby in Chile is powerful, leading many private and public leaders to refrain from criticizing Beijing (Han and Hanning 2022).

4 Telecommunications Sectors (5 Networks)

The expansion of Chinese 5G networks, particularly by Huawei, represents a significant shift in the global telecommunications landscape. Huawei's dominance in Africa, where it controls about 70% of the 4G network, is a strong foundation for its development of South Africa's first commercial 5G network. This move underscores the strategic importance of 5G technology in the digital revolution and indicates a major geopolitical shift, as China becomes a leading player in this sector.

In LAC Huawei's expansion is also noteworthy. Its plans to extend its cloud infrastructure across the region reveal the critical role of cloud services in the development and deployment of 5G networks. Already the fourth-largest cloud provider in LAC, Huawei is aiming to increase its market share, an ambition that is part of an even broader strategy to establish itself as a significant force in both cloud computing and 5G development.

These developments illustrate the varying depth of China's bilateral relations in diverse countries and locations within LAC. As the telecommunications landscape evolves, the influence of Chinese companies, particularly Huawei, is likely to reshape the dynamics of global digital connectivity and competition (Bartholomew 2020). From the fourth-largest cloud provider in the region Huawei is resolved to move up. The company has announced plans to extend its cloud infrastructure all over LAC, intimating the critical role of cloud services in the development of 5G as a universal connection platform (Bnamericas 2022c).

As in other sectors studied in this report, where China is present in LAC, the depth of bilateral relations varies by country and location. Overall, the data suggest that China's market penetration has come mostly via the Zhongxing Telecommunication Equipment Corporation (ZTE) and Huawei's involvement in 5G infrastructure development. This has varying political implications. Although the current condition of these projects cannot be known from the data, what is clear is that Huawei and ZTE are major players in LAC's telecommunications sector, leading construction of 5G infrastructure. The following discussion touches on some of the main advances identified. This is not an exhaustive account; the scarcity of verifiable empirical data and the rapid pace of China's progress in Latin America in almost every strategic aspect precludes that.

4.1 *Brazil*

The influence of China in Brazil, prominently led by Huawei, is extensive and diverse, spanning across telecommunications, manufacturing, mining, and digital entertainment. This influence was further solidified during Brazilian President Luiz Inacio Lula da Silva's state visit to Beijing in April 2023. During this visit, Lula defended Brazil's pursuit of Chinese communication and semiconductor technology, directly countering Western security concerns (Reuters 2023). His visit to Huawei's research center underscored Brazil's commitment to enhancing technological cooperation with China, notably in areas like 5G infrastructure—a field where Huawei has been actively involved since 2018 through an agreement with Brazil's Ministry of Science, Technology, Innovation, and Communication (MCTIC) (Chiplon 2018). Despite previous attempts in December 2020 by Brazil's government to legally obstruct Huawei's participation in Brazil's 5G network, the situation evolved. In February 2021, Communications Minister Fabio Faria met with Huawei executives to negotiate for Covid-19 vaccines. Following the successful vaccine negotiation, Faria announced the 5G auction rules, which permitted Huawei's participation, aligning with the broader context of Brazil-China cooperation as demonstrated by Lula's visit (ICPC 2023).

In 2022 it was reported that Huawei was planning to expand further into Brazil by concentrating on five key sectors: mining, manufacturing, agribusiness, logistics, and digital entertainment (Bnamericas 2022c). In the mining industry Huawei collaborates with Sul Americana de Metais and Senai Cimatec to develop 5G-driven autonomous operations and safety solutions (IM 2020). Huawei has set up an advanced 5G-run factory in São Paulo that uses AI, the cloud, and IoT technologies to transform manufacturing efficiency (Mello 2019). In collaboration with TIM, a major Brazilian telecommunications provider, Huawei has got a private 5G network operational inside Stellantis' automotive plant in Pernambuco, and is working on a smart city project in Curitiba (Huawei 2022).

In agribusiness Huawei is partnering with Claro Brasil to pioneer 5G pilot programs utilizing drone technology to enable precision farming (Bnamericas 2022c). The use of drones in agriculture can greatly benefit farmers. This technology enhances on-farm knowledge, leading to increased productivity and efficiency. It can also reduce the costs associated with labor, herbicides, pesticides, and fertilizers. Drones enable farmers to monitor their crops more closely and make more informed decisions about crop management, potentially leading to higher yields and more sustainable farming practices (FAO 2018).

However, the involvement of China via Huawei raises potential geopolitical concerns. One such concern is the possibility of Beijing gaining access to sensitive agricultural data through these drones. This data could include information on crop yields, pest and disease projections, and other key agricultural metrics. If accessed by China, this information could be used for their own political and economic advantage. For instance, knowledge about potential crop yields and pest infestations could influence global commodity prices. Beijing might use this information to make strategic decisions about purchasing agricultural products in advance, potentially at the expense of Brazil's interests.

There have been many other partnerships. In 2019 Huawei joined with Brazil's Telefónica, a subsidiary of Spanish Telefónica, to create a 4.5G network (Samena 2018). This joint venture was preparatory to the deployment of 5G technology in Brazil. It highlights the influence of Chinese tech companies in shaping the Brazilian telecoms landscape (Malena 2021).

Huawei and China Unicom have also partnered with CamTel to run the South Atlantic Inter Link (SAIL), a 6000-km submarine cable from Kribi Deep Sea Port in Cameroon to Fortaleza, Brazil (Bnamericas 2017). China Eximbank provided $85 million for this project, which according to the World Bank cost $136 million (Aiddata 2018). Completed in September 2018, this cable now offers the largest data transmission capacity between Africa and South America. The project is the first large-scale endeavor in the region led by a Chinese telecoms operator and an equipment manufacturer. It cements the pivotal role of Chinese firms in the development of Brazil's telecommunications infrastructure (Malena 2021).

4.2 Andean Countries

In Colombia Huawei has been involved in the development of 5G infrastructure as well. It was the first to deploy the 4G network in Colombia along with ZTE in 2016 and has been in talks over 5G implementation. As well, the Minister of Information Technologies and Communications stated that Colombia will not ban Huawei from the 5G auction. In May 2020 Telefónica's Movistar Colombia activated the first 5G pilot program with a 3.5 GHz band using Huawei technology. And in August 2019 ZTE signed a five-year 5G cooperation agreement with the Universidad Distrital Francisco José de Caldas to establish a joint 5G innovation research center in Colombia (ICPC 2023). In September 2022 Huawei announced that Colombia was to be the next major Latin American market with a dedicated physical location for its cloud infrastructure (Bnamericas 2022a).

In Ecuador Huawei and ZTE are both involved in the development of 5G infrastructure. In July 2019 Ecuadorian telecommunication company CNT launched its 5G network with the help of Huawei. In January 2020 it was reported that Claro, the leading mobile telecoms operator in Ecuador, would begin 5G technology tests in Quito in partnership with Huawei (ICPC 2023). In Ecuador also, Huawei and state-owned China National Electronics Import & Export Corporation (CEIEC) financed and built an emergency response network known as ECU-911 (Benitez 2019), consisting of 16 operational centers strategically placed throughout the country, equipped with a comprehensive system of 4,300 cameras in public spaces, which are closely monitored by security personnel (Rollet 2018).

Ostensibly, ECU-911 is purposed to respond swiftly to "emergencies". However, critics argue that these centers serve not only as a means of aiding those in distress, but also as an apparatus for domestic political surveillance (Benitez 2019). All video footage captured by the cameras is forwarded to the Ecuadorian security services, where it is handled using processes remarkably similar to what China employs to maintain social control over its own society (Ellis 2018). The extent to which Ecuadorian security officials share their data with their Chinese counterparts remains uncertain.

The PRC is also active in other Andean countries. In Venezuela ZTE collaborates with the Maduro government to operate a national identification system named the "Fatherland Card". This ID smartcard, connected to ZTE servers, gathers cardholders' data and could be misused to suppress dissent and gatekeep access to subsidized services (Malena 2021). In 2019 China gave 600 ZTE facial recognition surveillance devices to Bolivia as part of the BOL-110 security plan, which already included 2,000 similar devices and a central command hub in La Paz managed by CEIEC (Gobierno de Bolivia 2019). It cost more than $105 million. The Bolivian government is considering augmenting the system with 840 specialized cameras that automatically detect vehicle registration plates (Hernandez 2019).

4.3 Southern Cone

In Argentina ZTE has provided surveillance technology to help combat crime. In 2019 Jujuy province negotiated a package deal with ZTE that included security cameras, monitoring centers, emergency services, and telecommunications infrastructure. Jujuy is one of Argentina's more economically disadvantaged regions, with crime rates higher than the national average. It was reported that an Argentine government official claimed that local residents willingly acceded to the ubiquity of surveillance cameras as the trade-off for feeling safer (Hernandez 2019).

In 2017 Chile launched an ambitious venture to link up Latin America with Asia via a ground-breaking trans-Pacific submarine fiber-optical cable. This project received support from major Chinese tech giants Huawei Marine and China Unicom that aspire to bridge the geographical gap between Shanghai and Chile (Ramos Miranda 2019). Despite an initial preference for Huawei, unseen political forces soon imposed the decision to entrust execution of this monumental task to the Japanese company NEC instead (Zheng 2020).

In Chile, too, Huawei will construct a datacenter to enhance its cloud computing business, it recently announced. Its marketing director stated that Chile is notable for its significant adoption of cloud services, with transformation processes occurring at an accelerated pace. Yet it is not the only country in LAC experiencing such a trend. He provided no further details, but such a datacenter is consistent with what

Huawei is also doing in other parts of Latin America:—in Colombia, Brazil, Peru, and Mexico (Bnamericas 2022b).

Huawei's recent announcement to construct a datacenter in Chile signals a commitment to enhance its cloud computing business, echoing Chile's rapid adoption of cloud services and transformation trends seen in other Latin American countries (Bnamericas 2023). The expansion reflects Huawei's growing influence in the region. Yet the firm's expanding presence in Latin America is not without concerns; for example, subsidizing Latin American startups to use their cloud services might gain for Huawei unprecedented access to the region's emerging technological innovations. This raises critical questions about the depth and reach of the company's data centers, and the potential for Chinese companies to access the sensitive information and intellectual property of these startups.

5 Conclusion

Investigating China's expanding influence in Latin America, especially in the defense, aerospace, and telecommunications sectors, reveals a complex geopolitical landscape. Beijing's strategic footprint in these sectors carries profound implications for regional power dynamics, particularly in relation to the United States. LAC finds itself at a crossroads in navigating the opportunities and the challenges brought forth by Beijing's advent. Several strategic recommendations come to the fore in response to these developments.

Latin American countries must approach engagement with China from a well-thought-out strategic standpoint. While recognizing China's influence in critical sectors, they must also beware of overdependence on any single power, including the PRC. Striking a balance is essential to safeguard sovereignty while capitalizing on opportunities and managing risks. Prioritizing sovereignty should be the cornerstone of Latin American foreign policy and guide its decisions. Engagement with Beijing should be mutually beneficial and aligned with articulated national interests. Informed decision-making is paramount, underpinned by a comprehensive understanding of the geopolitical and economic implications.

The United States faces a distinctive set of challenges yet also opportunities resulting from China's growing influence in Latin America. It is imperative for U.S. policymakers to recalibrate their regional engagement strategy. The recalibration must transcend the short-sighted lenses of illegal immigration and drug trafficking, fostering a partnership rooted in mutual respect, shared interests, and a genuine commitment to the sustainable development of the region. Opportunities abound in areas such as technological innovation, educational cooperation, and support for sustainable economic practices. These offer the avenues for constructive engagement. Regional collaboration emerges as a unifying theme which ought to be emphasized. Latin American nations, along with the United States, should prioritize working together within the Western Hemisphere to create a cohesive bloc capable of negotiating effectively with global powers like China, tackling economic, technological, and security challenges with a unified strategy.

References

Aero-Naves (2023) Estados Unidos suministrará aviones y helicópteros a Uruguay. Aero Naves [Online]. Available: https://aero-naves.com/2023/02/07/estados-unidos-suministrara-aviones-y-helicopteros-a-uruguay/

Aiddata (2018) China Eximbank provides $85 million USD PBC loan for 5900 km South Atlantic Inter Link (SAIL). AidData—Research Lab at William & Mary [Online]. Available: https://china.aiddata.org/projects/58439/

Aiddata (2023) Chinese Government provides $50 million for construction of space station in Neuquén Province. AIDATA: A Research Lab at William & Mary [Online]. Available: https://china.aiddata.org/projects/52875/

Alden C, Mendez A (2023) China and Latin America: development, agency and geopolitics. Bloomsbury Academic, London, UK

Arostica Fernandez P (2022) Chile's once-pioneering relationship with China is turning into dependency. Available: https://merics.org/en/chiles-once-pioneering-relationship-china-turning-dependency

Barrett J, Ahlander J (2020) Exclusive: Swedish space company halts new business helping China operate satellites. Reuters [Online]. Available: https://www.reuters.com/article/us-china-space-australia-sweden-exclusiv/exclusive-swedish-space-company-halts-new-business-helping-china-operate-satellites-idUSKCN26C21L

Bartholomew (2020) China And 5G. Issues Sci Technol 36:50–57

Benitez G (2019) Implementing holistic security strategies to reduce crime and violence in Ecuador. KDI School of Public Policy and Management, Global Delivery Initiative [Online]. Available: https://archives.kdischool.ac.kr/handle/11125/41338. Accessed 04

BIS (2023) Commerce adds six to entity list for supporting PRC military modernization, intelligence, and reconnaissance activities. Bureau of industry and security: office of congressional and public affairs [Online]. Available: https://www.bis.doc.gov/index.php/documents/about-bis/newsroom/press-releases/3220-2023-02-10-bis-press-release-six-prc-entities-final-3/file

Bnamericas (2017) Huawei Marine to build South Atlantic Inter Link Cable System. Bnamericas [Online]. Available: https://www.bnamericas.com/en/news/huawei-marine-to-build-south-atlantic-inter-link-cable-system1

Bnamericas (2022a) Colombia is next in line for Huawei Cloud. Available: https://www.bnamericas.com/en/news/colombia-is-next-in-line-for-huawei-cloud

Bnamericas (2022b) Huawei aims to grow its LatAm cloud footprint. Bnamericas [Online]. Available: https://www.bnamericas.com/en/news/huawei-aims-to-grow-its-latam-cloud-footprint

Bnamericas (2022c) The industry segments Huawei is targeting in Latin America. Bnamericas [Online]. Available: https://www.bnamericas.com/en/news/the-industry-segments-huawei-is-targeting-in-latin-america

Bnamericas (2023) Huawei pledges new LatAm datacenters as cloud demand grows. Available: https://www.bnamericas.com/en/news/huawei-pledges-new-latam-datacenters-but-is-mum-on-details

Chin GT (2023) Lula's Return to China: Brazil between the Superpowers, Part I. Global Policy [Online]. Available: https://www.globalpolicyjournal.com/blog/11/04/2023/lulas-return-china-brazil-between-superpowers-part-i

Chiplon (2018) Huawei has won 25 5G contracts. This year's revenue will break the 100 billion dollar mark. Chiplon Microelectronics Limited—Beijing Public Security Bureau [Online]. Available: http://www.chiplon.cn/news/detail/id/153.

Control Ciudadano (2021) Fuerza Armada Nacional Bolivariana. Adquisición, recepción e incorporación de armamento y material militar. Período 2017–2021, Caracas, Asociacion Civil Control Ciudadano

CSET (2022) Chinese academy of sciences 2022 budget. Georgetown Center for Security and Emerging Technology [Online]. Available: https://cset.georgetown.edu/publication/chinese-academy-of-sciences-2022-budget/

CSIS (2020) How dominant is China in the global arms trade? China Power [Online]. Available: https://chinapower.csis.org/china-global-arms-trade/

Dialogo Americas (2022) China Cements Ties with Venezuela through Arms Sales. Dialogo [Online]. Available: https://dialogo-americas.com/articles/china-cements-ties-with-venezuela-through-arms-sales/#.ZGFaEHbMImE

Ellis ER (2016) Chinese engagement with Bolivia: resources, business opportunities, and strategic location. Air Space Power J 28

Ellis ER (2017) Testimony to the subcommittee on western hemisphere affairs. Foreign Affairs Committee: US House of Representatives [Online]. Available: https://docs.house.gov/meetings/FA/FA07/20170913/106398/HHRG-115-FA07-Wstate-EllisR-20170913.pdf

Ellis ER (2018) El apalancamiento de Ecuador sobre China para conseguir una vía alternativa de política y desarrollo. Air Space Power J 18–35

Ellis E (2022a) China's role in Latin America and the Caribbean. In: Evan Ellis R (ed) Statement before the Senate Foreign relations subcommittee on western hemisphere, transnational crime, civilian security, democracy, human rights, and global women's issues: a testimony. Washington, DC

Ellis E (2022b) Chinese security engagement in Latin America. In: CSIS (ed) Center for Strategic Studies & International Studies, Washington, DC

Ellis RE (2011) China-Latin America Military engagement: good will, good business, and strategic position. Strategic Studies Institute (SSI), Carlisle, PA

Elmer K (2019) Swedish defence agency warns satellite station could be serving Chinese military. South China Morning Post [Online]. Available: https://www.scmp.com/news/china/diplomacy/article/2182026/swedish-defence-agency-warns-satellite-station-could-be-serving

Eoportal (2013) NEE-01 Pegasus. Satellite Missions [Online]. Available: https://www.eoportal.org/satellite-missions/nee-01-pegasus#spacecraft

EXA (2023) NEE-02 KRYSAOR: 2do Satelite Ecuatoriano. Agencia Espacial Civil Ecuatoriana [Online]. Available: http://www.exa.ec/krysaor/

Fao (ed) (2018) E-Agriculture in action: drones for agriculture: food and agriculture organization of the United Nations

Forrest B (2020) China Arms Industry Ranks Second Behind U.S. Wall Street J [Online]. Available: https://www.wsj.com/articles/chinese-arms-industry-ranks-second-behind-u-s-report-says-11607295661

Forum China PLP (2020) Sino-Brazilian partnership to build satellite ground station. Supporting Office to the Permanent Secretariat to the Forum for Economic and Trade Co-operation between China and Portuguese-speaking Countries (Macao) [Online]. Available: https://www.forumchinaplp.org.mo/sino-brazilian-partnership-to-build-satellite-ground-station/

Franks J (2011) Chinese navy hospital ship visits Cuba, Caribbean. Reuters [Online]. Available: https://www.reuters.com/article/idINIndia-60058520111021

Funaiole M, Kim D, Hart B, Bermudez JS (2022) Eyes on the Skies: China's growing space footprint in South America. Center for strategic and international studies: hidden research issue [Online]. Available: https://features.csis.org/hiddenreach/china-ground-stations-space/

Garrison C (2019) China's military-run space station in Argentina is a 'black box'. Reuters [Online]. Available: https://www.reuters.com/article/us-space-argentina-china-insight-idUSKCN1PP0I2

Global Security (2020) Ecuador—Nanosatellites. Global Security, [Online]. Available: https://www.globalsecurity.org/space/world/ecuador/nanosatellites.htm?utm_content=cmp-true

Gobierno De Bolivia (2019) Presidente inaugura el sistema de seguridad ciudadana BOL-110, el más moderno de América Latina. Viceministerio de Comunicacion [Online]. Available: https://www.comunicacion.gob.bo/?q=20190821/27546.

Goñi U (2015) Argentinian congress approves deal with China on satellite space station. The Guardian [Online]. Available: https://www.theguardian.com/world/2015/feb/26/argentina-congress-china-satellite-space-station

Guevara I (2015) Peru receives Chinese missile launchers. Shephard News [Online]. Available: https://www.shephardmedia.com/news/landwarfareintl/peru-receives-chinese-missile-launchers/

Gurrola G (2018) China-Latin America Arms Sales: antagonizing the United States in the Western Hemisphere? Military Rev 124–132

Guzman J (2021) Programas espaciales en América Latina: Historia, operaciones actuales y cooperación futura. USAF J Am 3:22–42

Han B, Hanning S (2022) Beijing's global media influence 2022: Chile. Freedoom House [Online]. Available: https://freedomhouse.org/country/chile/beijings-global-media-influence/2022

Hermoso J, Fermín MV (2019) The Venezuela-China relationship, explained. The China Project

Hernandez C (2019) Bolivia invierte 105 millones de dólares en un sistema de seguridad ciudadana. Infodefensa [Online]. Available: https://www.infodefensa.com/texto-diario/mostrar/3128833/bolivia-invierte-105-millones-dolares-sistema-seguridad-ciudadana

Horta L (2008) In Uncle Sam's Bacyard: China's military influence in Latin America. Military Rev

Huawei (2022) TIM Brasil and Huawei Sign MoU to transform curitiba into the Country's First '5G City'. Huawei News & Events [Online]. Available: https://www.huawei.com/en/news/2022/3/mou-tim-5g-city-2022

ICPC (2023) Mapping China's Tech Giants. International Cyber Policy Centre [Online]. Available: https://chinatechmap.aspi.org.au/#/homepage/

IM (2020) Sul Americana de Metais signs 5G deal with Huawei to enable autonomous Block 8 iron ore mine in Brazil. International Mining [Online]. Available: https://im-mining.com/2020/07/09/sul-americana-de-metais-signs-5g-deal-huawei-enable-autonomous-block-8-iron-ore-mine-brazil/

Infoespacial (2014) Ecuador desarrollará su propio satélite con tecnología china. Infoespacial [Online]. Available: https://www.infoespacial.com/texto-diario/mostrar/3569902/ecuador-desarrollara-propio-satelite-tecnologia-china

Johnson R (2023) In Argentina's fighter competition, Washington and Beijing fight for regional influence. Breaking Defence [Online]. Available: https://breakingdefense.com/2023/09/in-argentinas-fighter-competition-washington-and-beijing-fight-for-regional-influence/

Johnson S (2005) Balancing China's growing influence in Latin America. The Heritage Foundation [Online]. Available: https://www.heritage.org/americas/report/balancing-chinas-growing-influence-latin-america

Klinger JM (2018) A brief history of outer space cooperation between Latin America and China. J Latin Am Geogr 17:46–83

Lazarus L, Berg RC (2022) Washington Must Respond to China's growing military presence in Latin America. The Diplomat [Online]. Available: https://foreignpolicy.com/2022/03/14/china-latin-america-military-pla-infrastructure-ports-colombia/

Li H (2007) China's growing interest in Latin America and its implications. J Strategic Stud 30:833–862

Malena J (2021) The extension of the digital silk road to Latin America: advantages and potential risks. Council on foreign relations [Online]. Available: https://www.cfr.org/blog/extension-digital-silk-road-latin-america-advantages-and-potential-risks

Mello G (2019) China's Huawei to invest $800 million in new Brazil factory. Reuters [Online]. Available: https://www.reuters.com/article/us-huawei-tech-brazil/chinas-huawei-to-invest-800-million-in-new-brazil-factory-idUSKCN1UZ1B6

Mendez A, Alden C (2021) China in panama: from peripheral diplomacy to grand strategy. Geopolitics 26:838–860

Mendez A (2017) Colombian agency and the making of US foreign policy: intervention by invitation. Abingdon, UK, Routledge

Nasa (2019) Tianlian 2–01. National Aeronautics and Space Administration [Online]. Available: https://nssdc.gsfc.nasa.gov/nmc/spacecraft/display.action?id=2019-017A

Nature (2022) China-Brazil Joint Laboratory for Space Weather (CBJLSW). Nature Index [Online]. Available: https://www.nature.com/nature-index/institution-outputs/brazil/china-brazil-joint-laboratory-for-space-weather-cbjlsw/6008f2210ab0d524b61f9b6d

NSSC (2014) China-Brazil Joint Laboratory for Space Weather. National Space Science Center, Chinese Academy of Sciences [Online]. Available: http://english.nssc.cas.cn/ic/ji/201410/t20141015_129395.html

Oliveira W (2023) Analysis: Beijing is and stays Brazil's main partner with new agreements worth US$10 billion. The Rio Times [Online]. Available: https://www.riotimesonline.com/brazil-news/new-multipolar-world-order/analysis-beijing-is-and-stays-brazils-main-partner-with-new-agreements-worth-us10-billion/

Paz GS (2006) Rising China's Offensive in Latin America and the U.S. reaction. Asian Perspect 30:95–112

Prensa Latina (2016) China is Interested in Increasing Military Ties with Bolivia. Newspaper of the Cuban province of Camagüey [Online]. Available: http://www.adelante.cu/index.php/en/noticias/world/8292-china-is-interested-in-increasing-military-ties-with-bolivia

RAND Corporation (2022) Assessing systemic strengths and vulnerabilities of China's defense industrial base. Rand Research Reports [Online]. Available: https://www.rand.org/pubs/research_reports/RRA930-1.html

Ramirez Aburto R (2023) Base China en Neuquén: en qué trabaja y cuáles son los planes de la estación espacial para el 2023. Río Negro [Online]. Available: https://www.rionegro.com.ar/sociedad/base-china-en-neuquen-en-que-trabaja-y-cuales-son-los-planes-de-la-estacion-espacial-para-el-2023-2727081/

Ramírez Bullón JE (2021) La adaptación del Perú ante el ascenso de China a inicios del siglo XXI: entre el pragmatismo y la aquiescencia. Agenda Internacional XXVIII:119–149

Ramos Miranda N (2019) Huawei wants to build first fiber-optic cable between South America and Asia. Reuters [Online]. Available: https://www.reuters.com/article/us-huawei-tech-chile/huawei-wants-to-build-first-fiber-optic-cable-between-south-america-and-asia-idUSKCN1VI2N2

Raquel A (2020) Brazilian Start-up ALYA NANOSATELLITES gains first partnership agreements. Space Watch Global, [Online]. Available: https://spacewatch.global/2020/08/brazilian-start-up-alya-nanosatellites-gains-first-partnership-agreements/

Redaccion El Universo (2016) The Chinese company CETC that sold radars demands $ 280 million to Ecuador. El Universo [Online]. Available: https://www.eluniverso.com/noticias/2016/11/07/nota/5892721/empresa-que-vendio-radares-pide-280-millones-estado/

Reuters (2013) China-Brazil satellite launch fails, likely fell back to Earth. Reuters [Online]. Available: https://www.reuters.com/article/us-space-china-brazil-satellite-idUKBRE9B90XK20131210

Reuters (2023) Lula courts Chinese tech for Brazil, brushes off prejudices. Available: https://www.reuters.com/technology/brazil-paves-way-semiconductor-cooperation-with-china-2023-04-14/

Rivas S (2022) Uruguay y el reemplazo de su aviación de combate. Pucará Defensa [Online]. Available: https://www.pucara.org/post/uruguay-y-el-reemplazo-de-su-aviaci%C3%B3n-de-combate

Rollet C (2018) Ecuador's all-seeing eye is made in China. Foreign Policy [Online]. Available: https://foreignpolicy.com/2018/08/09/ecuadors-all-seeing-eye-is-made-in-china/. Accessed 6 July 2019

Roy D (2022) China's growing influence in Latin America. Council on Foreign Relations [Online]. Available: https://www.cfr.org/backgrounder/china-influence-latin-america-argentina-brazil-venezuela-security-energy-bri

Saavedra G (2023) China pressures Argentina to build naval base. Dialogo Americas [Online]. Available: https://dialogo-americas.com/articles/china-pressures-argentina-to-build-naval-base/#.ZGFnKnbMImF
Samena (2018) Telefonica, Huawei ink '4.5G' agreement covering Bogota, Monterrey and Rio de Janeiro. SAMENA Telecommunications Council [Online]. Available: https://www.samenacouncil.org/samena_daily_news?news=67009
Sanders R (2023) F-16s for Argentina. Wilson Center [Online]. Available: https://www.wilsoncenter.org/blog-post/f-16s-argentina
Selligman L (2019) U.S. Military warns of threat from Chinese-run space station in Argentina. Foreign Policy [Online]. Available: https://foreignpolicy.com/2019/02/08/us-military-warns-of-threat-from-chinese-run-space-station-in-argentina/
Socialist China (2022) China and Cuba: an exemplar of solidarity and cooperation between socialist countries. Friends of Socialist China [Online]. Available: https://socialistchina.org/2022/11/30/china-and-cuba-an-exemplar-of-solidarity-and-cooperation-between-socialist-countries/.
Space Watch Global (2020) Venezuela's VENESAT-1 tumbling and out of service. Space News [Online]. Available: https://spacewatch.global/2020/03/venezuelas-venesat-1-tumbling-and-out-of-service/
SRE (2021) Signing of the Convention establishing ALCE, the Latin American and Caribbean Space Agency. Secretaria de Relaciones Exteriores de Mexico [Online]. Available: www.gob.mx/sre/en/articulos/signing-of-the-convention-establishing-alce-the-latin-american-and-caribbean-space-agency-283235?idiom=en
Stokes M, Alvarado G, Weinstein E, Easton I (2020) China's Space and Counterspace Capabilities and Activities. The U.S.-China Economic and Security Review Commissio*n* [Online]. Available: https://www.uscc.gov/sites/default/files/2020-05/China_Space_and_Counterspace_Activities.pdf
The State Council PRC (2022) White Paper: China's Space Program: A 2021 Perspective. People's Republic of China
U.S Department of State (20210 Table II.e Value of arms deliveries and total trade by country, 2009–2019. World military expenditures and arms transfers 2021 Edition [Online]. Available: https://www.state.gov/wp-content/uploads/2022/01/WMEAT-2021-Tables-II-IV-Arms-Transfer-Deliveries-2009-2019.xlsx
UEPB (2022a) UEPB e comitiva chinesa discutem ampliar cooperação na construção de Observatório Astronômico da Instituição. Universidade Estadual da Paraíba [Online]. Available: https://uepb.edu.br/uepb-e-comitiva-chinesa-discutem-ampliacao-de-cooperacao-na-construcao-de-observatorio-astronomico-da-instituicao/
UEPB (2022b) Universidade Estadual recebe terreno para construção de Observatório Astronômico em Juazeirinho. Universidade Estadual da Paraíba [Online]. Available: https://uepb.edu.br/universidade-estadual-recebe-terreno-para-construcao-de-observatorio-astronomico-em-juazeirinho/
UMSA (2013) Uso del satélite Túpac Katari como herramienta de investigación. Available: http://www.lacatedra.umsa.bo/web/la-catedra/2/-/asset_publisher/f55S/content/uso-del-satelite-tupac-katari-como-herramienta-de-investigacion
USCC (2018) China's engagement with Latin America and the caribbean. U.S.-China Economic and Security Review Commission, Washington, DC
Valiente I (2018) WMZ-551B1, los fallidos blindados argentinos comprados para Haití. Infodefensa [Online]. Available: https://www.infodefensa.com/texto-diario/mostrar/3072901/wmz-551b1-fallidos-blindados-argentinos-comprados-haiti
Watson P (2017) Perú investiga la adquisición de los lanzadores múltiples Norinco a China. Infodefensa [Online]. Available: https://www.infodefensa.com/texto-diario/mostrar/3079068/peru-investiga-adquisicion-lanzadores-multiples-norinco-china
Watson-Lynn E (2020) The gravity of China's space base in Argentina. Lowy Institute: The Interpreter [Online]. Available: https://www.lowyinstitute.org/the-interpreter/gravity-china-s-space-base-argentina

World Metereological Organization (2023) Observing Systems capability analysis and review tool. Available: https://space.oscar.wmo.int/satellites/view/vrss_2
Xia J (2022) China 5G: opportunities and challenges. Telecommun Policy 46:102295
Xinhua (2019) To build a powerful country in aerospace, Xi Jinping entrusted him earnestly. Xinhua News Agency [Online]. Available: http://www.xinhuanet.com/politics/xxjxs/2019-04/12/c_1124357478.htm
Zheng S (2020) China's Huawei loses out to Japan's NEC on Chile-Asia trans-Pacific cable project. South China Morning Post [Online]. Available: https://www.scmp.com/news/china/diplomacy/article/3095367/chinas-huawei-loses-out-japans-nec-chile-asia-trans-pacific

The Rise of China and De-Dollarization in a Multi-polar Digital World

Peter J. Middlebrook and Amba Tadaa

Abstract The Rise of China and De-Dollarization in a Multi-Polar Digital World," examines the transformative geopolitical shifts and monetary competition resulting from China's ascendance as a great power. It delves into the mechanisms and implications of de-dollarization, analyzing how digital currencies and financial technologies are reshaping global economic structures, within a broader context of great-power competition. The discussion emphasizes China's strategic initiatives to reduce dependence on the US dollar, fostering a multi-polar financial system. Key topics include the Belt and Road Initiative, the digital yuan, and China's role in international trade and finance. The chapter provides a comprehensive overview of the potential economic, political, and technological impacts of these changes, highlighting the increasing importance of digital innovation in global finance. It offers critical insights for policymakers, economists, and scholars interested in the dynamics of global power shifts and the future of international monetary systems.

US dollar hegemony has been the defining characteristic of the international financing order throughout the second half of the 20th Century; solidified by the Bretton Woods Agreement in 1944. With the U.S. trade war with China, U.S. sanctions on China and Russia, and with Brazil, Russia, India, China (BRIC) economic grouping actively working to substitute the dollar as the primary currency for international financial transactions, the primacy of the greenback in international markets is clearly being challenged. The concomitant decline in US unilateralism, US national debt exceeding US$ 34 trillion, global despite trade fragmentation, domestic challenges to raising the debt ceiling and the need to finance 750 military bases in more than 80 countries, makes for a somewhat perilous situation.

With both the BRIC (expanded to include Egypt, Ethiopia, Iran and the United Arab Emirates (UAE)), and Shanghai Cooperation Organization (SCO)

P. J. Middlebrook (✉) · A. Tadaa
Geopolicity, Dubai, UAE
e-mail: ceo@geopolicity.com

A. Tadaa
e-mail: atadaa@geopolicity.co

A. Farhadi et al. (eds.), *The Great Power Competition Volume 6*,
https://doi.org/10.1007/978-3-031-70767-4_10

members formally discussing de-dollarization in response to claims of Washington's weaponization of the US$, scaling-up local currency settlement has never been more in vogue. In such a context, and with members seeking to create a more equal global trading and financial system, trade settled in US dollars would *prima facie* appear to be in long term decline. While IMF data would appear to suggest challenges in de-coupling from the US dollar, the long view suggests that two major trading blocs have been formed—one monetized by US$ and one by RMB—changing the nature of Great Power Competition. Moreover, trends in cross-border payments show that fifteen years ago the RMB's share in cross-border transactions of Chinese non-bank entities was nearly zero; but by late 2023, it had reached around 50%. Conversely, the US$'s share declined from 80 per cent in 2010 to 50% in 2023.

Given the many obstacles to rapid de-dollarization and the continued need for stable trade currencies to peg value, decoupling will likely take decades not years. Even if (more likely when) black swan events such as US debt default, civil breakdown, world war, or the rapid adoption of central bank digital currencies occur, the outcome is uncertain.

In this context, this article explores what lessons can be learned from history in relation to the decline of monetary dominance in great power competition and what scenarios can be considered, as the basis for predicting the pace of de-dollarization by 2050? A review of the decline of the Roman and British empires suggest long term decline is inevitable, though exogenous shocks and domestic amplifiers in the US could either expedite or reverse long-term decline.

1 The Historic Importance of Monetary Dominance in Great Power Competition

Throughout history, and particularly in the context of Great Power Competition, Monetary dominance has played an instrumental role in the Hegemonic toolkit, enabling a great power to set the rules of the financial game. While currency dominance has been a hallmark of hegemonic powers throughout history, serving as both a symbol and a tool of their influence, the nature of monetary hegemony has evolved, shaped by technological and logistical advancements. At the same time, its role in asserting power, be it military, political, or economic, has remained—in some sense—constant.

In essence, from the Roman denarii to the U.S. dollar, the currency of a hegemonic power has always served as more than a medium of exchange; it has functioned as a tool of control.

The allure of a dominant currency is therefore not merely economic, as it extends into the realms of political, military, and cultural projection, though these specific benefits depend on the historical context in question. Direct economic gains from monetary order domination that may be achieved include:

- ***Seigniorage***: The issuing country gains from the difference between the face value of the currency and its production cost.
- ***Inflation Tax***: Issuing more currency can lead to inflation, effectively acting as a tax on holders of the currency while reducing the issuer's liabilities.
- ***Access to Cheap Credit***: The dominant currency is often held as reserves by foreign entities, who then invest it back, allowing the issuer to borrow at lower rates.
- ***Macroeconomic Flexibility***: The issuer has greater freedom in running deficits and implementing monetary policies, often without worrying about balance of payments, and at the same time the controlling authority can undertake capital tightening, increasing the costs of borrowing abroad.
- ***Global Pricing Power***: The issuing authority has far greater control over commodities such as oil and food, providing an additional layer of economic leverage.
- ***Monetary Policy Autonomy***: A dominant currency affords a nation greater policy autonomy, allowing it to set domestic monetary policy with fewer international constraints and to be less buffeted by exogenous shocks (thus the phrase when the United States sneezes, the world catches a cold).
- ***Financing Military Campaigns***: Armed with the capacity to mobilize debt, and where debt holders include major trading partners, military campaigns abroad can be financed unwittingly by international creditors.

Beyond the economic sphere, the benefits extend into the realms of geopolitics and soft power. The issuer gains significant financial power, without significant liabilities, which is translated into leverage and negotiating power on the international stage. The issuer can also influence other countries' monetary policies, impacting growth and the value of foreign currency receipts, and a country's political decisions. Additionally, having a dominant currency enhances the issuer's reputation and status globally, strengthening the potential for determining the terms of trade, for example. In time this can contribute to self-reinforcing cycle, where economic advantages feed geopolitical influence and vice versa.

However, achieving such monetary influence has certain prerequisites. From the perspective of the three basic functions of money; the currency in question must act a medium of exchange for international trade, a unit of account for most important goods worldwide, and a store of value for other economic actors. Here, the rise of the RMB and national currencies as units of account and exchange, clearly erode the position of the US dollar.

Other factors may also make universal adoption of a currency more likely, including (i) a large and robust domestic economy (ii) sound domestic macroeconomic policies to build trust in the currencies stability (iii) sustained gross domestic product and moderate inflation rates, which also instil confidence in the currency's stability (iv) centrality in both regional/international trade networks (v) ease of currency convertibility, and (vi) the size, depth, and openness of financial markets associated with the currency. What this means is that any monetary newcomer to the international stage must possess a broad set of political and economic capabilities to strike out alone and to influence the monetary preferences of trading partners.

2 Roman Empire Monetary Dominance

While the Roman Empire benefitted greatly from the dominance of its tri-metallic currency (consisting of gold coins (aurei), silver denarii, and copper or copper-alloyed denominations for small change), according to Meikle, technological and logistical constraints meant that the Roman state never fully transitioned to a fully cash-based taxation system (Meikle 2002). The primary roles of money in the Roman Empire were therefore practical: paying troops, exercising power, and displaying status. Monetary hegemony, while important, was a secondary consideration. Nonetheless, while it may not have been a system of 'monetary hegemony' in the modern sense, Roman currency hegemony was intrinsically tied to Rome's imperial political and military objectives.

The expansion of Roman coinage facilitated the growth of both common and foreign markets, thereby enhancing long-distance trade and regional specialization (Temin and Voth 2013). Widespread adoption of the Denarii also leant administrative convenience. Under Augustus Roman administrations were required to make some tax payments in Roman denarii, a policy aimed less at cultural imposition but for practical ease of administration (MacMullen 2000).

3 British Empire

It was not until the height of the British Empire that the world's first monetary hegemony can be said to have emerged. During this era, British trade, both exports and imports, made up a significant portion of global commerce, accounting for over 30% of the world's total trade in 1800, and 25% by 1850 (Imlah 1950).

The British monetary ascent was solidified in 1871 with the adoption of the gold standard, transforming London into the epicentre of global finance. The city became the leading supplier of both short-term and long-term credit, channelled abroad through its extensive financial facilities. The gold standard was instrumental for other financial centres like Paris and Berlin to attract business from London, as it represented creditworthiness and reduced transaction costs. Walter (1991) notes that 'sterling bills and short-term credits financed perhaps 60% of world trade' between 1870 and 1913, highlighting the pound's global reach.

Britain's net overseas assets rose from 7% of national wealth in 1850 to around 32% by 1913 (Edelstein 2004, p. 193), with this foreign investment deepening the use of sterling for international payments. By the late 19th Century, most loans were made in sterling, and most bill exchanges for international trade were drawn in London. These practices further enhanced the pound's value as a unit for storage and medium of exchange globally.

Britain's ability to run current account deficits through the issuance of sterling and its discount rate, endowed it with a unique advantage. The effects of the discount rate had a "controlling influence on Britain's balance of payments" allowing the Bank of England to exert unparalleled influence over international capital flows (Cleveland 1976, p. 17). This made British monetary hegemony resilient, as its gold reserves were insulated by the discount rate, and all foreign rates followed the British rate. Keynes famously wrote that London's sway on global credit conditions was so predominant that the Bank of England could be considered 'the conductor of the international orchestra' (Keynes 1930).

4 US Monetary Dominance in the 20th Century

The ascendance of the U.S. dollar as the world's leading reserve currency was a gradual process that gained momentum in the early 20th century. Although the United States' Gross Domestic Product had surpassed that of Great Britain prior to the First World War, the US dollar really did not challenge the British Pound's until after the 1920s.

The establishment of the Federal Reserve in 1913 was a pivotal moment which significantly bolstered the dollar's influence, coupled with the First World War itself, which further strengthened the dollar's standing. This was largely because Germany was required to borrow in dollars to pay for war reparations—with the proceeds of war reparations used by Britain and France to accelerate economic recovery.

During the interwar years, the U.S. dollar's influence continued to grow, fuelled by growth in U.S. exports, facilitating the dollar's wider use across European countries and accelerating its internationalization. By around 1925, the dollar had overtaken the pound as the world's number one reserve currency, although this status was briefly lost during the Great Depression (Eichengreen 2001). By 1945, however, the US Dollar was indisputably the world's global currency, holding approximately two-thirds of the world's gold reserves and over 50% of global industrial productivity (Eichengreen 2011). U.S. economic planning at the time emphasized the importance of an export surplus and aimed to establish a world economy based on free trade, devoid of bilateral payments and trading systems.

US interests were further cemented at the Bretton Woods Conference, in which the British delegation's proposal—led by Keynes—to create a new supranational currency, known as 'bankor', was rejected. Instead, the conference led to the establishment of a system where the dollar would be pegged to gold at a fixed price of US$ 35 per ounce, with other world currencies pegged to the dollar at fixed exchange rates. Keynes' endorsement of the General Agreement on Tariffs and Trade (GATT) further diminished British influence, as it led to the abandonment of the Imperial Preference system. This was accompanied with the creation of the International Monetary Fund

(IMF) and the World Bank, collectively known as the Bretton Woods System, which utilized financial instruments denominated in US dollars.

The Bretton Woods outcome, described by some as U.S. 'free trade imperialism' (Hudson 2003, p. 7), solidified the dollar's status as the uncontested reserve currency in the post-war era. In the post-war years, U.S. Monetary dominance was especially true as U.S. industrial products began to flood the global market. The death knoll to British monetary hegemony came during the Suez Crisis, in which U.S. demand the withdrawal of British troops from Egypt triggered a run on the pound and resulted in negotiated British withdrawal in exchange for U.S. support for securing IMF loans.

The dollar's hegemony was further secured because it could be converted into gold at any time at the fixed price of $35 per ounce. As European economies and Japan recovered in the latter half of the 1950s, they accumulated substantial dollar reserves. Concurrently, the U.S. balance of payments experienced a deficit, plunging to a $3.5 billion gap in 1958 and even larger deficits in subsequent years. The Bretton Woods system, however, allowed the U.S. to finance approximately 70% of its cumulative balance of payments deficits throughout the 1960s through gold demonetization and liability financing. This also enabled the U.S. to finance heavy overseas military expenditures and maintain flexibility in domestic economic policy (Gowa 1983, p. 63).

The Bretton Woods system began to show cracks as early as 1960 when, for the first time, dollar reserves held by non-residents surpassed U.S. gold reserves. This imbalance raised concerns about the dollar's devaluation and triggered a shift in discourse from a 'dollar shortage' to a 'dollar glut' (Cohen 2017). The U.S. faced its first significant dollar speculation attack in October 1960, leading to a run for gold and putting immense pressure on maintaining the $35 per ounce gold price (Eichengreen 2011).

To stabilize the situation, the Gold Pool was established in 1961 but was short-lived, collapsing in March 1968. The system's inherent flaws were already known, most notably articulated by Robert Triffin in his 1947 report to the Federal Reserve. The 'Triffin Dilemma' highlighted the paradoxical nature of the U.S. running balance of payments deficits to provide liquidity to the global economy, while also eroding confidence in the dollar (IMF 2001).

By 1971, with costs of the Vietnam War continuing to escalate and with increasing demands from foreign central banks to convert their dollar reserves into gold (Eichengreen 2011), President Richard Nixon 'closed the gold window', effectively ending the dollar's convertibility into gold. This marked the beginning of a new era of fiat money, with the U.S. dollar still at the forefront. The lifting of the dollar's peg to gold granted the U.S. full autonomy in issuing its currency and although initially dubbed a 'sick currency' (Silk 1973), the dollar has remained the dominant reserve currency ever since.

At the same time, a series of agreements known as the U.S.-Saudi Arabian Joint Economic Commission were being undertaken between 1972 and 1974. In essence, and in exchange for technical support and military assistance, Saudi Arabia agreed to accept only U.S. dollars for its oil, the world's most traded commodity (Bulent Gokay

2006). Given Saudi Arabia's unique position within OPEC as the 'swing producer', the agreement was universally adopted by OPEC.

This introduced the petro-dollar system, i.e. the denomination of critical commodities, most notably oil, in U.S. dollars (Liu 2002), cemented the U.S. dollar's role as the dominant global currency. This system ensured that the dollar remained the world's primary reserve currency, and enforced universality based on demand to purchase oil.

As long as OPEC oil was priced in dollars, the U.S. government enjoyed a "double loan" that allowed it to (i) print dollars to pay for oil without immediately offering goods and services in return, while also (ii) requiring other economies to acquire dollars to purchase for oil (Spiro 1999, p. 121).

By 1991, following the collapse of the Soviet Union and Japanese economic crisis, the U.S. had further solidified its position as the world's monetary hegemon. The dollar's functions as a medium of exchange, unit of account, and store of value were evident. For instance, by 1991, approximately 64% of the world's official foreign exchange reserves were held in U.S. dollars, and about 88 per cent of daily foreign exchange trades involved the U.S. dollar (Bordo et al. 2015). The dollar's international use conferred significant economic benefits to the United States, including seigniorage revenues from a growing demand for dollar reserves by central banks worldwide.

With the rise of US energy independence however has fundamentally undermined the future of the Petro-dollar. In May 2011, the US became a net exporter (See Chart 1) of refined petroleum products and by 2014 the US was the world's third-largest producer of crude oil, following Saudi Arabia and Russia, and the second-largest exporter of refined products, after Russia. Middle Eastern oil has therefore begun to look for other markets, particularly in the East, and with the bombing of the Nord Stream Pipeline in Europe on September 26, 2022, new energy corridors have opened up between the Middle East and Europe, supported by two EU Global Gateway Flagship projects and proposals to invest towards the India, Middle East and European Corridor (IMEC), supply European markets with green energy.

With Middle Eastern oil increasingly trading East, in 2018 the Chinese government issued the first long term oil trading contracts denominated in Petro-Yuan. The primary drivers of the emergence of the Petro-Yuan include Russian, Venezuelan and Iranian interest to avoid sanctions and escape from dollar hegemony, and Chinese interest to internationalize the yuan. As a result, as of 2024, approximately 20 per cent of global oil trade is now settled in national currencies, including the Chinese yuan, showing a slow trend towards rebalancing settlement away from the US dollar. The trend is likely to accelerate given Chinese, Saudi and UAE use of currency swap agreements, cross-border trade settlement arrangements, and collaboration in digital currency futures. The long-term cost of US energy self-sufficiency—at least in part—will be a further decline in the US dollar in international oil settlement, with a slow transition set to see the rightsizing of the US dollar in the decades ahead.

Despite all the noise about de-dollarization however, according to the IMF, the US dollar still accounts for over 80% of trade finance and nearly 60% of foreign currency reserves as of 2024, despite the gradual diversification of FX reserves away

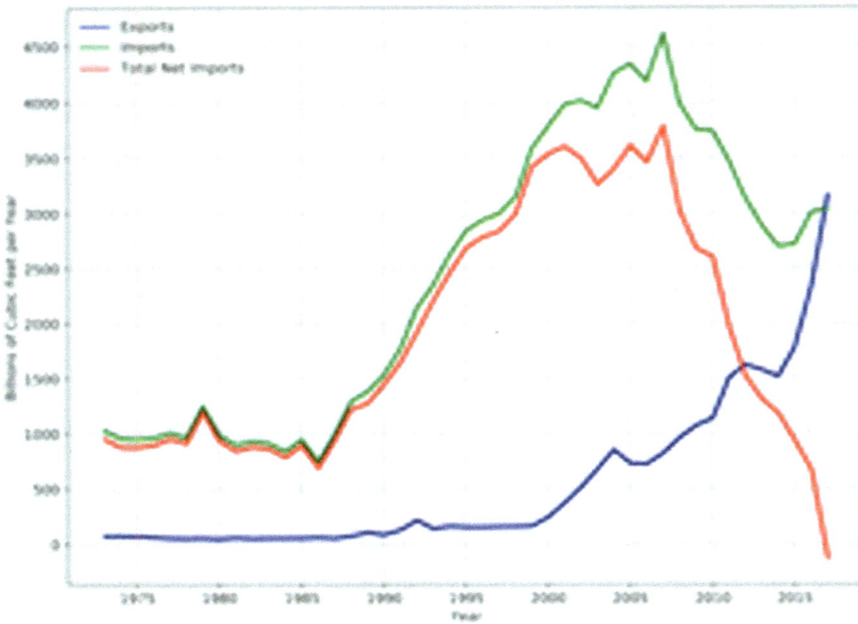

Chart 1 Trend of US net imports (imports minus exports) of natural gas, 1973–2017. Data from US EIA

from the dollar and partly into non-traditional reserve currencies (Stanford Institute for Economic Policy Research 2024) (Chart 2).

The outcome of the current war in Ukraine and the Middle East and the future pace of Israel's pivot to Asia all have implications for the pace of de-dollarization. For

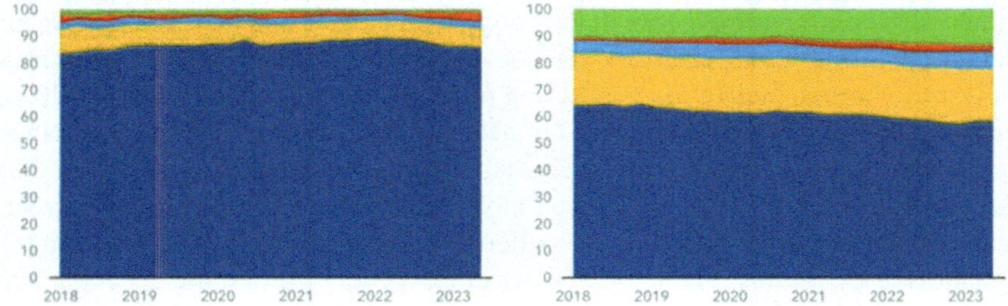

Chart 2 Dollar dominance in trade financing and foreign currency reserves. *Source* COFER, SWIFT and staff calculations.
Note Only includes trade finance covered by SWIFT (Code MT700 (Letters of credit)). Data as of 2023Q3 (FX reserves), 2023Q4 (Trade Finance). Data is exchange-rate adjusted

example, in recent years Israel has launched free trade negotiations with China, India, South Korea and Vietnam, defence tied with Thailand, security cooperation with South Korea and the Philippines and increased weapons sales to India and Vietnam (Witte 2022). Moreover, with the Abraham Accords signed in September 2020, establishing diplomatic relations between Israel, the UAE, Morocco and Bahrain and with the UAE and Israel having concluded a free trade agreement, the number of countries swinging to Asia can only seemingly increase. With US sanctions on China set to further ramp up, the war between the US$ and RMB now impacts foreign policy alignment and the future of trade relations.

5 Chinese Monetary Competition

The Renminbi (RMB) was first introduced by the People's Bank of China in 1948, following the establishment of the People's Republic of China in 1949, replacing several different currencies in circulation. Until 2025, the Yuan was pegged to the US$. In October 2016, the RNB became the first emerging market currency to be included in the IMF's special drawing rights basket, the basket of currencies used by the IMF as a reserve currency. By 2022, the RMB was the 5th most traded currency with a daily volume of 7%, behind Sterling, Japanese Yen, the EURO and US$.

The influence of the Chinese yuan in international finance has been steadily growing, particularly in cross-border transactions and trade finance within China-leaning countries. According to the IMF, the share of the yuan in trade finance for these countries has more than doubled since early 2022, and the use of the yuan in cross-border transactions by Chinese entities reached around 50% in late 2023. This increase is supported by the Cross-Border Interbank Payments System (CIPS) (Stanford Institute for Economic Policy Research 2024). For its part, the IMF have issued a litany of Working Papers on the rise of the Yaun, including Arslanalp et al.

A 2022 IMF Working Paper documents the decline in the US dollar's share of international reserves since the turn of the century (Arslanalp et al. 2022). This shift is attributed to active portfolio diversification by central bank reserve managers rather than changes in exchange rates, interest rates, or the actions of a few large central banks. Interestingly, the reduction in the dollar's share has not resulted in increased shares for other traditional reserve currencies like the Euro, Yen, or Pound Sterling. Instead, a significant portion of reserves has shifted towards the Chinese renminbi and currencies from smaller economies. The study highlights the increased market liquidity, better relative returns, and strategic reserve management practices as key factors driving this diversification. The findings suggest a gradual movement away from the dollar, a modest rise in the renminbi, and an enhanced role for nontraditional reserve currencies in the international financial system.

Monetary dominance continues to play a crucial role in China's broader economic competition strategy. By pushing for the international use of the renminbi (RMB) and expanding its financial institutions globally, China seeks to lessen its reliance on the U.S. dollar and counteract the effects of U.S. sanctions. This effort supports the Belt and Road Initiative (BRI), enabling more efficient trade and investment among participating nations. It also strengthens China's position in global economic governance, evidenced by its growing influence in organizations like the Asian Infrastructure Investment Bank (AIIB) and the BRIC New Development Bank. Ultimately, monetary dominance is key to China's ambition to reshape the global economic landscape in its favour, and with the growing membership and observership of the Shanghai Cooperation Organization (SCO) (founded some three months before the events of 911 to promote cooperation in various fields such as security, economy, culture, and education among its member states), inevitably linked to a de-dollarization agenda.

Through extensive financial contributions and proactive engagement, China has sought to enhance its voice and decision-making power within the IMF, reflecting its growing economic stature on the world stage. Additionally, China has played a central role in the BRIC grouping, advocating for greater focus on currency issues and challenging the dominance of Western currencies in global trade. China's leadership within BRIC has facilitated discussions on currency swaps and the establishment of the New Development Bank, providing alternative financing mechanisms to the existing Western-dominated institutions. Furthermore, China's initiative in founding the Asian Infrastructure Investment Bank (AIIB) showcases its commitment to reshaping the global financial architecture.

China began piloting its digital yuan (the Digital Currency Electronic Payment (DCEP), in 2020 with initial pilot programs conducted in several cities, including Shenzhen, Suzhou, Chengdu, and Xiong'an. In 2023, some three years later, a division of the New York Fed concluded digital-dollar test with banks to explore tokenized deposits. While digital currencies may not replace physical currency over the short term, with the 21st Century defined as the digital age, adoption soon or later of Central Bank Digital Currency (CBDC) is inevitable. The impact of digital currency on monetary dominance is unknown, though digitalization would likely make the informal economy more taxable, and it seems almost inevitable that issuance of the Yuan into the SCO trading block would lead to a precipitous downsizing of US dollar use in trade finance and exchange.

6 Chinese Ambitions in the Coming Decades

In an article in Foreign Policy entitled 'The Red Sea Crisis Proves China Was Ahead of the Curve',[1] Parag Khana explores the progressive nature of Chinese strategic investment, and the forward foreign policy impacts of the BRI (Khanna 2024). Khanna outlines the following Chinese ambitions for the coming decades:

- ***Expansion of the Belt and Road Initiative (BRI)***: China aims to solidify and expand its BRI, which has already seen about US$ 1 trillion invested in construction projects and nonfinancial investments in BRI member states. This initiative

[1] https://foreignpolicy.com/2024/01/20/url-red-sea-houthis-china-belt-road-suez-trade-corridors/.

seeks to create extensive trade networks by developing infrastructure such as railways, highways, and ports across Eurasia, the Indian Ocean, and even the Arctic Ocean.

- ***Mitigating Vulnerability to Maritime Chokepoints***: The BRI is partly a defensive strategy to reduce reliance on vulnerable maritime chokepoints like the Suez Canal and the Strait of Malacca. By developing alternative land and sea routes, China seeks to ensure more reliable and flexible supply chains, minimizing the impact of geopolitical and logistical disruptions.
- ***Increasing Global Connectivity***: China's infrastructure projects aim to boost global connectivity, facilitating smoother trade flows between Asia, Europe, and Africa. This includes expanding the number of trans-Eurasian rail routes and developing ports and corridors that enhance global trade reliability.
- ***Strategic and Economic Influence***: Through the BRI, China intends to extend its strategic and economic influence globally. This involves not only economic investments but also deepening strategic ties with various countries, which sometimes includes contentious practices like "debt-trap diplomacy."
- ***Response to Global Volatility***: China's approach emphasizes building more supply chains to address global volatility and disruptions. Khanna views this strategy as a way to stabilize global trade and mitigate the impacts of unpredictable events such as terrorism, geopolitical conflicts, and climate change.
- ***Climate Resilience and Migration***: Anticipating climate-induced migration, China plans to invest in infrastructure that supports population movements and urban development in climate-resilient areas. This includes constructing housing, healthcare facilities, and transportation networks necessary for large-scale human resettlement.
- ***Leadership in Infrastructure Development***: By spearheading large-scale infrastructure projects, China positions itself as a global leader in development, influencing global trade patterns and economic growth. This is also seen as a counter to Western-led initiatives and a way to secure China's place at the centre of global commerce.

While China faces demographic and debt challenges (at home and abroad), it also now faces the US as it moves to deepen sanctions and to repatriate industrial production out of China, into India and other markets with lower costs of Labour. The India, Middle East and European Corridor (IMEC) would—once complete—open up Indian markets through the Middle East into Europe—potentially undermining BRI strategic trade and investment interests. With IMEC—a planned economic corridor that aims to bolster economic development by fostering connectivity and economic integration from India to Europe through the United Arab Emirates, Saudi Arabia, Israel and Greece—delayed due to conflict in the Middle East however, a fully functional east west corridor seems unlikely any time soon.

7 Future Great Power and Monetary Dominance

Irrespective of black swan events, it seems almost inevitable that the SCO trading block and active efforts by its members to decouple from the greenback will further erode the dominance of the US dollar over the medium to longer term. China already has the 3rd largest bond market, is likely to be among the first to issue a central bank digital currency and it sits at the centre or the SCO and the BRIC economies. That said, China's currency policy has focused on making its exports significantly less expensive, and its imports more expensive, and the US through sanction is already offsetting the impact of the undervalued RMB, and this trend is likely to continue.

A 2023 report by Credit Suisse entitled 'The Future of the Monetary System—Leading perspectives to navigate the future' (Pozsar et al. 2023) provides the following insights on the future of the monetary system, reflecting the weaponization of the US dollar and the longer term implications for currencies in an increasingly fragment and multipolar trading and monetary system:

- ***USD-Centric Monetary System History***: The dominance of the US dollar has often been questioned due to perceptions of US monetary policy being either too loose or too tight.
- ***Macroeconomic Imbalances and Geopolitical Conflict***: US economic imbalances and geopolitical conflicts, like the Russia-Ukraine war and US-China tensions, could undermine trust in the US dollar.
- ***Foreign Currency Reserves***: The US dollar's share in global foreign exchange reserves has decreased, but significant reallocation to other currencies is limited.
- ***Evolution of the Monetary System***: No immediate replacement for the US dollar is in sight, but the global monetary system is gradually becoming more multipolar.

Several factors are likely to determine the outcome of the monetary competition between the US and China with emerging financial infrastructures (and digital banks) and standards for digital currencies providing the technological arena for monetary competition. As of the June 2024, CoinMarketCap includes some 2.4 million crypto currencies with a market cap of US$ 2.63 billion. Given that all crypto currencies are essentially global in nature, it is of course entirely possible that the ultimate winner in monetary competition may not be either the US or China, but instead the frictionless global market. With Bitcoin having a market cap of U$ 1.4 trillion, similar in value to the GDP of Indonesia in 2022, and with potential growth in value, virtual currencies appear to have a stronger comparative advantage into the future, as long as they can maintain the three primary characteristics of money (medium of exchange, unit of account and store of value).

As of June 2024, the combined market capitalization of the top ten market capitalized firms is approximately US$ 26.47 trillion, led by Microsoft at US$ 3,18 trillion. The wealthiest person as of June 2024 is Bernard Arnault, the CEO and chairman of LVMH with a net worth is estimated to be around $212.1 billion. In such a context, where transnational corporations and ultra-high net worth individuals (there are some 626,619 individuals with a wealth of more than US$ 30 million), the currency habits of private wealth (which could eventually be substantially invested in virtual global currencies) may play as big a role in determining who wins currency wars, rather than sovereign states per se.

Moreover, with political divisions emerging within the US, and with conflict breaking out in Europe, the Middle East, Africa and potentially around Taiwan and the Pacific Rim, and with global public debt at US$ 91 trillion or around 93 per cent of global GDP, the potential for conflict to determine winners and losers in the monetary arena should not be understated. Put on top of these growing confidence in digital currency and blockchain based immutability, the future of monetary policy and currency hegemony will have no historic point of reference.

What seems most likely is that an increasingly fragmented multi-polar currency order will emerge, following which slow consolidation will likely occur into major trade blocks. A tenfold increase in the value of Bitcoin would value the currency at around 14–16% of global GDP, equally the British Pound. Moreover, once dependence on oil for energy is broadsided by green alternative energy technology (perhaps by 2050), the idea of a Petro-dollar or Petro-Yuan may also become defunct. Such an eventuality would lead to a complex cocktail of traditional and virtual currencies operating interchangeably, fundamentally changing the role that Central Banks have played in monetary stability and prudence. Under such a scenario—and where the global market-place is the issuer of currency to a larger extent—however it is denominated—globalization will have finally given birth to a highly fragmented but potentially interoperable and interchangeable medium of exchange, unit of account and store of value, that would be definition need new oversight institutions, perhaps informed by AI technology, given the complexity of the transaction marketplace.

8 Conclusions—Recommendations for US Policy

A hegemonic monetary system (and the hegemony as a whole) arguably breaks down when the balance of the empire—broadly, the difference between revenues from booty, taxes and exports, and expenditures on the overall cost of empire plus imports—turns negative. This leads to what has been termed 'hegemonic shifts'—changes in the hierarchical structures of centres of surplus accumulation according to a continuous 'rise and fall' pattern.

What the analysis above makes clear is that for the US dollar to remain the predominant trade and exchange currency, its overall global standing as a political, military and economic hegemon would need to be consolidated. Currently, this appears unlikely unless political fractures and loss of social cohesion at home urgently addressed. Moreover, the fiscal regime that has led to extraordinary levels of national debt would need to be turned around, and a new foreign policy would be needed to address global concerns of dollar weaponization. Either way, in the digital era, the dollars future will need to be anchored firmly in the emerging global digital economy, and a future world of virtual monetary governance that still to even be conceptualized.

References

Arslanalp S, Eichengreen BJ, Simpson-Bell C (2022) The stealth erosion of dollar dominance: active diversifiers and the rise of nontraditional reserve currencies, IMF Working Paper, No. 2022/058. Available at: https://www.imf.org/en/Publications/WP/Issues/2022/03/24/The-Stealth-Erosion-of-Dollar-Dominance-Active-Diversifiers-and-the-Rise-of-Nontraditional-515240. Accessed 28 April 2024
Bordo MD, Humpage OF, Schwartz AJ (2015) Strained relations: US foreign-exchange operations and monetary policy in the twentieth century. University of Chicago Press, Chicago
Cleveland HV (1976) The international monetary system in the interwar period. In: Rowland BM (ed) Balance of power or hegemony: the interwar monetary system. New York University Press, New York, pp 1–60
Cohen BJ (2017) International political economy. Routledge, London
Edelstein M (2004) Foreign investment, accumulation and empire, 1860–1914. In: Floud R, Johnson P (eds) The Cambridge economic history of modern Britain, vol 2. Cambridge University Press, Cambridge, pp 190–226
Eichengreen B (2001) Capital account liberalization: what do cross-country studies tell us? World Bank Econ Rev 15(3):341–365
Eichengreen B (2011) Exorbitant privilege: the rise and fall of the dollar and the future of the international monetary system. Oxford University Press, Oxford
Gokay B (2006) Soviet Eastern policy and Turkey, 1920–1991: Soviet foreign policy, Turkey and communism. Routledge, London
Gowa J (1983) Closing the gold window: domestic politics and the end of Bretton Woods. Cornell University Press, Ithaca, New York, USA
Hudson M (2003) America's monetary imperialism. Global Dialogue 5(1/2):73
Imlah AH (1950) The terms of trade of the United Kingdom, 1798–1913. J Econ Hist 10(2):170–194. https://doi.org/10.1017/S0022050700064123
Keynes JM (1930) A treatise on money. Macmillan, London. [Two volumes]
Khanna P (2024) The red sea crisis proves china was ahead of the curve, Foreign Policy. Available at: https://foreignpolicy.com/2024/01/20/url-red-sea-houthis-china-belt-road-suez-trade-corridors/. Accessed 8 April 2024
Kindleberger C (1973) The world in depression, 1929–1939. University of California Press, Berkeley
Liu HCK (2002) US dollar hegemony has got to go, Asia Times, 11. Available at: [URL]. Accessed: Date
MacMullen R (2000) Romanization in the time of augustus. Yale University Press, New Haven
Meikle S (2002) Aristotle on money. In: Ingham G (ed) What is money? Routledge, London, pp 167–183
Pozsar Z, Adler O, Botteron M, Hechler-Fayd'herbe N (2023) The future of the monetary system. In: Adler O (ed). Credit Suisse Research Institute
Schwartz A (1996) The operation of the specie standard: Evidence for core and peripheral countries, 1880–1990. In: Bordo M (ed) (1999) The gold standard and related regimes: collected essays. Cambridge University Press, Cambridge
Silk L (1973) The Dollar's Tribulations, The New York Times. Available at: https://www.nytimes.com/1973/07/04/archives/the-dollars-tribulations.html. Accessed 8 June 2024
Spiro DE (1999) The hidden hand of American hegemony: petrodollar recycling and international markets. Cornell University Press, Ithaca
Stanford Institute for Economic Policy Research (2024) IMF's Gita Gopinath: Geopolitics and its impact on global trade and the dollar, SIEPR. Available at: https://siepr.stanford.edu/news/imfs-gita-gopinath-geopolitics-and-its-impact-global-trade-and-dollar. Accessed 12 May 2024
Temin P, Voth HJ (2013) Prometheus shackled: Goldsmith banks and England's financial revolution after 1700. Oxford University Press, USA, Oxford

Witte C (2022) Can Netanyahu prioritise Israel's Asia pivot? East Asia Forum. Available at: https://eastasiaforum.org/2022/12/22/can-netanyahu-prioritise-israels-asia-pivot/. Accessed 24 May 2024

Other Points of Reference

Books

"Exorbitant Privilege: The Rise and Fall of the Dollar and the Future of the International Monetary System" by Barry Eichengreen: Specifically, Chapters 3 and 4 offer a deep dive into how the U.S. dollar became dominant and what challenges it faces

"Currency Wars: The Making of the Next Global Crisis" by James Rickards: Focus particularly on Chapter 7, which discusses China's role in the currency war scenario

Academic Articles

"China's Competitive Dollar Policy" in the Review of International Economics: Focuses on China's strategic management of its currency

"Currency Wars, Coordination, and Capital Controls" in the International Journal of Central Banking: Discusses the mechanisms countries use to manage currency competition

"The Evolution of Reserve Currency Diversification: Changes in International Monetary Power?" in International Studies Quarterly: A useful piece on how reserve currency diversification influences global power dynamics

"Renminbi Internationalization: A Conflict of Statecrafts" in the China Quarterly: Focuses on China's approach to making its currency more globally accepted

Reports and Whitepapers

IMF Working Papers: Several papers on currency competition and its role in the international system

Peterson Institute for International Economics: Offers reports and policy briefs specifically focused on currency competition and U.S.-China relations

Bruegel Policy Contributions: Articles like "The international role of the euro: down but not out" provide a European perspective on the topic

Online Resources

Council on Foreign Relations: Articles like "The Big Shift: Challenge to the Dollar's Role" offer contemporary takes on currency competition

Bloomberg, Financial Times: These often have opinion pieces and analyses that touch on currency competition between the U.S. and China

Primary Documents

IMF Special Drawing Rights (SDR) Allocations and Composition: Provides hard data on reserve currencies

Federal Reserve or People's Bank of China Policy Papers: Often these will contain clues or explicit statements about currency policy

Whither the Belt and Road Initiative?

Lawrence C. Reardon

Abstract To lessen the impact of the 2007–2008 Global Economic Crisis, Chinese leaders invested large amounts of capital to build its domestic infrastructure and production capacity. Having saturated the market and created production overcapacity, General Secretary Xi Jinping readjusted the strategy in 2013 by promoting the Belt and Road Initiative (BRI). Xi sent Chinese delegations around the world who offered financial loans, engineering expertise, and skilled labor to fulfill the dreams of local politicians and businesspeople by constructing large infrastructure projects. China subsequently lent over $1 trillion to build 3,000 BRI projects, ranging from high-speed rail in Indonesia, a highway linking Montenegro and Serbia, and port facilities in Pakistan and Sri Lanka. China thus developed strong economic connections around the world, which secured markets for Chinese export goods to relieve Chinese domestic overcapacity and strategic materials to fuel China's production juggernaut. However, Chinese banks failed to evaluate these megaprojects, many of which had already been rejected by the "old era" multilateral aid programs such as the World Bank as financially unfeasible. BRI recipients were unable to meet their loan repayment schedules and requested loan adjustments and even loan forgiveness. Thus, in November 2021, Xi Jinping implemented an interim policy readjustment to scale down or halt BRI planned projects and to focus on "small yet smart" projects.

1 Introduction

Harkening back to 114 BC and the complex series of trade routes known as the Silk Road that connected the Chinese Han Dynasty with Central Asia, East Africa, and Southern Europe, Xi Jinping in 2013 announced his signature policy, the Belt and Road Initiative. Based on decades of Chinese experience building foreign aid projects in the developing world, Xi's vision has been realized with over $3.1 trillion in foreign currency reserves, of which China has lent $1.3 trillion to build over 3,000 infrastructure projects that link the People's Republic of China with many of the

L. C. Reardon (✉)
University of New Hampshire, Portsmouth, NH, USA
e-mail: chris.reardon@unh.edu

A. Farhadi et al. (eds.), *The Great Power Competition Volume 6*,
https://doi.org/10.1007/978-3-031-70767-4_11

medium and smaller economies around the world.[1] By signing a series of trade and investment agreements and acting as the lender of first resort to many developing economies, China has gained access to key natural resources necessary to fuel China's economic growth, has developed new markets to sell its goods, as well as fostered new allies who now prefer to align with China.[2] The success of the BRI also has laid the foundation for Xi's ambitious plan to promote the Global Development Initiative that challenges the older post-1945 era of global governance with a Sino-centric global economic regime.

On 17 October 2023, General Secretary Xi Jinping opened the Third Belt and Road Forum for International Cooperation with 23 world leaders and 10,000 representatives from 151 countries in attendance. After holding bilateral meetings with Vladimir Putin and leaders from Africa, South America, Central and Southeast Asia, Xi Jinping focused on the success of the BRI infrastructure projects around the world in his keynote address. While critical of the old global governance regime, which Xi described as promoting "ideological confrontation, geopolitical rivalry and bloc politics," Xi praised the signing of cooperation documents with 150 countries and 30 international organizations that had established "a global network of connectivity consisting of economic corridors, international transportation routes and information highways as well as railways, roads, airports, ports, pipelines and power grids." With Chinese knowledge and financial support, "new airports and harbors, smooth roads, and newly built industrial parks for business cooperation have created new economic corridors and new growth drivers and have become the trading routes and staging posts of the new era."[3]

During this 10-year period, developing economies warmly welcomed visiting Chinese BRI delegations. They offered financial loans, engineering expertise, and skilled labor to fulfill the dreams of local political and business leaders to build large infrastructure projects, many of which had been rejected by the "old era" multilateral aid programs such as the World Bank. While describing the BRI as having "gotten off the ground, grown rapidly and produced fruitful outcomes," Xi failed to mention the BRI's flagship infrastructure projects, many of which have encountered serious implementation problems. Perhaps the most famous is the $1.4 billion loan to build the Hambantota International Port in Sri Lanka, which China has assumed a 99 year lease in partial repayment of Sri Lankan BRI debts. Neither did Xi mention China's largest BRI recipient, Pakistan, which has borrowed $62 billion to build ports, railways, highways, and hydroelectric plants, of which 40% have encountered

[1] B. C. Parks., et. al., *Belt and Road Reboot: Beijing's Bid to De-Risk Its Global Infrastructure Initiative*. Williamsburg, VA: AidData at William & Mary (2023),1, 8.

[2] Tsubasa Suruga, "Majority of ASEAN people favor China over US, survey finds," *NikkeiAsia*, 2 April 2024, Majority of ASEAN people favor China over U.S., survey finds—Nikkei Asia.

[3] Xi Jinping, "Full text of Xi Jinping's keynote speech at 3rd Belt and Road Forum for Int'l Cooperation," delivered at the Opening Ceremony of the Third Belt and Road Forum on 18 October 2024, *China Daily*, 18 October 2024 Full text of Xi Jinping's keynote speech at 3rd Belt and Road Forum for Int'l Cooperation—Chinadaily.com.cn; China's Belt and Road Leaves Kenya With a Railroad to Nowhere—Bloomberg 18 July 2019.

significant implementation problems, some of which have been delayed indefinitely[4] (see below).

Starting around 2020 and COVID, Xi Jinping began to face a multitude of BRI recipients whose loans were becoming "debt traps," as Chinese state banks refused to take a "haircut" to accept a markdown of the loan value. A good example is the Chinese state-owned China Road and Bridge Corporation (CRBC), which won in 2014 a non-competitive bid to build the phase I of Kenya's Mombasa-Nairobi Standard Guage Railway. China's Export–Import Bank (Exim Bank) provided a $3.6 billion loan to finance the BRI project, which Kenya would repay by collecting cargo fees. However, from 2017 to 2020, the railroad ran a $200 million deficit as Kenya had yet to start phase II that would connect railway lines to Uganda and Rwanda. Businesspeople thus continued to rely on container trucks to ship their goods to Mombasa along an established highway. If the CRBC and Exim Bank had done their due diligence and carried out risk analyses and feasibility studies, they would have discovered that a 2013 World Bank study had concluded "that there is no economic or financial case" for building the railroad.[5]

Instead, Xi Jinping announced a readjustment of the BRI goals to emphasize "small yet smart" people-centered projects, to reduce the amount of loans to be distributed by China's development banks, as well as to highlight small-scale assistance and green development projects. While arguing the previous ten years had demonstrated that the BRI was "the right path forward," Xi warned about "headwinds and tailwinds. We need to stay focused on our goal, take result-oriented actions, persevere, and keep moving forward..." China has not eliminated the BRI program and has remained the world's largest official source of international development finance. Yet, since its peak in 2016 when it provided over $80 billion for BRI projects, China has greatly reduced its lending portfolio to focus on "small yet smart" projects.

To explain this readjustment, this chapter focuses on the impact of China's decision to invest large amounts of capital to lessen the domestic impact of the 2007–2008 Global Economic Crisis. After five years, Xi Jinping readjusted the strategy by announcing the BRI that would develop new overseas markets to relieve Chinese domestic overcapacity. However, during the implementation phase, the BRI encountered various problems that have diminished China's global reputation, including BRI recipients requesting loan bailouts or forgiveness. Thus, in November 2021, Xi Jinping proposed an interim policy readjustment to scale down the BRI, to halt future project plans, and to focus on "small yet smart" projects.

---

4 How China Got Sri Lanka to Cough Up a Port—The New York Times (nytimes.com) 25 June 2018; "Ten Years of China's Belt and Road: What has $1t achieved?" *Financial Times*, 22 October 2023 Ten years of China's Belt and Road: what has $1tn achieved? (ft.com).

5 George Downs, "Kenya's $3.8 Billion Chinese-built Railway Shows why Belt and Road Is Being Reworked," *Wall Street Journal*, 2023 Kenya's $3.8 Billion Chinese-Built Railway Shows Why Belt and Road Is Being Reworked - YouTube; Oscar Otele, "China's approach to development in Africa: A case study of Kenya's standard gauge railroad," *Council of Foreign Relations*, 2021 Otele_A Case Study of Kenya's Standard Gauge Railway.pdf (cfr.org).

2 Chinese Policy Cycles

This chapter adopts a policy cycle approach to explain Xi Jinping's revision of the BRI initiative. Western economists have used various cycle theories to analyze China's economic growth, including harvest cycles involving good and bad harvests and the demand for consumption goods affecting the business cycle.[6] Political scientists have emphasized the impact of Chinese elite conflict and opinion groups on policy cycles.[7] Building upon Premier Zhou Enlai's statement that China had experienced three policy stages since 1949 (recovery, development and readjustment),[8] the Chinese policy approach encompasses four phases: policy crisis; policy readjustment; policy innovation; and policy implementation.[9]

During the first policy phase [危机], policy elites use the threat of crisis to readjust policy and implement innovative measures to avoid total political or economic disorder that would threaten the Party's legitimacy.[10] A crisis can pivot favorably and result in an improved situation; just as easily, any crisis can devolve into utter chaos [乱].[11] For example, Chinese leaders used crisis to realize two of the most important changes in Chinese economic development policy during the 1970s.[12] Premier Zhou Enlai eliminated the Cultural Revolution's extreme autarchic economic policies following Marshall Lin Biao's attempted military putsch in 1971 to reintroduce the import substitution (IS) development strategy of the 1950s. By the late 1970s, Deng Xiaoping avoided economic chaos by eliminating most of Hua's expensive large-scale IS import projects and introducing a new outward oriented development strategy that relied on foreign trade and investment.

During the policy readjustment phase [调整], policy elites recognize that the country is facing a crisis and must adopt new emergency measures to address the

---

[6] Hiroyuki Imai, "Explaining China's Business Cycles," *The Developing Economies* XXIV, 2 (June 1996), 154–185.

[7] Richard Baum, "The Road To Tiananmen: Chinese Politics in the 1980s," in *The Politics of China, 1949–1989*, ed. by Roderick MacFarquhar (NY: Cambridge University Press, 1993), 340–471; Richard Lowenthal, "Development vs. Utopia in Communist Policy," in *Change in Communist Systems*, ed. by Chalmers Johnson (Stanford: Stanford University Press, 1970), 54.

[8] Zhou Enlai, "Guomin jingji fazhan de fangzhen he mubiao" [The Plans and Goals of National Economic Development], delivered during the discussion of the «Guanyu gonghui fazhan wenti (Chugao)» [Industrial Development Problems (Initial Draft)] on 23 August 1963, in Zhonggong Zhongyang Wenxian Yanjiushi, *Zhou Enlai Wenxuan* [A Selection of Zhou Enlai's Economic Works] (Beijing: Zhongyang Wenxian Chubanshe, 1993), 516–517.

[9] Lawrence C. Reardon, *Reluctant Dragon: Crisis cycles in Chinese foreign economic policy* (Seattle, WA: University of Washington Press, 2002), Chap. 1.

[10] Merilee Grindle and John Thomas, *Public Choices and Policy Change: The Political Economy of Reform in Developing Countries* (Baltimore: The Johns Hopkins University Press, 1991, Chap. 4; Kent E. Calder, *Crisis and Compensation* (Princeton: Princeton University Press, 1988), 20; 40.

[11] Solinger (1991).

[12] Reardon, *Reluctant Dragon*, Chaps. 5 and 7.

problem while the party state maintains control.[13] Besides resolving critical problems, elites also discuss new economic policies that can either expand existing policies or initiate new innovative approaches. This process can be prolonged as proponents of existing policies will often defend the status quo or offer alternative solutions before achieving elite consensus. Thus, when Premier Zhou Enlai and Liu Shaoqi promulgated emergency measures to roll back the Great Leap Forward (GLF) strategy in 1959, elites only fully implemented a readjustment after 1961 because of Mao Zedong's desire to defend his economic development strategy. In 1979 after rejecting Premier Hua Guofeng's large-scale IS development strategy, Li Xiannian announced that the readjustment period would take time to resolve the "proportional imbalances."[14] During the ensuing three years, a new consensus on development was transmitted to the Party's grass roots organizations to mobilize the party state.

During the third policy innovation phase [创新], the paramount leader introduces innovative economic policy options. Concurrent with the readjustment of Hua's large-scale import substitution program in 1979, the State Council issued document 1979.202 that expanded older export promotion policies first implemented in the early 1960s. The Central Committee also issued document 1979.50, which opened Guangdong and Fujian to the global economy and granted them greater foreign economic authority.[15]

The most complicated phase is the fourth and final phase of the Chinese policy cycle, policy implementation [实行] and readjustment [调整]. Policy implementation can go smoothly, or result in unforeseen problems, including political opposition, technological shortcomings, financial distress, or long-term economic disruption. To ensure smooth implementation, policy elites often mandate inspection visits and require the relevant ministries to submit reports. Elites thus determine the need to adopt policy readjustments to avoid problems that could impact on the preeminent leader's reputation and possible reinitiate a new Chinese policy cycle. Thus, after two years of sponsoring high-level policy review conferences in Beijing, the Central Committee issued an Emergency Circular in January 1982 that halted the expansion of Guangdong and Fujian's foreign trade rights, investigated local provincial leaders accused of economic malfeasance, and recentralized provincial economic controls.[16]

Thus when explaining Xi Jinping's readjustment of BRI policy, the Chinese policy cycle theory posits that Chinese elites injected large amounts of capital investment in China's domestic infrastructure to inoculate the economy against the 2007–2008 Global Economic Crisis. However, because the domestic investment strategy created production overcapacity, Xi Jinping readjusted the domestic strategy by announcing the Belt and Road Initiative in September 2013 during his visit to Kazakhstan. During

[13] Solinger, *From Lathes*, 9–16.

[14] William A. Joseph, *The Critique of Ultra-Leftism in China, 1958–1981* (Stanford: Stanford University Press, 1984), 140–150; 179–182.

[15] Reardon, *Reluctant Dragon,* Chap. 7.

[16] Lawrence C. Reardon, *A Third Way: The Origins of China's current economic development strategy* (Cambridge, MA: *Harvard University Asia Center,* 2021), Chap. 3.

the third phase, policy innovation, Xi Jinping introduced the concept of a Sino-centric global governance structure, in which China hoped to replace the Bretton Woods economic regime with China's global development initiative. However, during the implementation phase, the BRI has encountered various problems from poor construction, criticisms of debt traps, and requests for loan forgiveness. Thus, in November 2021, Xi Jinping proposed an interim policy readjustment to scale down the BRI, to halt future project plans and to focus on "small yet smart" projects.

3 Chinese Policy Cycles: Phases I–IV

Phase I: Crisis: The Global Economic Crisis (2007–2008)

The Global Economic Crisis of 2007–2008 was the most severe global recession since the Great Depression of the early 1930s.[17] Originating in the United States following the issuance of risky subprime mortgages that inflated the US housing market, investment and commercial banks suffered significant losses in mortgage-backed securities, including Lehman Brothers. This banking crisis threatened the entire global economy, which the International Monetary Fund predicted would lose over $2.8 trillion between 2007 and 2010.[18] Chinese elites were quite concerned about the impact of US economic problems as the Chinese outward oriented development strategy had depended on growing international demand for Chinese manufactured products. They thus allocated RMB 4 trillion ($585 billion) in stimulus financing to build infrastructure and develop specific industrial sectors to stimulate economic growth.

According to Barry Naughton, the Central Committee issued Document 2008.18 that promulgated 10 policies to guarantee domestic demand and growth.[19] By the fourth quarter of 2008, the party state invested RMB 100 billion in transportation and power infrastructure, rural village infrastructure, environmental investment, affordable housing, technological innovation, and health and education. In addition, the party state allocated RMB 20 billion to recover from the May 2008 Wenchuan, Sichuan earthquake for a total of RMB 120 billion. The National Development and Reform Commission (NDRC) held a special emergency meeting to determine the distribution of the RMB 100 billion with the focus on encouraging domestic production. Provincial governments viewed Document 2008.18 as a prime opportunity to list local projects in the national investment plan, which would be included in the NDRC block grants.

---

[17] Tambunan and Hamonangan (2011).

[18] Reuters, "FACTBOX-US, European bank writedowns, credit losses," 5 November 2009 FACTBOX-U.S., European bank writedowns, credit losses | Reuters.

[19] Naughton (2009).

Naughton argues that the party state eventually allocated over RMB 4 trillion to stimulate the economy, with a large majority devoted to transport and power infrastructure.[20] In addition, the NDRC allocated financial support to 10 industrial production sectors, including automobiles, steel, shipbuilding, industrial machinery and electronics and information technology. This included RMB 600 billion to develop the next Internet generation, RMB 1 billion to encourage owners to dispose of older cars to buy new vehicles, subsidies for household electric and electrical devices such as computers, cell phones and microwave ovens. By the end of 2009, the RMB 4 trillion stimulus package stabilize the Chinese economy and had fostered a growth from the first quarter GDP growth rate of 6.1% to 9.45 for 2009.[21]

In implementing Document 2008.18, central and local elites focus the largest percentage of investment in key infrastructure, especially developing the High-Speed Rail (HSR) Network and the national expressway system. During his visit to Japan in 1978, Deng Xiaoping described the Shinkansen as "faster than the wind" and a technology that China should adopt. Yet, the construction of China's first HSR was delayed until 2008, when the China Railway opened the Beijing-Tianjin HSR. With the infusion of infrastructure stimulus money, both the central and local governments strongly supported HSR construction during the next decade, which eventually became the world's largest system with over 28,000 miles. In June 1984, China built the Shenyang-Dalian highway, with new highways gradually being built across the country over the next decade. Starting in 2005, China's minister of transport called for the establishment of the 7918 network, which was composed of seven highways from Beijing to key population centers, nine highways running north to south, and 18 east to west highways to be completed by 2035. However, following the promulgation of Document 2008.18, the party state increased the highway allocation from RMB 100 billion between 2010 and 2020 to RMB 865.8 billion in 2009. This 765% increase in funds came from domestic bank loans, central and local government outlays, and vehicle taxes.

During this period, Chinese engineers and architects gained significant experience in dealing with various construction problems; the standardization of design reduced the overall cost, which was one-third lower than other countries.[22] Another crucial factor was China's migrant labor force, which had gained decades of experience building China's modern cities. In the early 1980s, labor recruiters working for private construction firms signed up rural farmers to join migrant construction work teams assigned to various infrastructure projects in the Shenzhen Special Economic Zone and subsequently moved on to build China's cities. They became temporary residents of the urban areas living in company provided housing, working for low wages, and would return home, especially during the Lunar New Year. They worked together as specialized teams, gained experience working on a specific task, and

[20] Naughton, 6.

[21] China GDP—Gross Domestic Product 2009 | countryeconomy.com.

[22] Lawrence et al. (2019).

learned to collaborate with other member of their team who were often from the same village speaking a similar Chinese dialect.[23]

These skills and methods would become especially important as Chinese construction teams and development bank managers searched the world after 2013 looking for potential infrastructure projects utilizing China's competitive advantage in design, skills, and labor.

Phase II: Policy Readjustment: Domestic Investment, 2011–2013

For three years following the implementation of Central Committee document 2008.18, the Chinese GDP growth rate rose from 9.55% in 2008 to 10.64% by 2010.[24] By investing in transportation, industries, housing, and disaster rebuilding, the stimulus shielded the Chinese economy from the detrimental effects of the Global Economic Crisis and helped to stabilize the global economy.

Yet after 2010, China's growth rate began dropping as the domestic economy, with its market forces constrained and dominated by the state-owned enterprises, suffered from overinvestment and overcapacity. Having absorbed the stimulus investment, domestic industries increasingly were unable to sell their goods because of excess domestic production capacity. Unable to earn capital to repay their creditors, local governments threw a financial lifeline to ailing companies as they defaulted on their loans. When some SOEs declared bankruptcy, the local governments adjudicated the bankruptcy procedures, which often restructured the debt in favor of the SOE.[25] The construction sector suffered similar problems with overcapacity as demand for new infrastructure and factories had waned by 2011, which also affect overproduction and excess capacity in China's iron and steel industries, cement, and flat glass industries.[26]

To resolve these problems, General Secretary Hu Jintao injected new capital that enabled Chinese industries to export their excess capacity. These funds also helped the state-owned and private enterprises to lower production costs and successfully compete for overseas contracts. The party state actively supported SOE and private company delegations during their visits to various economies throughout the world looking for contract opportunities. Thus, Hu Jintao not only was the originator of the BRI, but also the "China model of development," which is not only is characterized by its state-led, export-oriented growth model but also structural overcapacity. While its once abundant and cheap labor is no longer as important to the model, the party state more recently has concentrated support for certain key advance industrial sectors, such as electronic vehicles and solar cells. Once the domestic market is saturated with a particular product, the party state uses its export promotion policies and international market access to export its oversupply of goods at a cheaper price, thus

---

[23] Solinger (1999).

[24] China GDP Growth Rate 1961–2024 | MacroTrends.

[25] Xiaohuan (2024).

[26] Ping et al. (2019).

often undercutting foreign producers and creating a protectionist backlash abroad.[27] In recent times, China exported seven times the number of electronic vehicles in 2023 than in 2019, while China exported five times the number of solar cells in 2024 than in 2018. As one study warned, "These surges are potentially devastating to market-constrained producers in advanced economies."[28]

Thus, Hu Jintao's readjustment of the 2008 stimulus policy to become part of China's outward oriented development strategy became the economic foundation of the BRI program.[29] General Secretary Jiang Zemin had initially promoted the "Go Out policy" in December 1997 to encourage Chinese SOEs and private enterprises to rejuvenate domestic industry by going global and investing abroad.[30] Hu Jintao expanded upon Jiang's "Go Out policy" at the 17th Party Congress in 2007 by promoting the slogan, "Community of Common Destiny" to signal China's renewed interest to reengage the global economy. When promoting the export of China's overcapacity, Hu subsequently broadened the concept in his 2012 Report to the 18th Party Congress to emphasize global interconnectedness and cooperation among humanity to build a harmonious world of peace and prosperity.[31] This Chinese vision of international relations envisioned a new type of equitable partnership to promote global development and the pursuit of common interests. After becoming party general secretary in 2012, Xi Jinping adopted the slogan, which became the theoretical justification for the Belt and Road Initiative and the Global Development Initiative.[32]

Just as China's outward oriented development strategy adopted in the late 1980s had its origins in export promotion policies of the 1950 and 1960s, the BRI is built upon Chinese foreign aid schemes developed during the Cold War and the Sino-Soviet conflict, when the Chinese economy was barely above subsistence level. Prior to 2000, China financed over 4,000 projects in which China provided grants or interest-free loans, as well as taking responsibility for the background surveys, designing, and construction activities, including the equipment, building materials, technical personnel, and labor. To quote the State Council's foreign aid report of 2011, "China

[27] Mary Gallagher, "For China's Economy, Excess Capacity is a Feature, Not a Bug," *World Politics Review*, 2 April 2024 For China's Economy, Excess Capacity is a Feature, Not a Bug | WPR (worldpoliticsreview.com).

[28] Camille Boullenois, et al. *Overcapacity at the Gate,* Rhodium Group, 26 March 2024 Overcapacity-at-the-Gate.pdf.

[29] For more on China's adoption of an outward oriented development strategy, see Reardon, *A Third Way.* 1–41.

[30] Jiang Zemin, "实施 '引进来 '和'走出去 '相结合的开放战略" [Combining the Opening Strategy of 'Attracting Investment' and "going out'], 江泽民文选 [Collected Works of Jiang Zemin], vol. 2 (Beijing: People's Press, 2006), 91–4.

[31] Hu Jintao, "十八大报告全文英汉对照 Full text of Hu's report at 18th Party Congress," 12 December 2012, 十八大报告全文英汉对照 Full text of Hu's report at 18th Party Congress_时事英语_福建省外事办 (archive.org).

[32] Zhang (2018).

has made full use of the mature technologies and relative low cost of manpower to help other developing countries construct a host of infrastructure projects."[33]

Starting in 1954, China repaired the Vietnamese and North Korean bridges, roads, and port infrastructure damaged during their conflicts. Yet, the most publicized early foreign aid projects include the Karakoram Highway, which was started in 1962 and completed in 1982 that connects the Pakistani provinces of Punjab to the western Chinese province of Xinjiang. In the middle of the Cultural Revolution, Mao Zedong approved the Tanzania-Zambia Railway, which is a 1,160-mile track linking Dar Es Salaam with central Zambia. China provided a $406 m interest free loan which opened in 1975 after five years of construction. The Chinese actively promoted these two projects as a symbol of its solidarity with the developing world.

Guinea's experience of welcoming Chinese workers and expertise to harness its river system to generate energy is an excellent example of the evolution of Chinese foreign aid. First implemented during the Maoist period, China focused on small and relatively uncomplicate infrastructure projects financed with interest free loans. In 1963 while China still was recovering from the economic devastation of the Great Leap Forward and the Sino-Soviet Schism, Premier Zhou Enlai approved the building of the 2-megawatt (MW) Kinkon Falls hydroelectric plant in Guinea, which was its first power station since its 1958 independence. Under Hu Jintao in 2006, China's Hunan Construction Engineering Group upgraded the dam to 3.4 MW for $4 million.

Yet during its initial phase, the BRI undertook more grandiose, billion-dollar dams that would facilitate the extraction of China's needed raw materials. China's Exim Bank in 2018 provided a $1.3 billion credit to build the 450 MW Souapiti Hydropower Station. China International Water and Electric Corporation contracted to design and build the dam for $1.38 billion, using turbines from the Harbin Electric Group Motor Company, and utilizing the Third Hydropower Bureau for all civil engineering, grouting, and metal structures.[34] While the costs of building these hydropower projects had ballooned to over $1b under the BRI, these Chinese projects were warmly welcomed by the Guinean people. As the world's second largest exporter of bauxite, Guinea needed a more dependable source of energy for mining and exporting the ore. Guinea's largest trading partner, China, also depended on the ore to supply its aluminum oxide refiners. The Guinean government has considered these projects an important national achievement and has enshrined these Chinese foreign aid projects from both the 1960 and the BRI versions on Guinean currency.[35]

During the pre-2013 period, China also gained experience with export promotion policies, including interest free loans, concessional loans and investment banks which evolved into important players in the post-2013 BRI period. In 1993, China established the Foreign Aid Fund for Joint Ventures and Cooperative Projects that offered

---

[33] Information Office of the State Council The People's Republic of China, *China's Foreign Aid (2011)*, 9 August 2018 China's Foreign Aid (2011) (cidca.gov.cn).

[34] NS Energy, "Souapiti Hydropower Station," Souapiti Hydropower Station, Konkure River Basin, Republic of Guinea (nsenergybusiness.com).

[35] Austin Jordan, "Setting the Record Straight on China's Overseas Infrastructure Investments," 17 March 2023, Setting the Record Straight on China's Overseas Infrastructure Investments - Fairbank Center for Chinese Studies (harvard.edu).

interest free loans for 20 years to help small and medium sized Chinese companies to establish joint ventures and cooperative production in developing economies. In 2009, the state-funded Exim Bank of China provided concessional short-term and long-term loans to developing economies planning to improve their transport, communications, and electrical infrastructure. For instance, by the end of 2009, the Exim Bank had given RMB 256 billion in foreign aid, RMB 106 in grants, 76 billion in interest free loans, and 73 billion in concessional loans.[36] The State Council established the state-owned China Development Bank (CDB) in 1994 to promote domestic infrastructure investment, including for the Three Gorges Dam and Shanghai's Pudong Airport. Utilizing this accumulated experience, the CDB took a larger role in the BRI in 2017 by lending over $34 billion medium- and long-term loan interest loans to developing economies, which increased to $280 billion that was lent to 1,300 projects in 2023.[37] Based on the experience accumulated in managing these financial institutions, Xi Jinping proposed the Asian Infrastructure Investment Bank (AIIB) during his 2013 trip to Indonesia to inaugurate the BRI. China established the AIIB's headquarters in Beijing, whose membership grew to 109 members from the Global South and Europe with a total investment of $100 billion. Furthermore, the BRICS (Brazil, Russia, India, China, and South Africa) opened the AIIB counterpart, the New Development Bank (NDB) in Shanghai with $100 billion in initial authorized capital to invest in infrastructure projects among its shareholders.[38]

Phase III: Innovation: The Global Development Initiative, 2013–2021

While the BRI is an expansion of previous policies to support domestic infrastructure investment and foreign economic expansion, Xi Jinping introduced new innovations: the Global Development Initiative (GDI) announced in September 2021, the Global Security Initiative (GSI) of April 2022, and the Global Civilization Initiative (GCI) of March 2023. Xi stated that the purpose of the GDI was to relieve global poverty, poor health, and inequality by investing in basic infrastructure; the GSI promotes global peace, stability, and cooperation, while the GCI promotes a harmonious world of diverse cultures and a share humanity. Yet Xi also stated that his ultimate goal was to emerge as the new "global leader in terms of comprehensive national power and international influence."[39] Xi thus is challenging the post- 1945 economic, security,

---

[36] Information Office of the State Council The People's Republic of China, *China's Foreign Aid*.

[37] "CDB plays active role in boosting Belt, Road," *China Daily*, 17 October 2023, CDB News_CDB plays active role in boosting Belt, Road.

[38] Members of the Bank (aiib.org), 12 September 2023.; Shareholding—New Development Bank (ndb.int), 12 September 2023; Bas Hooijmaaijers, "Understanding success and failure in establishing new multilateral development banks: The SCO Development Bank, the NDB, and the AIIB," *Asian Perspective* 45, no. 2 (2021): 445–467.

[39] Xi Jinping, "Secure a Decisive Victory in Building a Moderately Prosperous Society in All Respects and Strive for the great Success of Socialism with Chinese Characteristics for a New Era," delivered on 18 October 2017 at the 19th CCP National Congress. Secure a Decisive Victory in Building a Moderately Prosperous (xinhuanet.com).

and cultural regimes by reforming the global governance structure to promote a more "fair and just system" for China and the developing economies.[40]

To transform the global economic regime, Xi's GDI utilizes the BRI to build a new Sino-centric global economic order.[41] Xi reminded the delegates to the March 2023 National People's Congress that the CCP worked "hard for a century to put an end to China's national humiliation," during which "China was reduced to a semi-feudal and semi-colonial society, when bullying by foreign powers and frequent wars tore the country apart and plunged the Chinese people into an abyss of Great suffering." To avoid such "national humiliation," Xi has used China's economic power to transform the current economic regimes dominated by the United States and the advanced industrialized economies.

In his 7 September 2013 speech to the assembled dignitaries in Astana, Kazakhstan, Xi explained that the origins of the BRI concept was the Chinese Han Dynasty's (206 BC to 220 AD) establishment of the Silk Road. This intricate network of trading routes linked the Han capital in Chang'an (today's Xi'an) to the Central Asian countries, India, Pakistan, and eventually Europe. One year later during his official 2014 visit to Indonesia, Xi invoked Admiral Zheng He who had undertaken seven voyages in the 1400 s to open a Maritime Silk route that linked Ming Dynasty China with Southeast and Central Asia as well as the countries surrounding the Indian Ocean. According to Xi, "the pioneers of the ancient silk routes won their place in history not as conquerors with warships, guns, horses, or swords. Rather, they are remembered as friendly emissaries leading camel caravans and sailing ships loaded with goods."[42] In this way, Xi proposed to establish a new Silk Road Economic Belt and the 21st Century Maritime Silk Road that would become the foundation of a new Sino-centric global economic regime.

Enshrining the BRI in the 2017 Chinese Constitution,[43] Xi mandated that the BRI would establish land and sea routes linking Chinese firms with the global economy, develop connectivity between areas, and promote an integrated global trading and financial system. Chinese multinational firms thus could gain a competitive advantage with the dominant Western multinationals.[44] According to a 2023 State Council White Paper, the BRI would break through the strictures of the old post-WWII Bretton Woods economic regime, in which "some countries, holding to the notion of might makes right, willfully engage in bullying, plundering and zero-sum competition."

---

[40] Ministry of Foreign Affairs, "Proposal of the People's Republic of China;" Stella Chen, "Community of Common Destiny for Mankind," 25 August 2021, Community of Common Destiny for Mankind - China Media Project. For similar arguments, see Cheng Yawen, "构建 "新三环":面对全面脱钩可能的中国选择"[Building the 'New Three Rings': China's options when possibly confronting complete decoupling], *Beijing Cultural Review*, 3, 6 (2022) 构建"新三环":面对全面脱钩可能的中国选择 | 文化纵横网 (21bcr.com).

[41] Lei (2022).

[42] Xi Jinping, "Building an Open, Inclusive and Interconnected World.".

[43] Jagannath P. Panda, "What the Inclusion of BRI in the Chinese Constitution Implies." *IDSA Comment* (2017). What the Inclusion of BRI in the Chinese Constitution Implies | Manohar Parrikar Institute for Defence Studies and Analyses (idsa.in).

[44] Altamira et al. (2024).

China would establish a new world order based on traditional Chinese cultural values that emphasize harmony and cultural diversity. This new Sino-centric global village, "where all human beings are one big family," would be led by the Chinese Communist Party, who would bring a good life to the Chinese people and also helping other peoples to prosper and contributing more to humanity."[45]

As the global economy becomes more dependent on China, Xi Jinping hopes to replace the Bretton Woods system, in which the US and others "have practiced unilateralism, protectionism and hegemonism, hampering economic globalization and threatening a global economic recession."[46] Xi Jinping proposed a new Sino-centric economic order composed of Chinese-based international financial, trade and development institutions utilizing the Chinese RMB. This new global order will be managed by a new cadre of Chinese bureaucrats working to transform existing global economic institutions,[47] who will be guided by "the Communist Party of China, [which] is a major political party with a global vision."[48]

Chinese leaders were especially motivated to build infrastructure projects in African and Asian countries with untapped mineral and agricultural wealth that could supply raw materials for China's industrial needs. Having accumulated three trillion dollars of foreign exchange reserves earned since China embraced outwardly-oriented development in the 1980s,[49] Xi Jinping announced a new investment initiative to finance infrastructure construction, such as railways, highways, bridges, port facilities, and power grids, especially in the developing world. Rejecting the Bretton Woods' development regime's strict loan eligibility criteria, Chinese banks generously supported the BRI projects by increasing Chinese bank loans from $390 billion in 2010 to $1.5 trillion in 2017. China became the global lender of "first resort," but also its largest debt collector.[50]

China thus has been challenging the "old era's" global economic regime by gaining friends and allies in the Global South and establishing the foundation of a new Sino-centric model of global governance. When Xi Jinping announced the GDI at the September 2021 UN General Assembly meeting, BRI recipients warmly welcomed China's new development initiative. Stating that "development holds the key to people's well-being," Xi argued for placing high priority on alleviating poverty, guaranteeing food security, overcoming COVID, accessing development funding,

[45] State Council Information Office of the People's Republic of China, "Full Text: A Global Community of Shared Future: China's Proposals and Actions," September 2023, Full Text: A Global Community of Shared Future: China's Proposals and Actions (mfa.gov.cn).

[46] State Council Information Office of the PRC, "The Belt and Road Initiative: A Key Pillar of the Global Community of Share Future," 10 October 2023, The Belt and Road Initiative: A Key Pillar of the Global Community of Shared Future_中华人民共和国国务院新闻办公室 (scio.gov.cn).

[47] Xie Zhihai," The rise of China and its growing role in international organizations." ICCS 現代中国学ジャーナル 4, no. 1 (2011): 85–96. Microsoft Word-85-96XIE.doc (aichi-u.ac.jp).

[48] State Council Information Office of the PRC, "The Belt and Road Initiative," 10 October 2023.

[49] Cabestan (2019).

[50] Keith Bradsher," China Invested $1 Trillion to Gain Global Influence. Can That Go On?" *New York Times*, 16 October 2023, What is China's Belt and Road Initiative and How Is It Changing? - The New York Times (nytimes.com).

and promoting green development.[51] To rally support, China established the Group of Friends of the Global Development Initiative, which held a ministerial meeting in New York City in September 2022 that was attended by ministers from 60 countries and ten international organizations.[52]

While China had begun to win allies from the Global South, Chinese leaders had been encountering implementation problems which had a negative impact on China's global reputation as the "lender of first resort." The numbers of global leaders attending the Third Belt and Road Forum for International Cooperation in October 2023 fell from 37 in 2019 to 23 in 2023, in part because of the growing concern about their growing debt burden. Xi Jinping thus undertook a midstream assessment and readjustment of the BRI strategy.[53]

Phase IV: BRI Implementation and "Small yet Smart" Readjustment

By 2016, China's annual overseas development commitments reached its peak at $87 billion, which was substantially larger than the World Bank's commitment of $34 billion that year. However, as demonstrated in Chart 1, China's commitments suffered a continuous drop, especially after the 2020 COVID pandemic. It thus became obvious that during the five-year review of the BRI program, Xi Jinping and the Chinese elite had reassessed the implementation of the BRI program and decided to implement a mid-course policy readjustment under the rubric, "small yet smart."

There are external and internal reasons for the BRI policy readjustment. China's global reputation had suffered as more developing countries encountered difficulties in repaying their BRI loans. When Sri Lanka could not meet its BRI debt obligation, the government struck a deal with their Chinese creditors of leasing the Chinese-built port of Hambantota to a Chinese shipping company for 99 years. While its loan was not forgiven, Sri Lanka also received $1.1 billion that is used to pay off its international creditors. Critics have described Hambantota as an example of a Chinese "debt trap," designed to gain access to a strategically important port on the Indian Ocean. However, the plan was the brainchild of the former Sri Lankan President Mahinda Rajapaksa and his family to develop this underutilized port located in his family's political district. Finally, the Sri Lankan Navy has established its southern command at the port and not the Chinese navy.[54]

---

[51] State Council, "Xi calls for bolstering confidence, jointly addressing global challenges at UNGA," 22 September 2021. Xi calls for bolstering confidence, jointly addressing global challenges at UNGA (www.gov.cn).

[52] Ministry of Foreign Affairs, "Press Statement of the Ministerial Meeting of the Group of Friends of the Global Development Initiative," 21 September 2022. Press Statement of the Ministerial Meeting of the Group of Friends of the Global Development Initiative (mfa.gov.cn).

[53] David Sacks, "China's Belt and Road Initiative Enters its Second Decade: Which Leaders went to Beijing to Celebrate with Xi Jinping?" *Asia Unbound,* Council on Foreign Relations, 17 October 2023 China's Belt and Road Initiative Enters Its Second Decade: Which Leaders Went to Beijing to Celebrate with Xi Jinping? | Council on Foreign Relations (cfr.org).

[54] Lee Jones and Shahar Hameiri, "Debunking the Myth of 'Debt-trap' Diplomacy," Chatham House, 19 August 2020 https://www.chathamhouse.org/2020/08/debunking-myth-debt-trap-diplomacy.

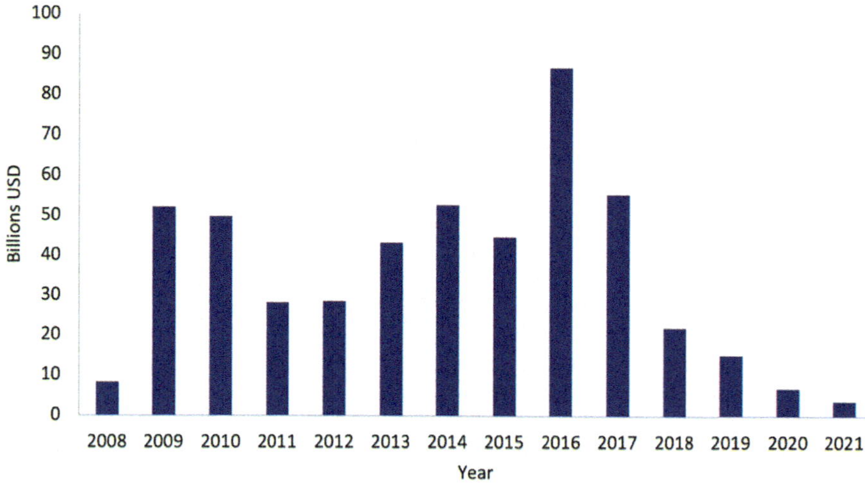

Chart 1 China's overseas development finance, 2008–2021. *Source* 10 Charts for the BRI at 10, Boston University Global Development Policy Center, 2023. 10 Charts for the BRI at 10 | Global Development Policy Center (bu.edu)

A second potential debt trap has been BRI loans to Laos. Governing one of the least developed countries in the world, the Laotian government warmly received Chinese BRI delegations promoting multibillion dollar projects to modernize this poor country of 7.5 million people. In 2022, Laos owed approximately $12.2 billion for Chinese BRI projects, which makes Laos as the country with the greatest exposure of public debt to China at 35% of its GDP.[55] Laos now must pay $1.2 billion per year for the next five years, which is 1/14th of its projected GDP. To avoid a default like Sri Lanka, China has approved a series of payment deferments, and $800 million in debt relief, as well as agreed to equity sharing arrangements. Laos thus borrowed $1.9 billion from China to build the $5.9 billion railway linking Kunming, China to the Laotian capital Vientiane, of which China owns 70%. China has established a joint venture with Électricité du Laos for 25 years to control the country's power grid, including hydropower.[56] In addition, the Chinese were granted land to build an export production zone in exchange for building a Laotian $100 million sports stadium.[57] While these two cases are extreme, BRI projects of other developing economies were facing various defaults and bad publicity, which forced China to restructure $78 billion of its external debt between 2020 and 2023. However, China

[55] Marwaan Macan-Markar, "Laos' debt pressure raises specter of a China vassal state,"*NikkeiAsia,* 6 September 2022, Laos' debt pressure raises specter of a China vassal state - Nikkei Asia.

[56] Kenya Akama, "China goes big on Laos power projects, boosting southeast Asian sway," *NikkeiAsia,* 12 March 2024, China goes big on Laos power projects, boosting Southeast Asian sway - Nikkei Asia.

[57] Alastair McCready," Laos debt at'critical level' with China payment still opaque," *NikkeiAsia,* 22 September 2023; Macan-Markar, "Laos".

was unwilling to "take a haircut" on its BRI loans, and has insisted on full debt repayment, especially among the less indebted borrowers.[58]

Internally, China lost nearly one fourth of its US dollar reserves following the 2015 Chinese stock market crash, which was initiated in the Shanghai stock exchange following a runup of stock values to over 150% between June 2014 and June 2015. When the stock market bubble burst in July 2015, the Shanghai bourse lost over 1/3rd of its value. A second round of stock selloffs occurred in early 2016 which was exacerbated by investor concerns over China's growing debt and weakened RMB. While Chinese growth rates were not adversely affected, there was an overall reassessment of domestic and international investments, resulting in policy banks reducing their overseas lending.[59]

The 2015 Shanghai stock market crash also brought renewed scrutiny on the BRI's financing structure. While having accumulated decades of experience dealing with foreign aid loans of millions of dollars, the BRI loans often reached billions of dollars. The state-owned CDB and Exim Bank were not answerable to shareholders, but to the Chinese party state. While the party state has emphasized the importance of accountability abroad with its Corporate Social Responsibility directives, its implementation has been influenced by CCP political objectives.[60] The banks thus were conflicted when lending projects involved Chinese state-owned enterprises. They have been hesitant to deny SOE loan applications, or write down concessional debt that would have a negative impact on the bank's balance sheet.

The eight big state-owned Chinese banks and SOEs had spent over ten years travelling the world investing in large infrastructure projects, resulting in the phenomenal growth of Chinese loans that totaled $390 billion in 2010 and escalated to $1.5 trillion in 2017. By 2023, many BRI loans' five-year grace periods had ended, with 55% of China's loans to developing economies in a repayment phase; at least 57 countries have encountered repayment difficulties.[61] The impact of COVID on global growth exacerbated a downturn in the global economy, forcing many leaders to ask for debt forgiveness to avoid defaulting on BRI loans. The deteriorating financial conditions thus forced Chinese banks to renegotiate BRI loans, including deferred payment schedules and reductions in interest rates.

---

[58] Matthew Mingey and Logan Wright, "China's External Debt Renegotiations after Zambia," 19 June 2023, Rhodium Group Research, https://rhg.com/research/chinas-external-debt-renegotiations-after-zambia/Also see Sebastian Horn, Bradley C. Parks, Carmen M. Reinhart, Christoph Trebesch, "China as an International Lender of Last Resort," *AIDDATA Working Paper* 124, Maryc 2023, China as an International Lender of Last Resort (aiddata.org).

[59] Nicolas Lardy, "False Alarm on a Crisis in China," *New York Times*, 26 August 2015 Opinion | False Alarm on a Crisis in China—The New York Times (nytimes.com); The big events that shook financial markets in 2016 (ft.com).

[60] Jie Zeng, "Corporate Social Responsibility in China" A tool of policy implementation," *Accounting, Economics, and Law: A Convivium*, De Gruyter (14 February 2024) Corporate Social Responsibility in China: A Tool of Policy Implementation (degruyter.com).

[61] Simone McCarthy, "Developing Countries Owe China at least $1.1 trillion-and the debts are due," *CNN*, 13 November 2023 Developing countries owe China at least $1.1 trillion—and the debts are due | CNN Business.

Chinese banks thus readjusted their loan criteria to allow for two types of borrowers: those who could pay, and those who were strategically important to China. BRI loans were given to countries who easily could repay their loans such as oil wealthy Angola and Venezuela, while loans to Africa were reduced. Pakistan is China's largest BRI recipient and perhaps the strategically most important recipient, with $62 billion in loans that finance 71 projects including the Gwadar Port, hydroelectric plants, highways, subways, fiber optic linkages, and rail links located in the China-Pakistan Economic Corridor (CPEC). Projects remain on hold, while others have run into serious construction problems, terrorist attacks, and corruption. Having readjusted the current CPEC plan to focus more on "small yet smart" projects and considering delaying Pakistan's debt repayment, Xi Jinping still hopes to develop a direct route from northwest Xinjiang Province to Gwadar's Indian Ocean port, which is to be transformed into a modern port including a free trade zone.[62] In a more direct policy readjustment, China's state-operated insurer, China Export and Credit Insurance Corporation, is refusing to approve any loan to a country that is seriously in debt.[63]

While state-owned construction companies continued to pursue contracts, poorer countries were struggling with repayments and Chinese banks were growing concerned about the runaway nature of the BRI approval process. Overall, there was a general lack of transparency, with too many actors involved in the BRI selection process, including the Chinese state-owned construction companies, state-owned banks, as well as politicians and businesspeople from the recipient country. The Chinese banks did an insufficient job evaluating the potential risk of investment, especially as the BRI was offering to build large scale infrastructure in worn-torn areas or countries rife with corruption. Furthermore, the banks failed to investigate why the World Bank and other development banks had already rejected many of these projects.

Xi Jinping formally announced the readjusted BRI commitment at the 10 year Anniversary forum in October 2023 where he presented an eight-point program. Xi's first commitment was to establish a network connecting Europe and Asia through improved transportation, communications, and digital infrastructure, including accelerating the completion of the China–Europe rail link. He pledged to establish a "silk road e-commerce zone," sign new free trade and investment treaties that would cancel restrictions to investments in China's manufacturing sectors and promote a global open economy. Xi's third point focuses on allotting $47.8 billion to the CDB and the Exim Bank to finance "small yet smart" projects. Xi's fourth point was to invest in green infrastructure, renewable energy projects and eco-conscious investments. Xi also required the BRI to emphasize investment in advanced research projects that can

---

[62] David Sacks, "The China-Pakistan Economic Corridor-Hard Reality Greets BRI's Signature Initiative," 30 March 2021, The China-Pakistan Economic Corridor—Hard Reality Greets BRI's Signature Initiative | Council on Foreign Relations (cfr.org).

[63] Chen Yunnan, Liu Zongyuan, "Hedging Belts, De-risking roads: Sinosure in China's overseas finance and the evolving international response," ODI, 14 December 2023 Hedging belts, de-risking roads: Sinosure in China's overseas finance and the evolving international response | ODI: Think change.

transfer technical knowledge and promote economic development. Xi's point six is to expand cultural and educational exchanges, while point seven is to promote accountability, transparency, and ethical conduct "to ensure that all BRI projects meet high standards." Xi's final eighth point is to establish BRI institutions and a BRI secretariat that would promulgate regulations and procedures for BRI countries.[64]

4 Conclusion

The Belt and Road Initiative is the latest iteration of the Chinese Communist Party's desire to win friends and influence people by building key infrastructure and raising the living standards of people living in the developing world. Older slogans encouraging the "proletariat of the world unite" has been replaced by Hu Jintao's "Community of Common Destiny" and the "China model of development." Interest-free loans of millions of dollars have been replaced by concessional loans worth billions of dollars. The inability of BRI recipients to repay their Chinese loans have not left their country in default, as this would have a detrimental impact on state-owned banks and enterprises. Instead, their repayment schedule is lengthened, or they are persuaded to sign over operation rights to run their facilities, as was the case of Hambantota, which has become infamous as a Chinese debt trap.

But beyond the sweet-sounding slogans and the ugliness of debt traps lies a fundamental Chinese policy process of crisis, readjustment, innovation, and implementation. The process reveals the autonomy of the Leninist system, in which the party state mobilizes its financial and political power to realize a political goal, as well as readjusts and augments policy without undue interference from internal actors. But it also shows the frailties of the system. During the policy implementation process, the party state lacks the capacity to control actors pursuing their self-interest or the interests of their bureaucracy. Beijing also has relied on an export development strategy that consistently results in overproduction and overcapacity, which is subsequently resolved by flooding international markets with cheaper goods.

In the end, analysts will continue to focus on China's desire to win friends in strategic places, and to establish debt traps to secure strategic assets. Instead, this chapter argues that the BRI is the product of an authoritarian regime that can learn from its mistakes as well as to help improve the living standards of people from the developing world.

The question remains how the world's second largest economy will resolve the high level of BRI debt owed to its policy banks and whether foreign economies will pressure China to alter its current export development policy based on overcapacity. Simultaneously, the party state is dealing with various domestic crisis ranging from the implosion of Chinese real estate to the high urban unemployment rate of Chinese

[64] Josephine Ma and Kawala Xie, "As it happened: Xi Jinping unveiled new action plan for Belt and Road Initiative, Vladimir Putin gave first international address since Ukraine war," *South China Morning Post,* 18 October 2023.

youth. While Xi Jinping had promoted the slogan that "the East is rising, and the West is declining" [东升西降],[65] the current global economic situation infers the inverse.

References

Altamira M, Fornes G, Mendez A (2024) Chinese institutions and international expansion within the Belt and road initiative: firm capabilities of Chinese companies in the European Union. Asia Pac Bus Rev 30(2):251–273

Cabestan J-P (2019) Beijing's 'Going Out' strategy and belt and road initiative in the Sahel: the case of China's growing presence in Niger. J Contemp China 28(118):592–661

Dorothy J (1999) Solinger, contesting Citizenship in Urban China: peasant migrants, the state, and the logic of the market. University of California Press, Berkeley, CA

Joseph WA (1984) The critique of ultra-leftism in China, 1958–1981. Stanford University Press, Stanford, pp 140–150; 179–182

Lawrence M, Bullock R, Liu Z (2019) China's high-speed rail development. World Bank Publications, Washington, DC

Lei W (2022) Global development initiative: an international public goods to promote common development. China Econ 17(4):2–25. 1700535743312.pdf (chinadaily.com.cn)

Naughton B (2009) Understanding the Chinese stimulus package. China Leadership Monitor 28(2):1–12

Ping L, Feitao J, Jianhai C (2019) Industrial overcapacity, and duplicate construction in China: reasons and solutions, vol 15. World Scientific, Singapore

Solinger D (1991) From lathes to looms: China's industrial policy in comparative perspective, 1979–1982. Stanford University Press, Stanford, pp 27; 30

Tambunan T, Hamonangan T (20211) The Impact of the 2008–2009 global economic crisis on a developing country's economy: studies from Indonesia. J Bus Econ 2(3):175 (175–197)

Xiaohuan L (2024) How China works: an introduction to China's State-led economic development. Springer Nature, NYC

Zhang D (2018) The concept of 'Community of Common Destiny' in China's diplomacy: meaning, motives and implications. Asia Pac Policy Stud 5(2):196–207

[65] Chris Buckley, "'The East is Rising': Xi maps out China's post-Covid Ascent," *New York Times*, 4 March 2021. "东升西降":习近平的后疫情时代中国崛起蓝图 - 纽约时报中文网(nytimes.com).

Identifying Risks from US China Economic Decoupling if Trump Wins

Leif Rosenberger

Abstract A recent New York Times poll has Donald Trump leading Joe Biden 48–43%. Trump recently said he would ratchet up the trade war even more against China if elected. Logistical plans and strategies have already begun for the US to economically decouple from China. Trump blamed China's exports for US unemployment. He promises to bring back US jobs. But there is nowhere to bring US manufacturing jobs back from. US manufacturing jobs were lost to technology, not trade. In the old days, manufacturing was done in-house, in one factory, under one roof, and in one country before being exported. Not anymore. Manufactured goods are now dispersed across different factories in different countries. Countries previously omitted from manufacturing can now manufacture intermediate components for the supply chain. This democratizes global production and creates Asian jobs. Trade wars destroy this economic democracy. The US trade war against China is a trade war against all Asian economies in this supply chain. By antagonizing them the US pushes them closer to China. When goods can't cross borders, armies will. Thus, economic interdependence and shared prosperity are strategically better ideas than economic coercion.

Keywords China · Trump · Economic interdependence · Decoupling · Trade war · Protectionism · Manufacturing · Supply chain · Technology · Exports · Democratization of global production

1 Introduction

As the 5 November 2024 US presidential election approaches, former US President Donald Trump prospects are on the rise. A 2 March 2024 New York Times (NYT)/ Siena College poll has Trump leading President Joe Biden 48–43%. That's the biggest lead that Trump has enjoyed in either a NYT/Siena or New York Times/CBS poll since Trump launched his first presidential campaign in 2015. If these trends continue,

L. Rosenberger (✉)
PKSOI, US Army War College, Carlisle, PA, USA
e-mail: leifrosy@yahoo.com

A. Farhadi et al. (eds.), *The Great Power Competition Volume 6*,
https://doi.org/10.1007/978-3-031-70767-4_13

Trump could actually get elected once again, notwithstanding his daunting legal problems.[1]

If Trump wins, how would that affect US China trade? Would Trump opt for US economic interdependence and shared prosperity with China? Or would Trump opt for economic decoupling from China and another trade war? As fate would have it, Trump has just responded to these questions for us.

On 11 March 2024 Trump told CNBC that he is ready to ratchet up the trade war even more against China. In a CNBC "Squawk Box" appearance on 11 March 2024 Trump escalated his trade war rhetoric, saying "I'm a big believer in tariffs," and indicated that he's likely to implement more duties on foreign goods should he win the election to a second term.

When Trump was president, his administration used delegated authorities under three trade laws to unilaterally levy tariffs without Congressional approval. President Biden continued Trump's trade war with China, with the current range of the tariffs on a wide variety of U.S. imports today between 10 and 25%.

Trump's tariffs would go a lot higher than Biden's current tariffs against China. In February 2024 the former president said he was considering a plan to impose tariffs of 60% or higher on Chinese goods as well as a blanket 10% tariff on all U.S. imports in his potential second term.

Trump's provocative statements are a green light to global logistics companies. In fact, these companies told CNBC that they have already started the planning for a potential Trump win in November and the strategies that would be needed to mitigate any additional tariffs.

2 Wartime Lessons Learned

The Treaty of Versailles after World War I provides a useful historical context. As cited earlier in the abstract, the allies assumed that an economically weak Germany could never again be a military threat. So, they opted for economic coercion against Germany. How did that work out? Don't worry, be happy? Guess again. Instead, this economic coercion led to German resentment, the rise of Adolf Hitler, territorial expansionism, and World War II. The 19th century French economist Frederick Bastiat puts it best, "When goods don't cross borders, armies will."

3 Economic Coercion

During his campaign for the US presidency, Donald Trump liked to wear a baseball cap that read, "Make America Great Again." He would tour the rust belt and promise the "forgotten" blue-collar workers that he would make their lives better.[2] These workers not only believed him—they helped to vote Donald Trump into office.

President Trump blamed China for "stealing" American jobs. In an ill-advised effort to create more US jobs, in July 2018 Trump announced $50 billion of tariffs on products that America imports from China. His 25% tax on Chinese imports—including steel, aluminum, washing machines and solar panels—was the most extensive trade protectionism in nearly a century.

China predictably retaliated in force, with a 25% tax on imports of US automobiles, Boeing aircraft and soybeans, and many other agricultural products. This didn't faze Trump; he just raised the stakes and began a cycle of mutual retaliations. For instance, on 7 September 2018, Trump threatened new tariffs of an additional $267 billion on US imports from China.[3]

4 Negative Impact on Americans

As expected, President Trump's high tariffs and China's retaliation hurt US workers across America. For instance, US car exporters in Southern states like Tennessee were hit hard, as were Boeing workers in the Northwest.[4]

In addition, Trump's trade war with China hurt American farmers in the Midwest who produced soybeans. They lost their export markets in China. China, the world's largest market for soybeans, then looked to Brazil to replace America as a supplier.[5]

To be fair, President Trump offered to compensate these Midwest soybean farmers. In fact, he proposed a $15 billion farm bailout. But this subsidy was a misguided attempt to cushion the damage and failed to address long-term risks. US soybean farmers feared that they could permanently lose their export market to China. Trump's subsidy could also trigger a challenge from the World Trade Organization (WTO).

US consumers were also hurt. The high tariffs were a tax hike on embattled American consumers. Previously, American consumers were free to buy the lowest-priced, best-quality products, which often meant buying products with final assembly in China. The trade war meant that Americans would be forced to buy inferior-quality, more expensive products from other countries. That unnecessary inflation was tantamount to a large tax increase and a lower standard of living for struggling US consumers.

5 Make America 1953 Again?

Trump's trade war with China played well with his political base, which longed for the days when those in the rust belt were doing better. To provide perspective, let's turn the clock back to the so-called "glory days" of US manufacturing. For instance, George Will notes that President Trump wanted to "make America 1953 again."

Trump said that America was "great" back then because it was "winning." America had a 30% share of global manufacturing. If this 30% market share of global manufacturing was all important to post-World War II US presidents, they could have tried to keep this manufacturing "success story" going into the future.[6]

But thankfully, back then, America had statesmen who had bigger things in mind—like internalizing lessons learned from the punitive Treaty of Versailles, which ended World War I. As cited in the introduction, keeping the German economy weak in hopes of avoiding a strong Germn military produced the opposite result. Economic coercion led to German resentment, the rise of a vindictive Adolf Hitler, German territorial expansionism and World War II. What is the take-away? The 19th century French economist Frederick Bastiat put it best: "When goods don't cross borders, armies will."[7]

6 Shared Prosperity

After World War II, the great French statesman Jean Monnet had a better idea than economic nationalism. Instead of economic coercion and endless war, Monnet argued for a European Coal and Steel Community (ECSC). These longstanding French and German enemies would bond and reduce the demand for violence. This web of economic interdependence fostered shared prosperity between France and Germany—not a bad way to orchestrate peace in Europe for 70 years. Economic interdependence and shared prosperity would spread and foster globalization in the post-Cold War era.

Toward this end, great US and European minds thought alike. One of the American heroes at this time was US Secretary of State George Marshall, who wisely created the Marshall Plan that helped Germany rebuild its manufacturing base. He had an implicit vision of "no (vindictive) victor, and no vanquished loser."

Instead of pursing a fool's errand of trying to keep Germany forever defeated and in ashes, Marshall (like Jean Monnet) fostered a web of economic interdependence and shared prosperity. Marshall also helped to turn a bitter enemy in World War II into a close friend and ally. The Marshall Plan enabled Germany to use its exports to earn hard currency to import American goods and services.

Just as George Marshall was an international hero, the scholars at the World Bank were equally impressive, connecting economics to global security. In their definitive 2003 study "Breaking the Conflict Trap: Civil War and Development Policy," the World Bank used statistical evidence to reveal a striking pattern: civil war did not occur randomly around the world.

Instead, civil war was heavily concentrated in the poorest countries. The important reason for this concentration was that poverty increases the likelihood of civil war. Thus, the World Bank study concluded that the root cause of conflict is the failure of economic development.[8]

More recently, scholars in the United Nations (UN) highlighted the importance of free trade to counter this rising demand for violence. The UN found that free trade across the world cut the extreme global poverty rate in half since 1990.

In contrast, Trump's protectionism and economic nationalism incited an unnecessary trade war that rolled back this progress in free trade. As a result, Trump needlessly reversed this decline in global poverty. This, in turn, provided fertile ground for an even greater rise in demand for violent extremism, which would put the brave men and women of the US military needlessly and repeatedly in harm's way.

7 US Protectionism is a Bad Idea

President Harry Truman once said, "the only thing new in the world is the history you don't know."[9] If we look at the past, we learn that protectionism was a bad idea then, and it was still a bad idea when President Trump opted for this kind of economic coercion against China in mid-2018.

Back in 1930, President Herbert Hoover's Smoot-Hawley Act's high tariffs achieved the glorious trade surplus that Donald Trump's economic nationalism calls "winning." But in economics, appearances can be deceiving, especially if you are obsessed with the wrong economic indicator. In 1930, shrinking the gains from trade by making the global economic pie smaller also produced a US unemployment rate of 25%—and the Great Depression.[10]

To be fair, Donald Trump and Herbert Hoover were not the only US presidents who tried to "bring back American jobs." But President Trump had no problem listening to ill-advised economists who dismissed lessons learned about protectionism from history and instead went with uninformed "gut instincts."

Unfortunately, each time an American president makes this painful mistake, protectionism backfires and worsens the overall US economy. For instance, when President Carter tried to protect US manufacturers by restricting imports of Japanese televisions, imports from South Korea and Taiwan increased. When imports from South Korea and Taiwan were restricted, manufacturers in Mexico and Singapore benefited.[11] When this replacement happened, American consumers lost because they had to settle for higher prices and worse quality. This unnecessary protectionism was not exactly life, liberty, and the pursuit of happiness!

The US middle class and other US businesses were also financially hurt by unhappy second- and third-order effects when the US government singled out one industry for special protection. In 2012, President Obama took credit for protecting over a thousand US jobs after there was a surge in Chinese tires. But this government interventionism cost $900,000 per job, paid for by American purchasers of then overpriced vehicles and tires. Meanwhile, imports of low-end tires from Thailand, Indonesia, Mexico and elsewhere largely replaced Chinese imports.[12]

To make matters worse, the Peterson Institute notes that this money needlessly taken out of the wallets of struggling US consumers also had other negative consequences. It reduced their spending on other retail goods, which, in turn, brought the net job loss from the "job-saving" tire tariffs to about 2,500. To add insult to injury, China retaliated with duties on US chicken parts, costing this US industry (which had done nothing wrong) $1 billion in lost sales.[13]

Moreover, the number of Americans hurt far outnumbered the number of people who benefitted from protectionism. For instance, Douglas Irwin from Dartmouth College asks: Is it fair to sharply increase the price of clothing for 45 million poor Americans just to save (a few) jobs in the US textile industry?[14]

It also made no sense for Trump to raise tariffs on imports and force American businesses and consumers to pay more and have less freedom to choose. Foreign competition keeps prices affordable for consumers and increases the quality of US-made goods.

8 Nowhere to Bring US Jobs Back From

Instead of learning about the negative aspects of protectionism and the upside of free trade, President Trump was stuck on blaming China for US unemployment in the rust belt and said that his trade war with China was designed to "bring back American jobs." The problem was that there was nowhere to bring the manufacturing jobs back from because the lion's share of US manufacturing jobs was lost to technology, not trade.[15] A Ball State University study notes that out of the 5.6 million US manufacturing jobs lost between 2000 and 2010, productivity improvements accounted for over 85% of the jobs that were lost.[16]

Take farming. Patricia Daly notes that half of all American jobs were in the agricultural sector in 1870. By 1980, only 4% of American workers were in agriculture. Thanks to stunning advances in agricultural technology, US agriculture is now a success story and certainly not what Trump would call a disaster. In many ways, America is now the breadbasket of the world.[17]

A similar pattern of rising productivity happened in US manufacturing. Thanks to technological advances, US manufacturing output was at an all-time high in 2015. Over the past three and a half decades, US manufacturers have been able to produce more goods than ever with seven million less people. American factories are now producing twice as much as they were in 1984 with one-third fewer workers.[18]

Thanks to equally amazing advances in manufacturing technology, US manufacturing became just as much a success story as US agriculture. And for the past 40 years, the US market share of global manufacturing has held at around 20%. All the while, American manufacturing has moved up the value-added ladder, thus fostering strategic stability and peace in the world by allowing developing countries to move into lower-end global manufacturing.

9 Dispersion of Global Production

Somewhere along the way, President Trump failed to understand the strategic importance of the transformation in global manufacturing. Back in the old days that President Trump says were "great," manufacturing was one-dimensional. Victor Fung notes that it was done "in-house" ... in one factory, under one roof, and in one country ... before that product was exported and sold in another country."[19]

But things changed, and products have become increasingly dispersed across different factories in different countries and thus globalized. Because of this dispersion of global production, countries previously omitted from manufacturing can now participate by performing just one or two pieces of the supply chain. Fung calls this the "democratization of the global production system."[20] Dispersion of global production also benefits American consumers because it reduces costs and makes products more affordable.

Notwithstanding covid challenges, we now live in a world where global supply chains involve goods repeatedly crossing open borders, thus making traditional trade metrics obsolete. In fact, two-thirds of international trade now takes place through these global value chains.[21] Therefore, when Trump criticizes China for its large trade surplus with the United States, there are simple reasons for this development. As it turns out, many Chinese exports are put together from parts produced elsewhere.[22]

With the dispersion of global production, China became a final assembly point for goods that the US previously imported from companies based in Japan, South Korea, and Taiwan. As the final leg of the supply chain shifted to China, US imports from China rose as the United States imported less from the rest of Asia. The US trade deficit with Asia did not significantly change until Trump's trade war. In other words, there is nothing evil about China's last leg in this dispersion of global production.

10 Trump's Trade War Against Asia

Not surprisingly, countries in ASEAN and elsewhere in Asia that were part of this supply chain with China were angry with Trump when he started his ill-advised trade war against China. Trump's trade war against China was also a trade war against other Asian countries which produce thousands of intermediate components for this supply chain they share with China. If we fast forward to 2024 and a possible Trump election, a deeper US economic decoupling with China would only make things worse. The US economic coercion and economic nationalism antagonizes former US partners and friends in ASEAN and the rest of Asia and pushes even more Asian countries into China's sphere of influence.

It's also important to understand that US and Chinese goods are often in different market segments—they therefore do not always compete head-to-head. So, when Trump's trade war increases the cost of US imports from China, the United States imports less from China and more from other, higher-priced Asian producers—not from producers in the United States. Not only is the result no net increase in US jobs, but it also means higher prices at Walmart for struggling US consumers. Thus, besides the imported inflation from the war in Ukraine, inflationary pressure that US consumers complain about today can also be traced back to the US-China trade war.

11 Manipulation of Exchange Rates?

Unfortunately, this inconvenient truth was not what President Trump told his base in the rust belt. Instead, he also accused China of stealing American jobs by manipulating its foreign exchange rate so that its export prices would be lower than US export prices. Is this argument persuasive? Let's check China's foreign exchange rate against the US dollar back then.

If China wanted such a low foreign exchange rate to underprice US products, why did its foreign exchange rate rise so much against the US dollar? This rate rose from 8 renminbi (RMB) to $1 in the mid-2000s to just 6.3 RMB to $1 in 2015.[23]

China's central bank depleted its foreign reserves numerous times to strengthen its foreign exchange rate. In fact, at the end of October 2016, its central bank's foreign reserves hit their lowest level since March 2011.

After President Trump came into office, the RMB did weaken. But that's because the Federal Reserve (or the Fed) had to raise interest rates to offset a super loose fiscal policy resulting from a huge tax cut. In other words, investors are chasing rising US interest rates, which strengthens the US dollar and weakens the RMB against the dollar.

12 Tax Cut, Fed Tightens, Strong Dollar, US Exports Too Costly

Instead of blaming China, President Trump should have blamed himself for policies which ended up strengthening the US dollar and therefore overpricing US exports. His huge $1.5 trillion tax cut increased public spending, a large percentage of which went into imports. As imports grew faster than exports, the trade deficit rose.

The US had to offset this deficit with a surplus in the capital account (financial flows) in the balance of payments. To attract capital inflow, the Fed had to tighten monetary policy by increasing interest rates. Rising interest rates strengthened the demand for a stronger US dollar, overpriced US exports, worsened US trade and current account deficits, and put Americans out of work in export industries.[24]

In addition, President Trump liked to justify his trade war against China by pointing to what he calls "America's $500 billion trade deficit with China." For starters, the US trade deficit with China in 2017 was only $375 billion.[25]

13 National Security Derisking

Second, the high US trade deficit with China also reflected national security realities regarding Taiwan and North Korea. US strategists had to forego trade advantages with China when they had bigger fish to fry. To help protect Taiwan's security, America placed an embargo on items which would give China's military advantages, even though this decision increased the US trade deficit with China. In addition, North Korean long-range missiles carrying nuclear weapons could blow up Seattle and other US cities unless something was done to curb North Korea's military buildup.

To avoid this existential threat to America, US strategists needed China to use its economic relationship with North Korea to dissuade North Korea from this unacceptable course of action. Finally, when the US launches a trade war against China with the hope of "bringing back" some US manufacturing jobs, China no longer has much incentive to put pressure on North Korea to curb its nuclear threat to Seattle.[26]

Third, President Trump fails to understand that this large trade deficit is what Nobel Laureate Paul Krugman calls a "statistical illusion." That's because (as stated earlier) over 50% of the value of these so-called "Chinese exports" to America are parts and components produced in places like South Korea, Taiwan, and Japan, even though China is the final assembler. As a result, over half the pain from Trump's tariffs has been felt by a wider group of countries, many of whom are close US allies and friends. And because of this dispersion of production, Trump actually had much less trade leverage over China than he imagined.[27]

In addition, over 20% of the goods produced in China and then exported from China to the US were created by foreign companies working in China.[28] President Trump says that these American companies doing foreign investment in China should return to the United States. But US businesses have significantly benefitted from large US investments in China. Over the years, American and other foreign companies have dominated China's high-tech and industrial exports.

Profits for these products are thus not China's alone. In the past, China only retained $4 out of the $300 final price for the iPods that Apple produced in China. US corporations benefit by recycling their profits back into the United States and other foreign companies. US consumers benefit by getting lower prices, higher quality, and greater variety.[29]

Instead of blaming China for America's high trade deficit with China, President Trump should blame himself. Trump's huge $1.5 trillion tax cut caused the budget deficit and national debt to soar. Trump's federal budget deficit headed back towards 7% of GDP over the next ten years.[30] That's over twice the 3% of GDP level which an EU Maastricht criterion says is financially unstable.

Similarly, the non-partisan US Congressional Budget Office (CBO) estimated back then that President Trump's public debt would hit 78% of GDP by the end of 2018, the highest level since 1950. The long-term projections were even worse. CBO said that the debt to GDP ratio was on track to hit 100% of GDP by 2030 and 152% by 2048.[31] That's a far cry from the 60% of GDP level which another EU Maastricht criterion says is financially unstable.

When President Trump says he wants to make America great again, he is half-right about one thing. In the 1950, 1960 and 1970s, America's national savings rate was about 8% of GDP. By international standards, that was still low, but it's a lot higher than it was when Trump left office. Unfortunately, the national savings rate fell sharply to just 2.4% of GDP in mid-2018. The problem was that President Trump was pursuing policies which were counterproductive. His 1.5% tax cut sent the already dangerously low national savings rate into negative territory.

The trade deficit that President Trump complains about exists because America is running a large deficit in national savings. The United States must import its savings from countries like China, Germany, and Japan, which have large surpluses in savings, because America lives beyond its means by consuming way too much, thus leaving it with a large deficit in savings. As long as the United States maintains its large savings deficit, eliminating a trade deficit with China simply turns into a larger trade deficit with other US trading partners.[32]

In this regard, President Trump's own domestic economic policies widened the US trade imbalance. He ran a loose fiscal policy that promised to generate rising budget deficits. The Fed said it would have to offset this loose fiscal policy with higher interest rates, which, in turn, strengthened the US dollar and increased the cost of US exports. More expensive exports worsened the overall US trade deficit and led to rising unemployment in US export industries.[33]

Trump officials were critical of Chinese economic initiatives in developing countries. They accused China of "predatory" lending policies. They said its five-year-old "Belt and Road Initiative" (BRI) was dangerous because it got developing countries hooked on excessive debt. That criticism is ill-advised. That US criticism of China is hypocritical.

The US should practice what it preaches. For instance, no country in the world is as highly indebted to China as the United States. The American government owes China over a trillion dollars in debt. And America's debt is becoming far worse because the Trump administration's tax cut for millionaires and billionaires increased the national debt by over a trillion dollars.[34] Perhaps they should have hired former Admiral Michael Mullen, Chairman of the US Joint Chiefs of Staff, who said, "the most significant threat to our (America's) national security is our debt."[35]

14 Conclusion

Earlier in this chapter we mentioned that President Trump could get elected once again. If so, Trump's economic strategy towards China would logically be an important issue. In February and March of 2024 Trump made it clear that he will ratchet up the trade war even more against China. Trump's legacy of economic coercion, economic nationalism, protectionism, and his trade war strongly suggest that these policies would continue.

Earlier in this chapter we learned that thousands of intermediate components for manufactured goods are produced in countries in ASEAN and the rest of Asia for

China's supply chain. Therefore, Trump's plan to ratchet up the trade war against China's exports is also a trade war against these countries. The trade war hurts exports from ASEAN and other countries in Asia. It kills their jobs. This approach antagonized these Asian countries when Trump was president and would antagonize these countries even more in second Trump presidency. This ill-advised US economic coercion is a bad idea. This strategy would push these Asian counties toward greater Chinese regional economic influence, isolate the US in Asia and undermine US national interests.

Notes

1. James Farrel, Trump Opens Largest-Ever Lead In New York Times Polling, Forbes, 2 March 2024.
2. See Ben Bradlee, Jr. The Forgotten: How the People of One Pennsylvania County Elected Donald Trump and Changed America, Little, Brown and Co., New York, NY, 2018.
3. David Dollar, "Unpacked: The US-China Trade War", Brookings, 12 July 2018.
4. Ibid.
5. Ibid.
6. George F. Will, "A Plan to Make America 1953 Again", The Washington Post, 28 December 2016.
7. David Henderson, Econ Lib, 29Aaugust 2013.
8. World Bank, Breaking the Conflict Trap: Civil War and Development Policy, Oxford University Press and the World Bank, Washington, DC, 2003.
9. Michael Beschloss, Presidential Courage, Simon and Schuster, New York, NY, p. 211.
10. See Robert Zoellick, "The Currency of Power: Want to Understand America's Place in the World? Write Economics Back into the Plan", Foreign Policy, 8 October 2012.
11. See Greg Ip, "When Presidents Defy Economic Gravity, Gravity Usually Wins", The Wall Street Journal, 7 December 2016.
12. Will, "A Plan to Make America 1953 Again".
13. Ibid.
14. Ibid.
15. See Knowledge at Wharton, "Can Trump – or Anyone Else – Bring Back American Manufacturing?", University of Pennsylvania, 30 November 2016; Claire Cain Miller, "The Long-term Jobs Killer Is Not China: It's Automation", The New York Times, 21 December 2016; Wolfgang Lehmacher, "Don't Blame China for Taking US Jobs", Fortune, 8 February 2017.
16. Will, "A Plan to Make America 1953 Again".
17. See Patricia A. Daly, "Agricultural Employment: Has the Decline Ended?" Monthly Labor Review, November 1981.
18. Drew Desilva, "Most Americans Unaware that as U.S. Manufacturing Jobs Have Disappeared, Output Has Grown", Pew Research Center, 25 July 2017.
19. Victor Fung, William Fung, and Yoram Wind, Compering in Flat World, Prentice Hall, New York, NY, 1979.
20. Ibid.
21. David Dollar and Zhi Wang, "Why a Trade War with China would Hurt the U.S. and Its Allies, Too, Order from Chaos", Brookings, 4 April 2018.
22. Paul Krugman, "Bumbling into a Trade War", The New York Times, 22 March 2018.
23. Kenneth Rapoza, "Is the Made in China Clothing Label a Thing of the Past?" Forbes, 11 October 2015.
24. See Barry Eichengreen, "Powerful Forces Will Lead to a Strong Dollar under Trump", Financial Times, 25 January 2017.

25. Krugman, op. cit.
26. Leif Rosenberger, APEU 2007, Volume 1, US Pacific Command, Camp Smith, Hawaii, 2007.
27. Krugman, op. cit.
28. Ibid.
29. Rosenberger, APEU 2007.
30. Stephen Roach is an economics professor at Yale University.
31. Economist Intelligence Unit, "Ballooning Deficits and Debt Cast a Long Shadow, United States", 5 July 2018.
32. Stephen Roach, "US-China Trade Deficit Is Set to Keep Rising", CNBC.com, 26 March 2018; and Stephen Roach, Project Syndicate, "The Current Account Counts", 27 August 2018.
33. Barry, Eichengreen, op. cit.
34. William Baldwin, "Tax Breaks for Musk, Bezos and Various Other Billionaires", Forbes, 20 December 2018.
35. CNN, Mullen: Debt is top national security threat, 27 August 2010.

China's Dominant Economic Influence: IMF and Asia Push Back Against Decoupling

Leif Rosenberger

Abstract The Biden administration was concerned about the wide gap between the dominance in China's economic influence in Asia and US marginal economic influence. Its US Indo-Pacific Strategy failed to close the gap. This chapter argues that this gap reflects China's shared prosperity in the region and US economic coercion. China benefits from longstanding co-production and economic interdependence with its Asian partners. China also bonds with them in regional trade diplomacy. In contrast, the US has disinterest in regional trade diplomacy. The US also misperceives "China's economic rise" as a threat. In fact, China's economy is too weak, not too strong. The US is making China's economic weakness even worse by doubling down with economic coercion. After World War I that kind of economic coercion led to German resentment, the rise of Hitler and World War II. Similarly, Harvard Professor Graham Allison says the US is destined for war against China. IMF recently criticized the US trade war against China as a financial risk to the global trading system and international economic stability. Similarly, Asian states are pushing back when the US puts "Cold War" pressure on them to "pick the US side" and decouple from their close economic ties with China. To increase US economic influence the US needs to promote US China shared prosperity and curb its US economic coercion against China and its Asian trading partners.

Keywords IMF · US · China · Greater China · Bangladesh · Garment industry · Asia · Europe · Decoupling · Risk · Instability · Economic coercion · Shared prosperity · Interdependence · Trade war · Regional trade diplomacy · Tariffs · Economic weakness · Push back · Jean Monnet

---

L. Rosenberger (✉)
PKSOI, US Army War College, Carlisle, PA, USA
e-mail: leifrosy@yahoo.com

A. Farhadi et al. (eds.), *The Great Power Competition Volume 6*,
https://doi.org/10.1007/978-3-031-70767-4_12

1 Introduction

The rise of China's economic influence in Southeast Asia and beyond has been a concern of the Biden administration ever since it came into office. But until recently, it was not clear that Team Biden fully appreciated the nature and extent of China's economic influence. Then a few things helped US situational awareness. For starters, the Economist Intelligence Unit (EIU) ranked Singapore as the country in the world with the best business climate for the next five years.[1] And when Singapore talks, Team Biden listens.

2 Wake Up Call

Second, the ISEAS-Yusof Ishak Institute, a prestigious Singapore-based think-tank, published its 2022 State of Southeast Asia report on 16 February 2022. This publication was based on a survey of over 1,000 individuals holding roles in sectors spanning business, government and academe. As the chart below shows:

- 77% of respondents said China was the country with the most economic influence in Southeast Asia.
- Only 10% said the US had the most economic influence in the region.[2] That was a wake-up call for Team Biden.

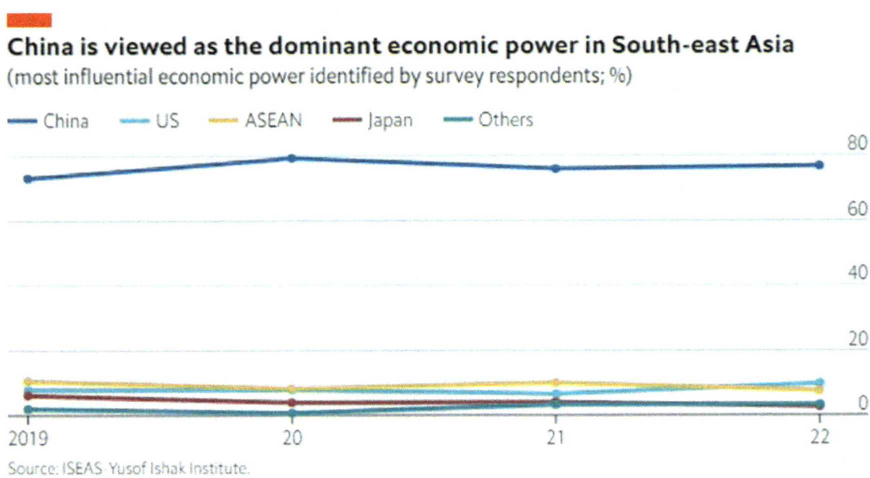

Unfortunately, the Singapore survey failed to address a number of the strategic questions. For instance, why is the gap between Chinese economic influence and US economic influence so large? What is China doing right? What is the US doing wrong? How can the US start to close this gap? This chapter attempts to answer these questions.

The military side of the strategy looks promising. Longstanding US military bases and military alliances throughout Asia arguably help to deter—and if necessary, counter—potential Chinese military aggression in the region.[3] Admittedly military power protects people. But it is not enough to win hearts and minds, nor does it close the huge gap in economic influence between the US and China.

The problem is the economic side of the Indo-Pacific Strategy is counter-productive. Not only does US economic coercion continue with China, the economic side of the strategy puts "Cold War" pressure on third countries in Asia to "pick the US side" between China and the US and then decouple from economic ties with China. In other words, US economic pressure on third countries in Asia antagonizes them and reduces US economic influence in Asia.

The US economic component of the strategy shows no signs of reengaging with Asia via existing trade agreements, such as the Comprehensive and Progressive TransPacific Partnership (CPTPP). Instead of pursuing economic integration and shared prosperity with China and other countries in ASEAN and elsewhere in the region, the US is reluctant to join regional trade frameworks. The US would rather seek to cultivate alternative economic relationships and lean on countries to reduce their supply-chain reliance on China. At the end of this chapter, we will show that economic strategists in each Asian economy would struggle to diversify their economic dependence significantly away from China.

3 Why US Economic Strategy Keeps Failing

To its credit, the US national security community realized that the economic component of the Indo-Pacific Strategy was not good enough. So, a few months later US economic strategists took another swing at an US economic strategy towards China and Asia. On 23 May 2022, the US national security community unveiled the Indo-Pacific Economic Framework for Prosperity (or IPEF). 13 other US friends joined IPEF hoping that it could somehow be a bridge to the more important regional trade organizations.[4]

Unfortunately, the IPEF framework was underwhelming given the economic stakes involved between the US and China (the two biggest economies in the world). IPEF did virtually nothing to start closing the economic gap in economic influence between China's dominance and the marginal US role. Instead, the economic strategy was prioritizing economic coercion and geopolitics in this strategy.

Lurking behind the framework is US domestic politics and US economic nationalism. Both US political parties are competing to see which one can be toughest on China's economy and third countries trading with China. Instead, what third countries wanted was US commitment to market access in the US and free trade. The US absence of free market access in the US curbed third country enthusiasm in Asia for regional trade diplomacy with the US in the future and thus reduced US economic influence.

4 Half the Battle: Show Up and Bond

IPEF especially contrasted with market pledges offered by Asia's existing mega trade agreements with which the US has avoided re-engaging. In this regard, US actor and director Woody Allen once said "showing up is half the battle." When the US fails to participate in regional trade diplomacy in Asia, it's virtually impossible for the US to bond with third countries in the region.

In contrast, China shows up and bonds. One reason China has so much economic influence is China's central role in the regional supply chain and its more proactive engagement in economic co-operation with members of the Association of South-East Asian Nations (ASEAN). A case in point was China's participation in the Regional Comprehensive Economic Partnership (RCEP) that started in January 2022.

This divergence in regional trade diplomacy between the US and China was also evident in the Comprehensive and Progressive Agreement on Trans-Pacific Partnership (CPATPP). At that time, China had a serious application to join CPATPP. Given the size and attractiveness of the Chinese market, China's application is promising. This cemented China's importance as a team player with third countries in regional economic networks.

In contrast, the US kept playing "absentee landlord" and did not even apply to join CPATPP. Thus, US failure to bond with third countries in regional trade diplomacy means it's increasingly difficult to close the large gap in economic influence between the US and China. Looking ahead, IPEF is unlikely to meaningfully restructure current Asian economic linkages. Thus, there is lack of an attractive US alternative to the existing China focused trade reality in Asia.

5 Derailing Trade Pillar of IPEF

To its credit, the US national security community realized by November 2023 that changes were needed in trade policy. If IPEF was to have any legitimacy in the international economic world, the 4 pillars of IPEF desperately needed to include a more credible trade pillar. Unfortunately, its plan to unveil a crucial part of this Indo-Pacific economic strategy was derailed by opposition from Congressional Democrats just as US President Biden prepared to meet Chinese President Xi Jinping in San Francisco on 15 November 2023.

The US intended to conclude critical elements of its Indo-Pacific Economic Framework trade pillar at the Asia–Pacific Economic Cooperation forum. It would have signaled to China that the US and its partners could finally co-operate on a strategy Washington hoped would counter China's dominant economic influence.

But the White House abruptly reversed course following pressure from influential Ohio Senator Sherrod Brown and other Democrats, who worried about the effect of a new trade deal on their election prospects in November 2024. One former US trade official cut to the chase: "We just handed China a double victory ... American politics is dysfunctional, and America has no economic agenda for Asia." In contrast to still more US protectionism from Team Biden, the US corporate leaders gave Xi Jinping a standing ovation when they met with Xi in San Francisco.[5]

Allies and partners were never that enthusiastic about the IPEF, partly because it does not offer any US market access, but they believed it was better than nothing. Some, including Japan, hoped that it would provide a bridge to the US joining the Comprehensive and Progressive Agreement for Trans-Pacific Partnership, a sweeping trade deal that was salvaged from the Trans-Pacific Partnership agreement that then President Donald Trump withdrew from in 2017.

Wendy Cutler, vice-president of the Asia Society Policy Institute, said that dropping the IPEF was a major setback. Cutler said that this mistake "shocked our trading partners, many of which have been lukewarm on IPEF from the get-go, but nevertheless were willing IPEF participants in hopes that this would lead to more US economic engagement."[6]

Officials from the US and its partners have stressed that they have made significant progress on the trade negotiations, but the prospect of a deal over the next year looks increasingly dire with US elections fast approaching in November 2024.

"It's hard to see how the politics that have guided the Biden administration's decision-making to date change as the US presidential election approaches," said Jake Colvin, president of the National Foreign Trade Council, who added that he hoped the progress to date could serve as a building block for the future.[7]

Not surprisingly, the first meetings of the 14 IPEF members from 8 to 9 September 2022 were fraught with difficulties right from the start. For starters, the meetings were held in the US rather than in Asia. That gave the signal to the Asians of top-down US economic coercion rather than shared prosperity among Asian countries.

Second, the US made it clear that China was excluded from IPEF and would never be a member of IPEF. Since most of the merchandise trade in Asia is with China, this insensitivity to the national interests of other Asian members of IPEF made it clear that the US had a US Cold War, zero sum geopolitical agenda which had little to do with shared prosperity in the region. Third, those IPEF countries which did not cave into the US top-down management could face the possibility of US economic pressure via tariffs, sanctions, export controls and other punitive economic measures.[8]

6 Historical Lessons Learned

But Chinese passion for regional trade diplomacy and US disinterest in regional trade diplomacy are not the only reasons why China has dominance in economic influence and the US has marginal economic influence. A historical analogy in Europe provides insights.

US economic problems can be traced back to the Treaty of Versailles after World War I. The US and its allies wanted to keep the German military weak. They assumed that if they could keep the German economy weak, it would be impossible for Germany to have a strong military. How did that work out? It led to German resentment, the rise of Adolph Hitler and World War II.

After World War II, the great French statesman Jean Monnet had a better idea. Monnet looked for common ground and social inclusion between the French and German people. In his strategic vision, French and German businessmen would work together and build a European Coal and Steel Community. Common bonds among businessmen would spread to the rest of the French and German populations, and help turn longstanding enemies into friends. That French empathy for Germany increased French influence with the German people and a French-German "special relationship."

Did Monnet appear naïve and hopelessly idealistic? Possibly. But appearances in economics can be deceiving. Monnet showed that a web of economic interdependence and social inclusion could lead to shared prosperity and a more durable peace than more short-sighted economic coercion. In other words, shared prosperity would reduce the demand for violence and thus lower mutual threat perceptions—not a bad way to keep the peace in Europe for 70 years.

Like Monnet, US Secretary of State George Marshall felt it was important to help Germany and the rest of Western Europe rebuild their factories and enjoy shared prosperity with America, rather than resorting to the ill-fated Clemenceau's economic coercion of Germany.

Now let's turn to China's economic trajectory. For four decades China's Deng Xiaoping's market reforms drove the greatest economic growth in the history of the world. US China economic interdependence was on a roll. Then Team Trump came along and opted to decouple from China. Team Biden felt it had to be even tougher on China.

So, imagine you are part of Team Biden. What's your mindset regarding China? You think about China like the allies thought about Germany in Versailles after World War I. I call this Versailles 2.0. You hope economic coercion will work against China better than it did against Germany. But hope is not a strategy.

Team Biden feels that China's economy is growing too strong. That enables China to have a strong military that is too much of a threat. So, you conclude that the US has to do everything possible to weaken China's economy. That means US economic coercion will make it more difficult for China to threaten the US and US allies and friends in the Indo-Pacific. In short, Team Biden will continue and even escalate Team Trump's trade war against China and reduce economic ties with China.

7 China's Economy is Now Too Weak

During the Cold War—long before Vladimir Putin became Russia's president—a Russian diplomat once told a US official that "we are going to do something terrible to you. We are going take the Russian threat away from you." Fast forward to present times. Financial Times (FT) journalist Martin Sandbu says it's time for the West to worry about "the other economic security threat from China." China's economy is "too weak, not too strong." [9] Unfortunately, Team Biden appears unaware of China's economic decline—or at least the strategic significance of China's economic decline.

If China's economy is struggling, how do we confirm this and measure the economic decline? Well, let's briefly start with China's economic rise. China's economy had an amazing run. After China's economy stagnated under Mao in the 1960 and 1970s, Deng Xiaoping's free market reforms in the late 1970s paved the way for four decades of amazing economic growth. In fact, Ruchir Sharma—Head of Rockefeller Capital Management's International Business—notes that no nation had ever risen so far, so fast. China's share of the global economy rose nearly tenfold from below 2% in 1990 to 18.4% in 2021.[10]

But as frequently happens in this world, all good things often come to an end. The biggest global economic story for the past half century shows strong signs that China's amazing run is over. Sharma's research shows that China's share of the world economy shrank from 18.4% in 2021 to just 17% in 2023. That two-year drop of 1.4% is the largest since the 1960s.

Sharma's research also shows that China's real long-term potential growth rate—the sum of new workers entering the labor force and output per worker—is only about 2.5%. This low figure reflects the negative demographic realities in China. The ongoing baby bust in China has already lowered its share of the world's working age population from a peak of 24–19%, and it is expected to fall to 10% over the next 35 years.

Sharma states that these forces are slowing growth in productivity, measured as output per worker. This combination—fewer workers, and anemic growth in output per worker—will make it difficult in the extreme for China to start winning back its previous share of the global economy. With a shrinking share of the world's workers, a smaller share of growth is almost certain.

So, we have measured demographic reasons for the decline in China's economic growth decline. But let's drill down a little more and explore other reasons for China's decline in economic growth. The root causes of China's economic decline can be traced back to the global financial crisis of 2007–2009, which shook China's confidence in the Western-style financial system.

8 Beijing's Economic Control

After the crisis, China's government sector started to control more and more of China's economy. Starting in 2008 the government started to fund infrastructure investment and a housing boom. But these debt-financed investments also planted the seeds for the debt problems Chinese developers and local governments are currently facing. Real estate investors used empty buildings as collateral to build more and more buildings.

Even before COVID shutdowns, China's growth began to stall due to a deadly combination of demographic headwinds, the deterioration of the real estate sector, and the re-emergence of state-led economic policymaking. University of Pennsylvania Professor Hanming Fang notes three ways the Chinese government began to play a larger role in the economy starting in the 2010s.

- Beijing began controlling small stakes in the private sector.
- Beijing took a larger role in corporate governance.
- And Beijing began to tighten regulations in Chinese domestic businesses.[11]

Hanming Fang's research also shows that China's increasingly prescriptive industrial policy had limited success in promoting productivity. Beijing's subsidies failed to result in productivity-enhancing breakthroughs. Instead of government money being used to stimulate corporate innovation or tax incentives for R&D, government money was wasted on unnecessary administrative expenditures. There was also little evidence that the Chinese government had consistently picked winners when allocating subsidies.[12]

From 1979 to 2008, the US and China enjoyed shared prosperity. Although diplomatic tensions took place now and then, our case study on Greater China later in this chapter will highlight pervasive shared prosperity among Mainland China, Taiwan and Hong Kong.

But today, US China relations are at its lowest point in the last 40 years. The change became evident in President Obama's "Pivot to Asia" in 2011 and intensified with President Trump's anti-China campaign rhetoric and the trade war initiated in 2018 and which President Biden in turn has continued and escalated.

9 National Security Concerns

Hanming Fang notes three areas where the economic ties between the US and China are now intertwined with national security concerns.

- The U.S. has instituted restrictions on both outbound investment in Chinese firms and inbound capital flows from China to protect against intellectual property theft from China.
- The total bilateral foreign direct investment (FDI) between the US and China fell to $15.9 billion in 2020 amid pandemic-related disruptions and rising tensions in the U.S.-China relationship.
- These restrictions, along with U.S. embargoes of key technological goods like semiconductors and smartphone processors, have further hampered growth of Chinese firms.[13]

In many ways, the US is trying to extricate itself from the constraints imposed by economic interdependence with China.[14] For instance, Ashley Tellis—a senior scholar from Carnegie Endowment for Peace—notes that US decoupling from China would create separate global production chains that have little or no reliance on Chinese materials, technology, or production facilities. Tellis concludes that the endgame would include independent production networks. Each network would create separate supply chains and possibly different technical standards, business models, and legal and regulatory frameworks.[15]

Keith Johnson and Robbie Gramer—analysts from Foreign Policy Online—note that the last big economic split brought on two wars and a depression. They point out that this kind of US China economic restructuring is unprecedented in peacetime.[16] In the past, war has been the main precipitant for radical dislocations in commercial exchanges across national borders. If Columbia Professor Robert Jervis was here, he would remind us that the US is sending the wrong signal to China. Beijing could easily misperceive that US signal and now assume the US is mobilizing for war. That's why restoring US China military communications back in December 2023 and US National Security Advisor Jake Sullivan's de-conflicting false signals helps to stabilize US China stability.[17]

But it's not just Chinese misperceptions of US wartime intent that is at stake here. Ashley Tellis underscores that decoupling involves shattering the global trading system. That's because the US has previously been the principal underwriter of the open trading system. Since decoupling would be a major US shift toward protectionism, decoupling would amount to a fundamental rewriting of the rules of the game.[18]

Former MIT Professor Charles Kindleberger's hegemonic stability theory is relevant here. The theory says that there is a relationship between trade openness and hegemonic power. Kindleberger emphasizes that open trading systems survive only if there exists a hegemonic power (like the US) to bear the costs of upholding them.[19]

Ashley Tellis argues that what's at stake is more than just how decoupling structurally disrupts trade flows which in turn undermine the postwar US vision of global economic integration. In addition, I would argue that decoupling also threatens shared prosperity, which China has used to boost China's economic influence.[20]

Victor Fung—Chairman of the Asia Global Institute at the University of Hong Kong—notes that back in the old days, manufacturing was one-dimensional. As stated in my other chapter on Team Trump's approach in this book, manufacturing was done "in-house" … in one factory, under one roof, and in one country … before that product was exported and sold in another country.[21]

But things changed, and products became increasingly dispersed across different factories in different countries and thus globalized. Because of this dispersion of global production, countries previously omitted from manufacturing can now participate by performing just one or two intermediate components of the supply chain. Fung calls this the "democratization of production." Dispersion also benefits American consumers because it reduces costs and makes products more affordable for US consumers.[22]

10 Greater China—Mini-Case Study

Does Victor Fung just have a vision about the democratization of production and do economic interdependence and shared prosperity really matter in terms of economic performance? I researched this question and found strong evidence that the economic interdependence and shared prosperity were impressive. The following case study below provides hard data from the initial growth of Greater China—consisting of Mainland China, Taiwan and Hong Kong—after Deng Xiaoping's reforms in China (circa 1980 until 2007). My purpose is to document the impressive economic and business performance of Greater China. In short, shared prosperity worked back then and still works.

The point is China did not achieve its economic success alone. China increased its economic influence by sharing its wealth with others in a positive sum game. Greater China was an economic powerhouse that Mainland China, Hong Kong, and Taiwan all bought into.

In fact, Greater China was one of the biggest economic mergers in history. The concept of a Greater China started developing in the 1980s with the mass influx of manufacturers from Hong Kong, and then Taiwan, to the Southern China coast soon after Beijing opened its doors.

11 Industrial Unification

Starting with China's entry into WTO in 2002, the industrial unification of the three economies accelerated and spread deeply into the mainland. Political barriers to investment in strategic industries such as semiconductors, oil, and banking began to crumble. Infrastructure systems were fused. Despite occasional tensions over political and security issues, Beijing, Taiwan and Hong Kong were collaborating on economic affairs.

12 Economic Convergence

There were numerous manifestations of this accelerated convergence. For instance, the number of Taiwan officially living on the mainland swelled from 300,000 to 500,000. Several thousand Taiwan students were pursuing degrees in China rather than at US universities. China overtook the U.S. as Taiwan's biggest export market in 2002.

13 IT Ties That Bind

In addition, China passed Taiwan as the world's No. 3 maker of information-technology (IT) products. But many of the benefits flowed to Taiwan companies, which controlled 70% of the output. China overtook Taiwan as a maker of desktop PCs, optical drives, and liquid crystal displays. The share of Taiwan notebook PCs made in China jumped from 4 to 30% from 2001 to 2007. Hong Kong's role as an international capital window for China grew. The Bank of China (Hong Kong) placed its initial stock offering in Hong Kong rather than New York. Dozens of other Chinese companies were listed on Hong Kong's new second board as startups.[23]

14 Greater China: Export Powerhouse

The impact of Greater China as an exporter was felt around the world. Greater China's share of total world exports went from 6.9 to 9.6% in just four years (1998–2001) and surpassed Japan by 2001.

By 2007, Greater China's share of total world exports reached 13.7%. Also, by that time, as the Mainland and Taiwan opened their markets to comply with WTO commitments, Greater China surpassed $2 trillion in exports and imports, excluding the trade between the three economies.

That made Greater China nearly twice as big as Japan as a trading power and two-thirds the size of the US. Measured in terms of purchasing power, Greater China soon overtook the EU, with a combined gross domestic product of $12 trillion.

By that yardstick, it became nearly triple the size of Japan. The market capitalization of the stock bourses of Hong Kong, China, and Taiwan challenged Tokyo's $2.1 trillion by that time.

15 Overseas Chinese

This concept of Greater China was expanded even further if we consider the huge populations of overseas Chinese who still considered the mainland home and increasingly returned to start businesses there. Of the more than 400,000 Chinese who studied overseas, some 140,000 moved back to the mainland, bringing with them advanced degrees, Silicon Valley experience, and venture capital. Shenzhen Mayor Yu Youjun reported that his city already had 300 companies run by returnees. Over 1,000 overseas-educated Chinese moved there each year.

16 Commercial Hub of Northeast Asia

Stretch the notion a bit more and it was easy to view Mainland China as the hub of North Asia. What was emerging was an integrated Chinese network, where dozens of growth zones essentially rented themselves out to foreign investors. An estimated 100,000 South Koreans lived in cities on the Yellow Sea such as Qingdao and Tianjin. Some 4,000 Japanese companies had operations near Dalian. Singaporean companies also helped the cause and were a big source of capital.

17 Architects: Chinese Entrepreneurs

The real architects of this growing economic colossus were the entrepreneurs of Geater China and the officialdom of cities such as Shanghai, Chongqing, Shenzhen, Dongguan, and Hong Kong. They spent two decades creating networks of economic influence and putting down roots—advantages that were extremely difficult for other foreigners to match. Equally important, these players were providing China with the managerial and financial expertise needed to compete in the world. In short, they all bought into Mainland China's vision of shared prosperity.

18 Why Was This Process Speeding Up?

A number of factors were driving the process: The biggest driver was China's economic reforms, now anchored by its entry into the WTO. Another was the ever-deeper integration of Hong Kong since the departure of the British in 1997. Security regulators and monetary authorities from Hong Kong and Beijing worked closely with each other. Calmer relations across the Taiwan Strait were also helping.

19 Corporate Cooperation

Meanwhile, mainland companies such as Bank of China, which used to be shadowy presences in Hong Kong, started to use Hong Kong as a base to learn international management standards. China's foreign trade ministry began bringing together delegations of private mainland entrepreneurs interested in opening Hong Kong offices.

20 Mainland Magnet for Hong Kong Residents

After years of trying to keep its distance from the Mainland, Hong Kong was also looking to the mainland as a savior. A few years back, people on Hong Kong were willing to commute by hydrofoil or air-conditioned bus to factories in the Pearl River Delta. But back then, few Hong Kong residents wanted to live on the mainland. Then they started to fight for jobs there. Thousands flocked to mainland job fairs.

21 China's Dependence on Taiwan and Hong Kong Talent

Talent from Taiwan and Hong Kong was pivotal to China's ability to maintain annual growth rates of at least 7%. It played a key behind-the-scenes role at advanced plants and research labs provided by multinationals such as Cisco Systems, Ford Motor, Nokia, Sony, and Motorola. All of the Chinese plants were built with Taiwan managers. Royal Philips Electronics cut its Taiwan workforce from 12,000 to 4,000, as it shifted production to China.

22 China and Taiwan

Why did Taiwan businessmen invest on the mainland? Taiwan's GDP growth slowed (from almost 7% in 1997 to less than 4% in 2001). In addition, Taiwan's unemployment rose from less than 3% in 2000 to over 4% in 2001). Therefore, Taiwan's corporations saw business opportunities drying up in Taiwan and starting to grow on the mainland. Taiwan's business community poured capital into the mainland to take advantage of the rising opportunities to make money there. Taiwan's pledged cumulative investment increased almost tenfold between 1991 and 2000.

23 Magnetic Attraction

In addition to slow GDP growth and rising unemployment in Taiwan, the rise in Taiwan's FDI on the mainland reflected the magnetic effect of the Chinese mainland market (which was characterized by the cheaper cost of land and labor). Numerous incentives were taken by the Chinese government. Social and economic interactions between Taiwan and mainland China took off in the late 1980s. Before then, there were almost no direct contacts between citizens of the two countries.

24 Visits Soar

Taiwan then announced in late 1987 that citizens would be allowed to visit relatives on the mainland for humanitarian purposes. Once the door was opened, visitors poured across the strait "like waters rushing through a breached dam."[24] Within about a year, 200,000 islanders had traveled to China. By the end of 1989, the number was near 1 million. By October 2002 Taipei's Mainland Affairs Council (MAC) reported that there were over 27 million visits to the mainland. In the year 2002 alone, the figure was over 3 million. At the same time, by the 1980s Taiwan's prosperity drove up wages. Taiwan's wages were so high that it was losing competitiveness in its labor-intensive production sector.

25 Mainland's Attraction to Taiwan's Businessmen

With the opening of the mainland, many Taiwan business people saw it as a prime location for new factory sites. In Taiwan business eyes, the Chinese mainland enjoyed almost unlimited low-cost labor. In addition, the language and cultural ties were obviously very strong. Finally, pragmatic provincial leaders on the mainland offered substantial incentives for Taiwan's investment in export industries.

26 China's Coastal Development Strategy

In the meantime, China in the late 1980s embarked on a "coastal development strategy" aimed at attracting the light and labor-intensive industries that were being priced out of Hong Kong and Taiwan. China also began to emulate the export-led industrialization of the East Asian capitalist nations.

27 Confluence

The result was a confluence of fundamental economic changes in Taiwan as well as the new Chinese economic strategy. This confluence produced tremendous growth in cross-strait economic interactions.

28 Taiwan Exports to China Soar

While the Taiwan and Chinese numbers differed, the trend lines showed a similar upward trajectory. Taiwan's exports to China grew nearly 6 times between 1987 and 1991 (from $1.2 billion to $6.9 billion). Between 1992 and 1996 Taiwan's exports to China more than doubled again (to $19.1 billion).

29 Shared Prosperity Resists Political Turmoil

This upward trend in cross-strait trade and investment was so strong and resilient that it shrugged off political turbulence, including the Tiananmen Square massacre of 1989 and the Taiwan Strait crisis of 1995–1996. Taiwan's exports to China remained steady at $21.2 billion from 1997 to 1999, despite the Asian financial crisis of 1997–1998 and Lee Teng-Hui's 1999 declaration of "two-state theory." Nor did the presidential election of Chen Shui-bian change this economic growth trend. In 2000, Taiwan's exports to China reached $26.1 billion, while the bilateral trade volume jumped to $32.3 billion. In 2002 Taiwan's exports to China increased to $33.05 billion, while the bilateral trade volume jumped to $41 billion.

30 Trade Up

Statistics from China's Ministry of Commerce suggest that in the 1st quarter of 2003, cross-strait trade volume increased by 39% to reach $12.12 billion. According to Taiwan's Board of Trade (BOFT), from January to March 2003 the cross-strait trade volume was $10.06 billion, a 33.2% increase over the same period in 2002, which in turn was 16.4% of Taiwan's total trade volume. The value of Taiwan's exports to China in the same period was $7.8 billion, a 29.9% increase from 2002. The value of Taiwan's imports from China was valued at $2.26 billion, a 45.6% increase, for a total $5.53 billion trade surplus with China.

31 Taiwan's Trade Surplus

Similarly, after 1995 the mainland became the single most important source for Taiwan's foreign trade surplus. In 2002, Taiwan's trade surplus with China was $25.3 billion, while its total trade surplus that year was $18.05 billion. That said, Taiwan's balance of trade would be in deep deficit if it were not for its large trade surplus with China.[25] On balance, China and Taiwan both benefited from their trading relationship and economic interdependence.

32 Investment Drives Trade

Taiwan's trade with the mainland was for the most part investment driven. That's because Taiwan investors on the mainland mainly depended on the island for their supplies of machinery equipment, spare parts, and certain raw materials. According to Taiwan BOFT, 54% of the materials and 75% of the machinery and equipment needed by Taiwan's businesses in China were imported from Taiwan.

33 Growth of Taiwan's FDI

Prior to 1987 there was virtually no Taiwan investment on the mainland. But then things changed. The growth of Taiwan investment on the mainland was particularly impressive. Following Taiwan's decades-long martial law in July 1987, Taipei relaxed foreign exchange controls. Adventurous Taiwan businessmen then began to invest in the coastal cities on the mainland, especially in the city of Xiamen, located directly across the Taiwan Strait.

34 The Rise of Taiwan's FDI

The initial level of Taiwan's investment on the mainland was actually quite moderate. But after 1990, Taiwan's investment sharply increased when new capital poured into real estate along the coastal cities of Eastern China. Between 1991 and 1995, there were 1,254 investment applications for the mainland. Taiwan statistics show that there was a total value of $56.45 billion investment applications that were approved while Beijing's statistics show that 31,780 Taiwan investment applications were approved for a total value of $114.27 billion.

35 Taiwan's # 1 FDI Destination: China

In addition, the mainland became the most important outlet for Taiwan's overseas investment. The mainland accounted for 44.53% of the island's total investment abroad from 1991 to January 2003. The mainland far exceeded Taiwan's 2nd country of investment, the U.S. (11.25% from 1969 to January 2003).

36 The Changing Nature of Taiwan's Investment

Then Taiwan's investments on the mainland started to change in a number of different ways: For instance, the nature of Taiwan's foreign investment became more long term. Taiwan's investors moved from joint ventures to solely owned enterprises. Taiwan investors also began to build their own factories in China. The structure of Taiwan's business ventures on the mainland was also upgraded—from simple assembly to upstream heavy and more capital-intensive or high-tech production.

By mid-to-late 1990s, the mix of Taiwan's investment on mainland China began to shift—from predominantly small business in labor-intensive exports to much larger businesses seeking to penetrate the Chinese market in heavy industry (e.g. Formosa Plastics) and consumer goods (e.g. President Enterprises). There was also a geographic spread of Taiwan investment, from an initial concentration in Fujian and Guangdong provinces, then to Shanghai, and finally to almost all regions of China.

37 Trade and Investment

By 2002 this economic integration began to generate another round of economic interactions between Taiwan and China. Cross-strait trade jumped that year by 36.9%, from $29.9 billion to $41 billion. Taiwan's FDI on mainland soared. Taiwan's investment surged even more dramatically, up to $20 billion in 2000–2002, reaching an estimated cumulative level of $80 billion—$100 billion.[26]

38 Taiwan Communities on the Mainland

These strong economic ties also augmented informal people-to-people diplomacy and created substantial Taiwan communities in many mainland cities. There were an estimated 400,000 to 700,000 Taiwan nationals living and working on the mainland, concentrated in the high technology and high economic growth areas of Shanghai and Shenzhen. Some sources say the number of Taiwan living in Shanghai alone was about 300,000.

39 IT Migration to the Mainland

2000 was a benchmark year for Taiwan investment on the mainland. Taiwan investment on the mainland changed from low and medium tech sectors into high-tech sectors. Furthermore, investment in services showed for the first time that Taiwan businesses finally began to view mainland China as a major export market in itself.

40 Taiwan's IT Production on Mainland

For year 2000, Taiwan-owned production lines manufactured an astonishing 72.8% of the total $25.535 billion production value of China's information technology (IT) hardware sector.[27] By the end of 2000, up to 48% of desk top computers exported by Taiwan were manufactured in mainland plants. Plants in Taiwan turned out 18%. The remaining 34% were produced in other parts of the world.

41 Taiwan's Production Items on the Mainland

In 2000 Taiwan-operated plants on Mainland China assembled 6.5% of Taiwan's notebook output, double the 1999 figure. About 56% of Taiwan's motherboard makers also had production lines on the mainland, which by the end of 2000 produced over 50% of Taiwan's total motherboard shipments. Of Taiwan's computer monitor producers, 58% had major production lines on the mainland. 74% of Taiwan's CD-ROM drive makers moved to mainland China. And 88% of scanner manufacturers moved to the mainland by the end of 2000.

42 Mini-Case Study: The Global Garment Industry

The economic trend lines cited above illustrated China's success in developing shared prosperity with two economies that had similar Chinese language and culture. But what about when China goes head-to-head with a country like Bangladesh in the competitive garment industry? How did China develop shared prosperity and economic influence with Bangladesh?

Let's turn the clock back to 2005 and look at over a decade (2005–2016) of Bangladeshi efforts to compete with China in the cut-throat global market industry. The garment trade in Bangladesh was possible because of the 30-year-old Multi-Fiber Arrangement (MFA) which gave it and other poor countries preferential access to important markets in the European Union (EU) and America. So, when the MFA rules governing garment exports to rich countries were changed at the start of 2005, Bangladesh faced a time bomb. It feared huge jobs losses because of the potential collapse of its garment industry.[28]

Why so devastating? The Bangladesh garment industry anchored the economy and sustained millions of families. Bangladesh received 75% of its hard-currency earnings from garment exports. Garment factories employed as many as two million workers or half the nation's industrial workforce. About 85% of these workers were women.

43 The China Price

Bangladeshi garment makers were afraid they wouldn't be able to compete against China. Why? For a six-year period—1998 to 2004—China faced price deflation in textiles and other low-end manufacturing. China's low price seemed more and more unbeatable.

Some people said, "Don't worry, it will take China a long time to capture new market shares after MFA expires on 01 January 2005." Well, guess again. China enjoyed virtually overnight success. In the first quarter of 2005, its cotton shirt sales in the US market rose 1,250% from the same period in 2004. Its cotton trouser sales in the US market rose 1,500% from the same period in 2004.

44 Bangladesh on the Ropes?

Bangladesh appeared to be on the verge of losing a key industry. The garment industry in Bangladesh previously accounted for 80% of its exports and lifted 13% of the country's poor households out of poverty.

The World Bank estimated that China would corner the market by controlling 50% of global garment exports by 2010. Why? The Chinese garment makers enjoyed vast economies of scale; a deep pool of cheap labor easily bent to the political will; and fully integrated cotton, textile and garment industries.

China's growing market share in garments did not bode well for Bangladesh once the remaining quota barriers disappeared. At first glance, it appeared that the end of MFA would sharply diminish the demand for garments stitched in Bangladesh.

Unemployment in Bangladesh could soar. Most experts predicted that the country would lose one million garment jobs. Also at risk were the jobs of many of the 15 million people who worked in related industries, such as button-makers and truckers. For them, there was no state social security plan and no unemployment benefits.

45 Possible Violent Extremism

To make matters worse, Bangladesh—a poor and overpopulated country with rising income disparity—was always fertile ground for violent extremism. But all of a sudden, the extremists had a major issue they could exploit. Violent extremists blamed the United States for globalization and the end of MFA, and they planned to exploit this ill-advised but pervasive perception of social and economic injustice. The concern was that these young men might not find jobs outside Bangladesh. These angry young men might get politicized and fall for an extreme version of violent extremism. Over time, this could tip the scale of a relatively moderate Muslim country and turn it into a more extreme one.

46 The Great Debate

This connection between the economy and violence contributed to a "great debate" in the country. Given the specter of rising unemployment, would terrorism break out in Bangladesh? One "supply" school of thought said that the Bangladeshi government had beefed up its "capacity" (or supply of intelligence and counter-violence) to combat terrorism. A competing "demand" school of thought said that capacity building—while necessary—was not enough. This demand school said that the socio-economic demand for violence was rising. Frustrated people were at risk of buying the propaganda lines of violent extremists.

Who was right? Was the terrorist threat up or down? In August 2005, the worst fears of the demand school played out. Five hundred bombings occurred in Bangladesh in just one month.

47 Reducing the Demand for Violence

But at a time when things looked especially bleak, things were changing in the Chinese countryside that would reverberate in every nook and cranny in the international garment industry. These developments would also reduce the demand for violent extremism in Bangladesh.

Social unrest was rising in the Chinese countryside. In 2005, there were 87,000 public disturbances in China. That's a 13% rise from 2004. Frustrated Chinese farmers were pouring into Chinese coastal cities looking for work in the garment factories. An oversupply of Chinese workers in the cities pushed Chinese wages below those of Bangladesh. Most of these ex-farm workers stayed in the cities. If the great American singer Eddie Cantor were still alive and visited China at this time, he might have sung, "How you gonna keep 'em down on the farm after they've seen Shanghai?".

Then everything changed. In September 2005—one month after 500 bombings in Bangladesh—Beijing announced that the agricultural tax—which dates back over 2,000 years—would be eliminated. What was the impact of abolishing the agricultural tax in China? At first glance, eliminating this seemed like a good idea to Beijing. It was more affordable for Chinese farmers to stay on the farm, now that their taxes were gone. This also reduced at least some of the social unrest in the countryside.

But be careful what you wish for. There were unintended consequences for Beijing. Improving life in the rural areas of China meant fewer migrant workers leaving the countryside and looking for work in urban garment factories along the PRC coast.

So, imagine you were running a garment factory in Shanghai. You all of a sudden had a shortfall of workers knocking on your door. How were you going to attract more workers? You had to boost wages. And in 2005, there was a double-digit rise in Chinese wages. It didn't just happen in the textile industry. It happened across the board in Chinese low-end manufacturing. As a result, its manufacturing competitiveness fell in 2005. It was no longer the low-cost producer.

48 Godsend for Bangladesh

Who benefited from rising prices in China? It was a godsend for the garment factories in Bangladesh. Back then, garment workers in Dhaka earned 39 US cents an hour, while the hourly wage for sewing and stitching in coastal China rose to 88 cents. In Bangladesh, orders started to soar at its garment factories. Thanks to rising prices, Bangladeshi garment exports rose by $500 million in 2005 compared to those in 2004. The UN fear of a million Bangladeshi garment workers being laid off never happened.

Bangladesh nearly doubled its garment exports from 2004 to 2009.[29] China's garment workers made about $117 to $147 a month in 2010, while Bangladeshi workers made about half that. That spurred production in Bangladesh. The production of Walmart and Liz Claiborne products in Bangladesh rose 20% in 2010, while China's production for the same two companies fell 5%.[30]

If we fast forward to 2015, we discover that China's rising production costs (particularly, for labor and land) allowed Bangladesh and other rival producers to eat into China's market share. In fact, China's wages in the garment industry doubled, while those in Bangladesh remained relatively flat.[31]

Over the first eight months of 2015, US garment imports from China fell by 7.7% year on year. Those who most clearly benefitted from China's decline were Bangladesh, Vietnam and India. EU imports from China fell by 9.7% year on year from January–July 2016, while those of Bangladesh and Vietnam surged by 6.3 and 8.8%, respectively.

49 China's Support for Port Expansion

News flash. This doesn't sound like China cheated. Far from it. In fact, China deserves three cheers for fostering shared prosperity with Bangladesh. In the past, Bangladesh suffered from unreliable energy utilities and inadequate logistics infrastructure.[32] Beijing was committed to helping Bangladesh overcome these challenges. China's government contributed to Bangladesh's ability to cope with the rapid growth in shipment volumes at its ports by providing Bangladesh the financing for port and power infrastructure upgrades (i.e. more port capacity) under its "One Belt, One Road" initiative (BRI).[33]

The point was that China believed in shared prosperity in the region, even when its own percentage of the garment pie declined. BRI is all about a rising tide lifting all ships of state. This Chinese positive sum game and support for shared prosperity stood in sharp contrast to Trump's zero-sum game of "America First" economic nationalism. China's economic influence rose, US economic influence fell.

Now, imagine you were a recruiter for a violent extremist group in Bangladesh. Bashing globalization lost its punch. China's shared prosperity reduced the demand for violence. Jobs and incomes at Bangladesh textile factories rose. Garment workers felt that globalization was wonderful. The free market marginalized you. The demand for violent extremism fell. Nobody believed your propaganda.

50 Take Aways

To sum up, the economic merger and cumulative economic success of Greater China plus China's support for port expansion in Bangladesh are historical examples of China's longstanding support for shared prosperity.

51 Decoupling

Now let's fast forward to today. The discussion of Greater China's shared prosperity and China's dominant economic influence sets the stage for why most economies in Asia in June of 2024 are pushing back on US efforts to decouple Asian economic ties with China. Tellis provides us with an overview of six US decoupling related policies. Then we look at US attempts to persuade third countries in Asia to decouple from China in Asia. NBR research explains why each Asian country with economic ties to China is, on balance, pushing back against decoupling from China, notwithstanding security ties between the US and its Asian partners.[34]

Tellis builds on Victor Fung's insights. He argues that decoupling would also threaten the greatest previous achievements of globalization. Decoupling would subvert the increases in what he calls allocative efficiency and by extension the improvements in aggregate growth and welfare. Decoupling would also undermine technological innovation, which Tellis says has raised incomes and lifted millions out of poverty in developing countries as well as increased the overall prosperity of the developed world.[35]

Now let's turn to Europe. The Russian invasion of Ukraine pushed the EU away from economic interdependence with Russia and more towards decoupling from Russia. The costs of being deeply dependent on Russian energy became too burdensome.[36] As a result, Europe finally bought into the US argument for decoupling.

The energy war pushed the US and Europe—previously the greatest champions of globalization—together on a new course that reflects a marked suspicion of free trade and deeper international integration. Tellis persuasively argues that the loss of policy support for open-ended expansion of globalization is a significant change from the optimism that prevailed after the end of the Cold War. He says US and

European movements away from globalization represent "a clear—and potentially dangerous—break from the past when the US generally supported free trade during the postwar era."[37]

For most of this post World War II period, the US viewed China's economic rise as a blessing rather than a threat. The US and China enjoyed shared prosperity. But President Trump flipped the script. He went back to the economic coercion strategy the allies opted for and which failed so miserably after World War I and hoped for a different result this time. I call this Versailles 2.0.[38]

President Trump launched a trade war against China and its trade and production partners in Asia. President Trump blamed US unemployment on Chinese exports despite a Ball State study that showed 85% of the US job losses were due to technology rather than trade. But in US domestic politics, both US political parties are trying to be tougher on China than the other party.

52 US Decoupling Policies

Now that we've looked at Team Biden's misperceptions of China's economy and the doubling down on it with an economic coercion strategy, let's now turn to a breakdown of Team Biden's ill-advised efforts to decouple the US and its allies away from connections to China's economy. Tellis gives us his persuasive take on six decoupling policies in this regard. These include: Team Biden's escalated trade war, US restrictions, Asian push back on incoming Chinese foreign direct investment (FDI) into US, prohibition of outgoing US FDI coming into China, rising subsidies for US industrial policy, and friend-shoring. To be fair, some of Team Biden's policies are useful and protect US national security. That said, too many of Team Biden's policies are aimed at weakening China's economy and are arguably destabilizing.[39]

53 The Trade War Continues

When President Trump started the trade war with China, former US Vice President Joe Biden was critical of America First and US protectionism. Then when Joe Biden became president, he not only retained all of Trump's expansive tariffs on China, he escalated the trade war. For instance, on 14 May 2024 US President Joe Biden sharply increased US tariffs on Chinese electric vehicles, solar products, semiconductors, medical supplies and other goods.

54 IMF Criticizes US Economic Coercion

Two days later, IMF pushed back. IMF publicly knocked President Biden's extraordinarily high tariffs as a risk to world trade and the global economy. IMF spokesperson Julie Kozack openly criticized President Biden's trade restrictions for distorting trade and investment, fragmenting supply chains and triggering Chinese retaliatory actions. She said that "fragmentation of this type can be very costly for the global economy."[40]

With respect to the US tariffs, she said the IMF view is that "the U.S. would be "better served by maintaining open trade policies that have been vital to US economic performance" in the past. Kozack also said "We also encourage the U.S. and China to work together toward a solution that addresses the underlying concerns that have exacerbated trade tensions between the two countries." In short, IMF confirmed our recommendation for US shared prosperity rather than US economic coercion.[41]

The Trump tariffs were intended to initially correct the U.S. trade deficit with Beijing and eventually rectify China's structural distortions through the planned "phase two" trade negotiations. But as I say in the other chapter of this book, US trade deficits reflect loose US fiscal policies, high interest rates and an overvalued US dollar to offset them. That's why high US tariffs have been ineffective in lowering the bilateral US trade deficit that has risen sharply while Donald Trump and Joe Biden have been president.

Since the Biden administration has also not demonstrated any interest in pursuing trade negotiations with China to address its structural distortions, Tellis persuasively argues that the value of preserving the inherited tariffs is questionable—except for making domestic political points.[42]

In contrast, Tells notes that Team Biden's approach to sanctions on Chinese apps and companies has been more sensible. The earlier bans on Chinese apps were revoked, and the Commerce Department now has the authority to monitor and appropriately control any software applications that may threaten US national security.[43]

Similarly, the Pentagon has arguably made the right decision to transfer oversight of Chinese military companies from the Department of Defense to the Treasury's Office of Foreign Assets Control so that the sanctions on these entities may be administered more effectively.[44]

55 New US Restrictions

US restrictions are also a mixed bag.

- On the one hand, the IMF has identified 3,000 US global trade restrictions in 2023, up from 1,000 in 2019. IMF notes that in a worst-case scenario of severe fragmentation into geopolitical blocs, these US restrictions could reduce global economic output by some 7%, equivalent to removing the GDP of Japan and Germany combined.[45]

- On the other hand, Team Biden has correctly come up with responsible new restrictions in the area of national security. The US has done its due diligence in regards to the export of advanced integrated circuits, electronic components containing advanced integrated circuits, semiconductor manufacturing equipment, and related software and technology to China.

Tellis correctly points out that the expansion of US export controls and the creation of new rules that constrain third-country activities pertaining to China's semiconductor and supercomputing manufacturing capabilities are intended to limit the threat that Beijing could pose to U.S. national security. This is especially important when it comes to the development of advanced weapons.[46]

In this regard, it's important to understand that technology is central to great-power competition we discuss in this book in a way that industrialization was during the twentieth century. And so, at least at first glance, Team Biden's strategy is reasonable and has been defended by National Security Advisor Jake Sullivan as exemplifying the "small yard and high fence" approach that restricts trade only in regard to a small set of foundational capabilities.[47]

That said, Tellis reminds us that the gambit is not without risks. For instance, what if future US controls in other arenas—such as quantum technologies, artificial intelligence, biotechnology and biomanufacturing, advanced telecommunications, and advanced materials—are identified as critical to the evolving U.S.-China competition?

These items could easily expand Jake Sullivan's "small yard" in ways that disfigure it beyond recognition. And what if this were to occur through either bureaucratic momentum or leadership indiscipline—which often accompany such policy shifts once they are initiated? Tellis notes that the US ambition of maintaining "as large of a lead as possible" over China could itself be subverted. Tellis argues that sustaining technological supremacy when international knowledge flows of different kinds are constricted could prove to be daunting in an era of deep economic interdependence.[48]

56 Push Back on Incoming Chinese FDI into US

Third, as Hanming Fang mentioned earlier, the U.S. has instituted restrictions on both outbound investment in Chinese firms and inbound capital flows from China. In this regard, Tellis notes that Team Biden has continued with Team Trump's policy of intensively scrutinizing Chinese foreign direct investment into the US. The review authority of the Committee on Foreign Investment in the US (CFIUS) was expanded in 2018. The intent is to prevent Beijing from acquiring US firms with advanced technology. CFIUS is especially anxious to prevent Beijing from acquiring US military related US companies.[49]

Beijing has tried to circumvent US export controls pertaining to advanced technology by instead acquiring the U.S. companies that create them. CFIUS is also on the lookout for Chinese attempts to acquire properties located within a certain proximity of important national security installations.[50]

57 Prohibition of U.S. FDI into China

Tellis adds that Team Biden has also tried to prohibit certain outward U.S. investments in semiconductors and microelectronics, quantum information technologies, and artificial intelligence capabilities going to China. The intent is to prevent the technologies relevant to military, intelligence, surveillance, or cyberspace operations, getting into the hands of Chinese nationals and in turn aid the growth of China's technological capabilities.[51]

58 Rising Subsidies for US Industrial Policy

One of the most controversial policies of Team Biden has been its industrial policy. Tellis notes that the industrial policy was born of the conviction that the free-market solutions that drove globalization in previous years failed and what was needed was a massive provision of government funds.[52]

While these de facto efforts are well meaning, Tellis rightly argues that the dangers are not trivial. For instance, industrial policies have failed in the past. That's because it's difficult for Washington bureaucrats to connect labor, capital, and innovation. Moreover, Tellis adds that producing such success might end up being even more dangerous if it leads to the large-scale subversion of markets within the U.S. economy as a result of increased state dominance.[53]

There are also a number of other problems to consider. First, one of the biggest reasons China's economy is in decline is because they have abandoned much of the free market at the time of the global financial crisis and pursued their own industrial policy.

In addition, the US spent many years helping Eastern European economies make the transition away from industrial policies to the free market. Now Team Biden wants to abandon the free market and let the government call the shots. Why is this a problem?

The free market produces signals that tell a private company how best to use its resources. But unless these resources are privately owned, those signals will be ignored by socialistic, government dominated companies. In addition, when the government owns an enterprise, the mangers of the enterprise face no financial discipline because a soft state budget can always bail out sick and inefficient companies. In contrast, private entrepreneurs have an obvious reason not only to preserve the value of their assets, but to enhance their value.[54]

In addition, private companies need to rely on loans from commercial banks, not the government. Instead of the government increasing subsidies to companies, these companies need to be freed from government subsidies. Finally, those companies that are chronically unprofitable and cannot attract commercial loans need to go bankrupt so that factors of production can be reallocated to those more resilient companies that would benefit more from productive resources.[55] Workers should go to private companies that have a more promising free market future.

59 Friend-Shoring

Partly in an effort to mitigate the risks I have just mentioned regarding its industrial policy, Team Biden has declared its support for "friendshoring." Tellis says friend-shoring focuses on shifting existing supply chains (or parts thereof) that are currently anchored in China to countries that are either political allies or strategic partners of the United States.[56]

But the danger is that Team Biden is considering making decisions mainly for political reasons. That raises the question: When it comes to country evaluation and selection, can US multinational companies make better decisions than US government officials?[57]

Yes. Most US multinational companies already have competent organizations in charge of country evaluation and selection. If so, how does this market-led methodology work?

Because companies seldom have sufficient resources to exploit all opportunities, free market managers face two major considerations: Which markets should they serve? And where should they locate the production to serve those markets?

The business research team at BYU notes that company decisions on market and production location tend to be highly interdependent. That's because companies often need to serve markets from local production and because they like to use existing production capacity.[58]

Scanning techniques help managers consider alternatives that might otherwise get overlooked. Scanning techniques also help reduce the number of feasibility studies to a few of the ones that look most promising. Since each company has unique competitive capabilities and objectives, the factors affecting the choice of operating location will be different for each company.

That said, there are similar country comparative indicators when seeking advantages from foreign sales or foreign assets. Four broad categories of risk that companies may consider are political, foreign exchange, natural disaster, and competitive.[59]

When allocating resources among countries, companies need to consider a number of issues. These include how to treat reinvestments and divestments, the interdependence of operations in different countries, and whether they should follow diversification versus concentration strategies.

Companies may reduce the risk of foreign liability by moving first to countries more similar to their home countries. Alternatively, they may contract with experienced companies to handle operations for them, limit the resources they commit to foreign operations, and delay entry to many countries until they are operating successfully in one or a few. Companies also must develop location strategies for new investments and devise means of deemphasizing certain areas and divesting if necessary.[60]

Thus, most multinational corporations have this kind of logical, coherent market led decision-making for country evaluation and selection. In contrast, Team Biden's friendshoring is political and is vulnerable to crony capitalism abroad. Without market analysis Team Biden's friend-shoring is potentially dangerous and therefore arguably not needed.

60 Asian Push Back on Decoupling from China

Despite pressure from the US to decouple from China, research at the National Bureau of Asian Research (NBR) notes that Asian states have pushed back from decoupling from China. In other words, despite political and national security ties to the US, the value of Asian states maintaining strong trade and production ties with China has not diminished.

Tellis persuasively argues that that China is simply too big and too important, either as a supplier or as a market, for Asian states to cut ties with China. As a result, diversification from China may be the best approach. But even this solution has inherent limits where merchandise trade is concerned. That's because Tellis notes that China:

- Maintains its manufacturing leadership,
- Remains a vital supplier of many commodities or raw materials for which there are no economical substitutes,
- And, thanks to its size and growing prosperity, offers huge markets for goods produced by others at a time when there are few comparable alternatives.[61]

61 Country by Country Push Back and Why

With this in mind, excellent research by scholars connected to NBR provides a useful breakdown for a number of Asian countries that have economic ties with China:

Japan

Given Japan's strong military ties to the US, some observers argue that Japan might want to bandwagon with the US and totally decouple from China. However, Kristin Vekasi persuasively argues that despite Tokyo's strong desire to limit economic linkages with Beijing because of the latter's growing assertiveness, Japanese businesses still view China as indispensable to their economic viability.[62]

In an era where Japan's ubiquitous industrial presence is manifested not simply through branded products but by myriad internal (and invisible) components, the importance of supply chains that traverse China, among other countries, has only increased. Despite Japanese efforts to diversify its trade with Southeast Asia, Kristin Vekasi notes that its trade with China has increased. Japan simply cannot—and does not seek to—sever commercial ties with China.[63]

South Korea

South Korea is between a rock and a hard place when it comes to its close military ties with the US and the economic success it deeply owes to South Korea's connection to China, its largest export market. Yul Sohn and Hyo-young Lee argue that the integration between China and South Korea is particularly pronounced in the production of semiconductors—the latter's single-largest export item to China.

Yul Sohn and Hyo-young Lee also note that South Korea remains highly dependent on China for raw materials used in the production of other manufactured goods. They argue that it's especially difficult for South Korea to extricate itself from China, notwithstanding US pressure to decouple from China. Decoupling from China would also be difficult because South Korean exports to Southeast Asia—which also depend on Chinese intermediate products—are themselves rising.[64]

Decoupling from China would also be difficult because the structure of Chinese–South Korean trade is now essentially intra-industry trade. Yul Sohn and Hyo-young Lee conclude that any consequential decoupling from China is likely to lie beyond reach for some time to come.[65]

Taiwan

Given US military support for Taiwan and Taiwan's desire to be more politically independent of China, one might think that Taiwan would be especially anxious to decouple from China. But Taiwan's longstanding economic interdependence and shared prosperity with the mainland (cited earlier in our case study) cause Taiwan to push back against US efforts for it to decouple from China.

Syaru Shirley Lin notes that Taiwan's most significant contribution to the international economy—semiconductor manufacturing—is inextricably dependent on its commercial and technological integration with China.[66]

Syaru Shirley Lin also points out that the imperative for Taiwanese ICT firms—which contribute around 40% of the nation's GDP—to remain ensconced in China has only intensified. Taiwan's trade with China constitutes over 40% of its total trade, with both its bilateral trade and its trade surplus increasing to record levels.[67]

Admittedly, the dangers of excessive Taiwan economic dependence on China are well understood in Taiwan. The Taiwanese government has prevented advanced semiconductor manufacturing from moving to China and has acquiesced to U.S. pressure to create new semiconductor manufacturing facilities in the United States.

But Syaru Shirley Lin persuasively argues that the survival of the Taiwanese ICT sector, to include advanced semiconductor fabrication facilities on the island, simply depends on its continued presence in China. That's because Taiwanese manufacturers have no readily available alternative to economic ties to China. The fruits of globalization have thus made China the vital gateway for Taiwan even when the latter seeks to expand trade with wider global markets.[68]

India

Unlike the countries cited above, India's economic connection is relatively new. In the aftermath of its 1991 economic reforms, however, India's trading links with the world, and especially with the Asian economies, have expanded impressively.

Gulshan Sachdeva notes that one outcome of this new orientation has been growing Sino-Indian trade, with China catapulting to within the top two spots on the Indian trading roster. That said, political tensions have impelled India to pursue decoupling from China. In fact, New Delhi may be the only outlier of the Asian economies we discuss in the Indo-Pacific region that is deliberately attempting to limit economic connectivity with China.[69]

For all these efforts, however, Gulshan Sachdeva argues that autonomy from China is nowhere in sight. Many of the heavily promoted "Make in India" projects, including marquee initiatives such as Apple's manufacture of iPhones in India, still rely heavily on imported Chinese components, as does the otherwise successful Indian pharmaceutical industry, which cannot operate without active pharmaceutical ingredients sourced from China. Furthermore, the economics of much of India's import-substituting manufacturing are controversial, and the ambition to entice production exiting from China has not yet borne significant fruit.[70]

Southeast Asia

Southeast Asia remains a champion of globalization. The region's economic success with globalization is closely connected to international interdependence. As a result, Southeast Asian states are determined to protect their structure of trade. Vikram Nehru argues that this structure has witnessed the region's progressive integration into the China-centered production system.[71]

Frankly, the economic officials in the region are concerned that US China decoupling could destroy its shared prosperity with China. Consequently, Southeast Asian states have refused to become party to any policies aimed at eroding their economic integration with China. The dangers of further tariff wars, fragmenting standards, and technological cleavages are viewed as forcing unpalatable choices upon the regional states.[72]

Australia

Australia's economy has been deeply connected to China for many years. Darren Lim, Benjamin Herscovitch, and Victor Ferguson note that this connection was a natural outcome for an open economy that relied heavily on trade—especially exports of agricultural goods, minerals, and energy—for its growth.

China's demand for exactly these inputs made the country an important partner for Australia. Accordingly, Canberra sought to protect commercial ties with China, which had become Australia's largest two-way trading partner and an important source of foreign investment. Despite recent political and security tensions with China, Australia still remains economically intertwined with China.[73]

In fact, as the chart below shows, Australia's trade with China surged to record levels in 2023. Australia's total trade with China reached A$219bn (US$145bn) in 2023, the highest-ever level and up from A$168bn in 2019.[74]

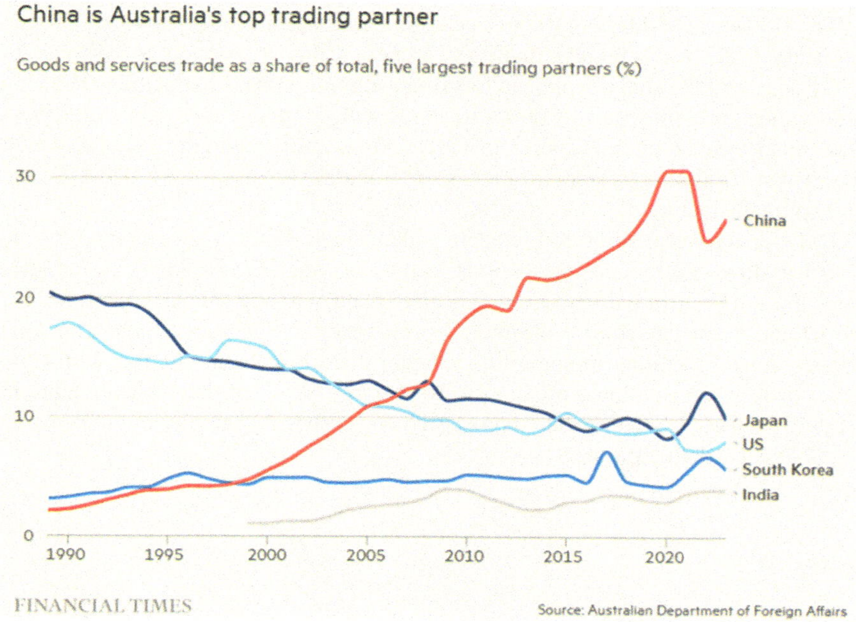

Lim, Herscovitch, and Ferguson argue that "the economic opportunities presented by China's internal market in terms of size and price premiums remain unparalleled. They conclude that the potential risk of politically motivated disruption recurring in the future does not justify the economic costs of a permanent exit from Australia's trading relationship with China."[75]

62 Conclusion

Based on US alarm over China's dominance in economic influence, the US created an Indo-Pacific Strategy. The US did not understand why China's economic influence was so dominant. The study used two mini-case studies to illustrate how China's shared prosperity promoted its economic influence. The first case study looked at Greater China. The second case study looked at Bangladesh and the competitive garment industry. This Bangladesh case study also explained how China's shared prosperity reduced the demand for violent extremism.

The study also argued that China's success reflected its longstanding democratization of Asian production, economic interdependence and shared prosperity.

Asians previously omitted from manufacturing benefitted by producing intermediate components of production. China also bonded with Asian partners in regional trade diplomacy. The study argued that the US economic influence was marginal due to its disinterest in regional trade diplomacy.

The US also kept thinking China's economic rise was a threat. The chapter argued that Team Biden needs better situation awareness. China's economy is too weak, not too strong. And the study explained lots of reasons why this is happening. The US is making this problem worse by doubling down with economic coercion. After World War I that kind of economic coercion led to German resentment, the rise of Hitler and World War II. Similarly, Harvard Professor Graham Allison says the US is destined for war against China.

IMF persuasively criticized President Biden's high tariffs as a financial risk to the global trading system and international economic stability. Similarly, Asian states were spot on to push back when the US put "Cold War" pressure on them to "pick the US side" and decouple from their close economic ties with China. The Asian states confirm the larger theme in this chapter that close economic ties with China are critical to their national prosperity. To increase US economic influence the US needs to promote US China shared prosperity and curb US economic coercion against China and its trading partners.

Notes

1. Economist Intelligence Unit, (EIU), Singapore retains its lead in business environment rankings, October 25, 2022.
2. EIU, Survey shows more divergent ASEAN views on US-China rivalry, February 21, 2022.
3. EIU, The US plans a long-term challenge to Chinese influence, February 16, 2022.
4. EIU, Indo-Pacific Economic Framework falls short, May 24, 2022.
5. Demetri Sevastopulo and Alex Rogers, Joe Biden halts plan for Indo-Pacific trade deal after opposition from Democrats, FT, November 14, 2023.
6. Ibid.
7. Ibid.
8. EIU, The future of the IPEF remains fraught with challenges, September 13, 2022.
9. Martin Sandbu, The other economic security threat from China, FT, 25 May 2023.
10. Ruchir Sharma, China's Rise is Reversing, FT, November 19th, 2023.
11. Hanming Fang, "Where Is China's Economy Headed?" by Hanming Fang in the Aspen Economic Strategy Group's 2023 policy volume, Building a More Resilient US Economy, Aspen, Colorado, 2023.
12. Ibid.
13. Ibid.
14. See Keith Johnson and Robbie Gramer, "The Great Decoupling," Foreign Policy, May 14, 2020, https:// foreignpolicy.com/2020/05/14/china-us-pandemic-economy-tensions-trump-coronavirus-covid new-cold-war-economics-the-great-decoupling.
15. Ashley J. Tellis, Interdependence Imperiled? Economic Decoupling in an Era of Strategic Competition, from Ashley J. Tellis et al., edited, Reshaping Economic Interdependence in the Indo-Pacific, National Bureau of Asian Research (NBR), Seattle, 2023, p. 14, and "Economic Decoupling: Our New Reality?" American Chamber of Commerce in France, December 2021.
16. Johnson and Gramer, "The Great Decoupling" and "Decoupling—Severed Ties and Patchwork Globalisation," European Union Chamber of Commerce and MERICS, 2021.

17. Courney Kube, US and Chinese military leaders finally speak to each other, NBC News,21 December 2023.
18. Ashley Tellis, op. cit.
19. Charles Kindleberger, The World in Depression, 1929–1939 (Berkeley: University of California Press, 1973).
20. Ashley Tellis, op. cit.
21. Victor Fung, William Fung, and Yoram Wind, Compering in Flat World, Prentice Hall, New York, NY, 1979.
22. Ibid.
23. The data for this Greater China case study comes from Taiwan, Hong Kong and Mainland China sources. I will also mention other sources as we go along. From time to time there are discrepancies between Taiwan and Beijing's statistics. For instance, there are occasional discrepancies from the indirect trade and investment between Taiwan and the mainland. Since most Taiwanese businessmen made their investment on the mainland through Hong Kong and another third country, their records of investment are generally excluded from the official Taiwan statistics, while they are included in Beijing's statistics.
24. See Denny Roy, Taiwan, A Political History," Cornell University Press, 2003, p. 150.
25. See Chu-yuan Cheng, Economic Relations across the Taiwan Straits: Mutual Dependence and conflicts," in Winston L. Yang and Deborah A. Brown, ed., Across the Taiwan Strait: Exchanges, Conflicts and negotiations (New York: Center for Asian Studies, St. John's University, 1999); and Board of Foreign Trade, Ministry of Economic Affairs.
26. The cumulative statistics on Taiwanese investment on the mainland vary. For example, nongovernment sources estimate that Taiwan actually invested well over $140 billion on the mainland.
27. See John Thacik, "Taiwan Dependence: The Strategic Dimension of Cross-Strait Trade and Investment," in Andrew Scobell, ed., The Costs of Conflict: The Impact on China of a Future War, Strategic Studies Institute, October 2001, pp. 35–62.
28. The Economist, "Garments in Bangladesh – Knitting Pretty: The Clothing Business Is Flourishing Despite Chinese Competition", 16 August 2007.
29. Vikas Bajaj, "Bangladesh", The New York Times, 16 July 2010.
30. Vikas Bajaj, "With Low Pay, Moves in on China", The New York Times, 16 July 2010.
31. Kenneth Rapoza, "Is the Made in China Clothing Label a Thing of the Past?", Forbes, 11 October 2015.
32. Ibid.
33. EIU, "Cutting into China's Apparel Market", 18 October 2016.
34. Ashley J. Tellis, Interdependence Imperiled? Economic Decoupling in an Era of Strategic Competition, from Ashley J. Tellis et al., edited, Reshaping Economic Interdependence in the Indo-Pacific, National Bureau of Asian Research (NBR), Seattle, 2023, p. 14, and "Economic Decoupling: Our New Reality?" American Chamber of Commerce in France, December 2021.
35. Ibid.
36. See my study on the energy war between Russia and Europe in Rosenberger, L. (2023). The Energy War: How Europe Turned the Tables on Russia. In: Farhadi, A., Grzegorzewski, M., Masys, A.J. (eds) The Great Power Competition Volume 5. Springer, Cham. https://doi.org/10.1007/978-3-031-40451-1_13.
37. Ashley Tellis, op. cit. and Pinelopi K. Goldberg and Tristan Reed, "Is the Global Economy Deglobalizing? And If So, Why? And What Is Next?" Brookings Institution, Brookings Papers on Economic Activity, Spring 2023.
38. See my other chapter in this volume: Identifying Risks from US China Decoupling If Trump Wins.
39. Ashley Tellis, Interdependence Imperiled?, op. cit.
40. David Lawder, IMF: U.S. should keep open trade, work with China to resolve disputes, Reuters, 16 May 2024 and Eric Martin, IMF Knocks Biden's China Tariffs as Risk to US, World Growth, Bloomberg, 16 May 2024.
41. Ibid.

42. Ashley J. Tellis, Interdependence Imperiled? Economic decoupling in an era of Strategic Competition, Reshaping economic interdependence in the Indo-Pacific, edited by Ashley Tellis, et al., Seattle, The national Bureau of Asian Research, 2023, p. 19.
43. Ibid.
44. Brian J. Egan et al., "China Faces Existing and Expanded U.S. Restrictions on Trade, Investment and Technology," Skadden, January 19, 2022.
45. David Lawder, IMF: U.S. should keep open trade, work with China to resolve disputes, Reuters, 16 May 2024 and Eric Martin, IMF Knocks Biden's China Tariffs as Risk to US, World Growth, Bloomberg, 16 May 2024.
46. See "Biden Administration Restricts U.S. Exports of Advanced Computing and Semiconductor Manufacturing Equipment, Software, and Technology to China," Dorsey, November 28, 2022.
47. "Remarks by National Security Advisor Jake Sullivan on Renewing American Economic Leadership at the Brookings Institution," White House, April 27, 2023.
48. See Hugo Meijer, Trading with the Enemy: The Making of U.S. Export Control Policy toward the People's Republic of China (New York: Oxford University Press, 2016).
49. Martin Chorzempa, "U.S. Security Scrutiny of Foreign Investment Rises, but So Does Foreign Investment," Peterson Institute for International Economics, September 1, 2022.
50. Christian C. Davis et al., "CFIUS Expands Jurisdiction Over—and States Restrict Chinese Investment in—Real Estate," Akin, May 19, 2023.
51. Karen Freifeld, "Biden Order Curbing Investment to China Expected Next Week—Sources," Reuters, August 4, 2023.
52. Ibid.
53. "The Biden Administration Embraces Place-Based Industrial Policy," Economist, July 27, 2023.
54. See Leif Rosenberger, Economic Transition in Eastern Europe: Paying the Price for Freedom, East European Quarterly, XXVI, No. 3, September 1992.
55. Ibid.
56. "Remarks by Secretary of the Treasury Janet L. Yellen on Way Forward for the Global Economy," U.S. Department of the Treasury, April 13, 2022.
57. The difficulties of engineering successful friendshoring are clearly explicated in "Globalization 2.0: Can the U.S. and EU Really 'Friendshore' Away from China?" Allianz, October 5, 2022.
58. This section draws from the longstanding BYU research team at BYU. For updated information, see John D. Daniels, Lee H. Radebaugh, Daniels P. Sullivan and Reid W. Click, International Business: Environments and Operations, 17th Edition, Pearson, New York, 2021.
59. Ibid.
60. Ibid.
61. Yukon Huang and Genevieve Slosberg, "China's Response to the U.S. Trade War," China Leadership Monitor, June 1, 2 023, 8.
62. Kristin Vekasi, Japan's Approach to Economic Security and Regional Integration, NBR, 2023.
63. Ibid.
64. Yul Sohn and Hyo-young Lee, South Korea's Economic Statecraft: Between Interdependence and National Security South Korean Perspective, NBR, 2023.
65. Ibid.
66. Taiwan: Walking the Tightrope between the United States and China, NBR, 2023.
67. Ibid.
68. Ibid.
69. Gulshan Sachdeva, India's Reluctant Participation in the Evolving Indo-Pacific Economic Architecture, NBR, 2023.
70. Ibid.
71. Vikram Nehru, US-China Economic Rivalry in Southeast Asia and its Implications, NBR, 2023.
72. Ibid.
73. Darren Lim, Benjamin Herscovitch, and Victor Ferguson, Australia's Reassessment of Economic Interdependence with China, Australia's Perspective, NBR, 2023.

74. Nic Fildes, Australia's trade with China surges to record level after tariffs lifted, FT, 17 June 2024.
75. Lim et al., op. cit.

China and Afghanistan: Navigating Geoeconomic and Geopolitical Pathways at a Time of Great Power Competition

Omar Samad

Abstract The expanding "odd couple" bilateral relationship between China and Afghanistan, neighbors bound by a sliver of geography and limited historical ties, finds itself at a critical juncture following the NATO withdrawal from Afghanistan in 2021. Not only does it coincide with the rise of great power economic competition and a multipolar world order, but also with a tilt in some countries toward authoritarianism, rise of ethno-religious tensions, various forms of terrorism and rivalry for access to raw materials. Relations between China and impoverished Afghanistan—sitting on trillions of dollars of untapped minerals—will be watched closely by serious strategists, especially in America. China considers Afghanistan as a potential market and pit stop among many for its landmark Belt and Road Initiative (BRI) projects, offering access to valuable resources and a strategic land route to South-Central Asia and beyond. This paper explores the following key questions: (1) What are China's vital national security, geopolitical and geoeconomic interests and priorities regarding Afghanistan since the return of the Taliban to power in 2021? To what extent does the BRI/CPEC play a role? (2) Unrecognized as a de jure state, can the de facto Taliban government maintain equilibrium in its relations with great powers? What could change that formula? (3) China and the US say that they do not see Afghanistan as a space for great power competition. Is that practical and what could change that? The analysis will aim to contribute to a nuanced understanding of the challenges and opportunities both China and the United States face as global economic giants in navigating a region caught between economic aspirations and geopolitical complexities. This chapter will briefly look at the trajectory and history of China-Afghanistan relations, current Chinese interests in Afghanistan, Afghan interests in China, the role of Afghanistan in the great power competition agenda, and the various dimensions (security, geo-strategic, geo-economic, regional, and bilateral) of the Sino-U.S. interactions.

Keywords China · Afghanistan · Belt and Road Initiative (BRI) · CPEC · Geoeconomics · Geopolitics · Taliban · Islamic Emirate of Afghanistan (IEA) ·

O. Samad (✉)
Atlantic Council, Fairfax, VA, USA
e-mail: omarsamad@gmail.com

A. Farhadi et al. (eds.), *The Great Power Competition Volume 6*,
https://doi.org/10.1007/978-3-031-70767-4_14

Great power competition · South-Central Asia · Sanctions · Terrorism · Shanghai Cooperation Organization (SCO)

1 Introduction

On the surface, it can be described as an "odd couple" bilateral relationship, but China and Afghanistan, neighbors bound by a sliver of geography and limited historical ties, find themselves at a critical juncture following the NATO withdrawal from Afghanistan in 2021, coinciding with the rise of great power economic competition that may contribute to the emergence of a multipolar world. Some experts believe that a bipolar order is already in place. Furthermore, the specter of authoritarianism, the expansion of ethno-religious tensions in some parts of the world and rivalry for access to raw materials, are contributing to a new phase of competition in the years to come. Relations between economic powerhouse China and Afghanistan, one of the world's poorest (yet rich with untapped minerals) and most war-torn countries, will be watched closely by strategists and regional experts.

The tumultuous events unfolding in Afghanistan since state collapse in 2021, exacerbated by the Taliban's second takeover, the withdrawal of US forces, and the ensuing humanitarian crisis, have shifted geo-political alignments, and raised new questions for traditional aid donors and other regional stakeholders, including China. This chapter examines the complex relationship between China and Afghanistan, arguing that it is defined by national interest driven by the interplay of security concerns, geoeconomics and geopolitics. While some analysts see the interactions as an emerging dimension of great power rivalries, others are less scrupulous and attribute them to risks and opportunities at the periphery and regional level.

Regardless, one question that keeps strategists preoccupied is whether economic bloc competition over resources, access and geoeconomic foot-printing will impact geopolitical dynamics and rivalries or vice versa?

On the one hand, China views the war-torn nation as a potential market and pit stop for its Belt and Road Initiative (BRI) projects, offering access to valuable resources and a strategic land route to South-Central Asia, Iran, the Caucasus, and the Gulf region. However, achieving these goals hinges on political stability and security in Afghanistan, a challenge underscored by sanctions imposed by the United Nations and lack of international recognition of the Taliban's de facto regime, also known as the Islamic Emirate (IEA).

On the other hand, geopolitical considerations shape China's steady engagement with Kabul. The rise of the Taliban raises old concerns about separatism, extremism, and potential spillover effects on China's own security. China must also navigate the complex web of regional and global powers with competing interests in Afghanistan, including Western powers, Russia, India, Türkiye, and Iran, among others. This paper explores the following key questions:

1. What are China's vital national security, geopolitical and geoeconomic interests and priorities regarding Afghanistan since the return of the Taliban to power in 2021? To what extent does the BRI/CPEC play a role?
2. Unrecognized as a de jure state, can the de facto Taliban government maintain equilibrium in its relations with great powers? What could change that formula?
3. China and the US say that they do not see Afghanistan as a space for great power competition. Is that practical and what could change that?

By addressing these questions, this paper sheds light on the intricate dynamics shaping Sino-Afghan relations in a challenging neighborhood. It also looks at these bilateral and regional dynamics in the framework of East–West or great power rivalries at a time when tensions are high over conflict between Russia and Ukraine, the Israeli-Palestinian conflict and the latest clashes between Israel and Iran. These unfolding developments will undoubtedly have some bearing on state-to-state relationships, alignments, and future trends in the region and beyond.

The analysis will aim to contribute to a nuanced understanding of the challenges and opportunities both China and the United States face as global economic giants in navigating a region caught between economic aspirations and geopolitical complexities.

For Afghanistan, given its vulnerabilities, China could either become a regional anchor or an inevitable burden as the country puts behind almost five decades of conflict and instability, and attempts to come out of isolation while balancing its foreign, security and economic interests.

This chapter will briefly look at the trajectory and history of China-Afghanistan relations, current Chinese interests in Afghanistan, Afghan interests in China, the role of Afghanistan in the great power competition agenda, and the various dimensions (security, geo-strategic, geo-economic, regional, and bilateral) of the Sino-U.S. interactions.

2 A Brief History of Sino-Afghan Relations

2.1 Pre-revolution China

Dating back to the Han dynasty, earlier contacts between traders in today's Afghanistan and Chinese merchants were established as part of the lucrative Silk Road. The Hans protected parts of the Silk Road trade through Central Asia and today's Northern Afghanistan (Dai 1966). In pre-Islamic times, parts of today's Afghanistan were followers of Buddhism. Chinese records refer to Kabul as "Kao-fu," a flourishing urban center situated in the Hindu Kush mountains, strategically positioned along the trade routes connecting Central Asia with India (Dai 1966). A Buddhist monk from what is now Afghanistan passed through China in 2 BC and converted the first Chinese to Buddhism (Dai 1966). Buddhist sites in

Northern and Central Afghanistan, especially in Bamyan, attracted many pilgrims form China. Chinese accounts credit the first glassmaking in China to Afghan artisans, dating back to 424–428 AD (Dai 1966).

In the 7th century, under the Tang dynasty, Chinese rulers used military means to expand their sphere of influence to parts of Central Asia and beyond, encompassing protectorates in Soghd, Ferghana, Tashkent, Samarkand, Bukhara, Herat, Balkh, and Kabul (Dai 1966). Indian historian K. P. S. Menon writes those Chinese rulers "performed the remarkable feat of sending an army of 100,000 men which marched up the Pamirs from Kashgar and then occupied the Hunza Valley, part of today's Pakistan" (Menon 1958 as cited in Dai 1966). These conquests ushered a period of prosperity in China as the Silk Road gained prominence across a vast region stretching from China to the Mideast and beyond.

This Westward expansion into a vast region called Takharistan occurred during the defeat of Western Turk and Sassanid dynasties at the hand of Arabs (Dai 1966). As empires came and went, including the Arabs and later the Moguls, the geopolitics of the region kept shifting as new land and maritime trading routes emerged between East and West. The arrival of Western colonial powers in Asia changed the landscape to this day.

The first signs of tension between Chinese and Afghan rulers emerged in the 18th century during the Qing dynasty. Tensions flared up in the greater Turkistan region, including today's Xinjiang region, between the emperor and neighboring Muslim rulers (Schluessel; UBC 2024). Further South, the powerful Pashtun ruler Ahamad Shah Durrani (also known as Ahmad Shah Abdali), dissatisfied with Chinese expansion into parts of Central Asia, dispatched an envoy to the Kazakh Sultanate to pledge protection to Kazakh and Kyrgyz Emirs against Chinese threats (Newby 2005). In 1762, a year after having defeated the Maratha rulers at Panipat in Northern India, an emboldened Durrani demanded Chinese withdrawal from Muslim lands in a letter to the Qing court (Newby 2005). He also dispatched troops to Khokand and Tashkent as preventive measures in case of a surprise attack by Qing rulers. At one point, Afghan and Chinese forces were facing each other not far from Kashgar (Newby 2005) (Image 1).

Despite tensions, a full-blown conflict between Afghanistan and China never erupted. In a diplomatic shift, an envoy of Afghan representatives presented Emperor Qianlong with four magnificent horses as a gift in 1763 (Newby 2005). These horses later became the subject of Giuseppe Castiglione's renowned series of paintings, "Four Afghan Steeds." However, the envoy's refusal to perform the *kowtow*, a customary bow symbolizing respect, left a poor impression on Qianlong (Newby 2005). The breach of manners, along with the Qing court's perception of Afghan diplomats as troublemakers, led to discouragement of further envoys from Ahmad Shah's court (Newby 2005). Qianlong further expressed his disapproval the Durrani conquest of India in a letter to Ahmad Shah, asserting his own authority as the

Image 1 Lingkunbai, from the Four Afghan Steeds, gifted to Qianlong (Ragazzi 2014)

ruler who oversees all within and beyond the empire, dispensing justice accordingly (Newby 2005). Finally, Ahmad Shah backed the Khan of Bukhara to invade Badakhshan, and killed the ruler who had betrayed the Muslim population to the Qing. Not wanting to confront the Afghan ruler, the Chinese did not intervene nor try to save their protectorate (Newby 2005; Ragazzi 2014).

2.2 *Post-revolution China*

Thereafter, relations remained almost dormant for many years until the 20th century. Taking China's side after the Japanese attacks during WWII, Afghanistan became one of the first countries to recognize Communist China after the 1949 revolution (Dai 1966). Diplomatic relations were established thereafter as veteran diplomat Abdul Samad (the author's grandfather) was appointed as first Afghan ambassador to Beijing in 1955.

Chinese Premier Zhou Enlai (middle) welcomes Afghan ministerial delegation to Beijing, 1956. First Afghan Ambassador Abdul Samad is on his left. (courtesy of Embassy of China in Kabul)

Premier Zhou Enlai was the first PRC leader to visit Afghanistan in 1957. Both Prime Minister Mohammad Daoud (1957) and King Zahir Shah (1964) visited China, however they were unable to attract Chinese support for the Pashtunistan cause against Pakistan (Dai 1966). The Sino-Soviet split and rivalry was also viewed negatively by both countries as Kabul had established close ties to Moscow by the late 1950s. Nonetheless, a treaty of friendship and non-aggression, and a trade agreement were signed between the two countries in 1960.

King Mohammed Zahir Shah and the visiting Afghan delegation with Mao Zedong, Liu Shaoqi, Zhou Enlai and the Chinese leadership in Beijing on November 1, 1964. On 22 November 1963 Beijing and Kabul signed the Boundary Treaty. This treaty settled the territorial dispute over the Afghan-controlled Wakhan on the border between Badakhshan Province in Afghanistan and the Xinjiang Uyghur Autonomous Region in China

Beijing did not change its position vis-à-vis Kabul as former Premier prince Daoud staged a bloodless coup against his cousin Zahir Shah in 1973 and established a republican system. However, the ensuing bloody 1978 military coup d'etat in Kabul that in turn overthrew Daoud and brought two pro-Moscow factions to power, rang alarm bells in Beijing. Watching Afghanistan slide into civil war, Beijing was caught off guard when the USSR deployed military forces invaded Afghanistan in December

1979. Within days, the Chinese government denounced the Soviet military invasion and refused to acknowledge the Soviet-backed government. Diplomatic relations were suspended, degrading the Chinese embassy to a representative office that dealt solely with consular and visa issues.

To make matters worse, the pro-Soviet government in Kabul supported China's rival, Vietnam, and blamed Beijing for supporting Afghan anticommunist Mujahedeen militants. China decided to actively back the Afghan fighters headquartered in Pakistan and Iran, ramping up their military presence near the Xinjiang border crossing (Starr 2004). Ironically, both China and the U.S. covertly supplied weapons through Pakistan to the anti-Soviet Afghans. Both countries outsourced the management of the arms and money pipeline to Pakistan's powerful military leaders.

As China increasingly saw the Soviet presence in Afghanistan as a threat, it pushed for Soviet withdrawal as one of the conditions for any détente in Sino-Soviet relations (Qaiser 1987). This period also saw some unexpected rapprochement between Washington and Beijing on select issue, including Afghanistan. The Soviet military withdrawal in 1989, and the ensuing breakup of the Soviet Union, the collapse of the Warsaw Pact, as well as the reunification of Germany and regime changes in Eastern Europe, not only paved the way for new relations with Moscow but also, as part of a new world order, resulted in the emergence of America as the sole superpower. This new reality, coupled with drastic changes taking place across China, accompanied by an economic boom, opened a new chapter in great power relations as China not only strived to compete globally, but also put the U.S. on notice about its own strategic imperatives such as Hong Kong, Taiwan and threat perceptions along its borders.

In the 1990s, as Afghanistan experienced continued civil strife, the lesser-known yet more conservative Mujahedeen-turned-Taliban with sanctuaries in Pakistan, gained the upper hand. However, Beijing adopted a more hands-off approach as it suspected Islamic ties developing between the Talibs, who were hosting militant groups such as al Qaida, and members of East Turkistan Islamic separatist movement who had traveled to Afghanistan via Pakistan. Fearing chaos spilling across China's Western borders along the Xinjiang province, China sent an envoy to Kandahar to meet with the reclusive Taliban leader, Mullah Mohammad Omar. The Chinese diplomat was the first senior foreign official from a non-Muslim country to meet him (Singh 2016).

2.3 Post 9/11 Relations

The al Qaida attacks on September 11, 2001, on the U.S. changed the world and international dynamics altogether. China supported NATO efforts to pursue al Qaida and attended the Bonn Conference on the future of Afghanistan as a regional stakeholder. Relations between the new interim Afghan government and China were reestablished and China, mindful of American and Western military presence in the country, started to pursue a low profile yet engaging policy towards its Western neighbor. The focus initially shifted on targeted assistance for reconstruction and some capacity building projects, but over time turned toward potential for investment in resources and mining. By 2020, China became Afghanistan's top trading partner (OEC 2024).

Watching the Taliban insurgency grow over time, the trajectory of Western military engagement in Afghanistan, the decision to end combat missions by 2015 and to maintain a smaller footprint in the country, prompted the Chinese to initiate backchannel contacts with the Taliban, most probably facilitated by Pakistani intermediaries.

Relations between 2015 and the Afghan "republic's" collapse in August 2021 saw an odd mixture of multilateral attempts through trilateral and regional platforms, such as the Shanghai Cooperation Organization, The Heart of Asia—Istanbul initiative and others, to seek a durable solution to Afghanistan's growing instability and uncertain future as Taliban strength grew (Heart of Asia 2024).

Other contributing factors that undermined the trillion-dollar Western efforts to rebuild Afghanistan and prompted regional countries—including China—to adopt more hedging tactics, included systemic corruption, failures in implementing rule of law and democracy-building, mounting civilian and military losses, and bad governance. These setbacks created deep mistrust of Ashraf Ghani's kleptocratic administration, adding war fatigue and a changing global agenda to the mix, especially as al Qaida became a lesser threat following Osama bin Laden's killing by US special forces in Abbottabad, Pakistan in 2011.

Meanwhile, China continued to court Afghanistan for mineral contracts, including the famed copper mine in Aynak, oil explorations in the North and, more recently, showing interest in lithium deposits. None of the larger scale projects were kicked off because of insecurity, corruption and infrastructure-building costs.

Avoiding any type of military involvement, during this critical period, China played a wait-and-see game as the winds shifted and the U.S. started formal talks with the Taliban on withdrawal and a transition to an interim period through talks in Doha, Qatar in 2018. China was not consulted by the U.S. during this period as much as other countries, except through the *Moscow Format* meetings on Afghanistan (Zakharov 2023) but maintained close watch on developments that eventually led to an agreement in 2020 that was signed between the US and the Taliban, assuring NATO withdrawal by 2021 and a pledge by the Afghan rebels to prevent any terrorist activities from Afghan soil targeting American and allied interests (Maizland 2020). China welcomed the decision but remained concerned about the fallout. For their part, the Taliban, mindful of East–West tensions, fostered closer relations with the Chinese, while adopting a balanced foreign policy posture to prevent any future entanglements.

Following the collapse of Kabul into Taliban hands on Aug. 15, 2021, China adhered to the international consensus formed at the United Nations to not recognize the new rulers, who raised the Islamic Emirate of Afghanistan (IEA) flag, as the new de jure government. However, in September 2023, China's new envoy to Kabul presented his credentials to the IEA Premier Mullah Hassan Akhund. By December, the Chinese accepted to recognize Bilal Karimi as the new Emirate Ambassador to Beijing. Despite the de facto status, China went a step further and in January 2024, Karimi was allowed to present his Letters to President Xi Jinpin (Dawi 2024).

This was a first for the Taliban since returning to power in 2021. Although "the Chinese Ministry of Foreign Affairs was quick to issue a statement, clarifying that the acceptance of diplomatic credentials did not signal Beijing's official recognition of Afghanistan's current rulers." It sent a clear message to all other parties that China was only one beat away from formal diplomatic recognition of the ostracized Afghan regime. Washington's reaction was muted and dismissive. Others in the region have thus far adopted their won bilateral relationship based on their interests short of recognition.

In Dec. 2023, China and Russia broke ranks and abstained on a vote at the UN Security Council for Resolution-2721 that called for the appointment of a Special Envoy for Afghanistan, responsible for coming up with a new roadmap and addressing a host of issues that have prevented recognition of the Afghan regime by the UN and imposed severe sanctions after the collapse of the Republic (United Nation 2023). Chief among international criticism are Taliban restrictions on girls' education, female employment bans and suspension of intra-Afghan talks. At a UN-sponsored stakeholders conference held in Doha in Feb. 2024 (United Nation 2024), international rifts within the community of nations on Afghanistan were more evident as the IEA, mindful of its stature, did not attend as an Afghan side, and China joined by Russia and Iran, supported Kabul's stance. It is expected that the UNSC would reconsider a new envoy appointment, but it is not yet clear how Kabul and some of its regional backers might react.

3 Sino-Afghan Vital National Security, Geopolitical and Geoeconomic Interests and Priorities

According to Zhao Hong, Afghanistan linked to the China-Pakistan Economic Corridor (CPEC), can help China implement its 'March West' strategy to 'expand economic and strategic influence in Central Asia, the Middle East, and beyond (2013). On the other hand, regional instability can threaten China's Belt and Road Initiative (BRI) plans that include CPEC as one pathway. Security and stability are essential elements of Chinese growth and implementation of the BRI.

3.1 The Security Paradigm

Contrary to common perceptions, China's primary interest in, and concern about Afghanistan, are stability and counter militancy. For now, expanding trade, business and investment opportunities come a close second. This scenario might change rapidly, when these concerns are satisfactorily alleviated, and stepping away the position adopted by many Western donors, could pave the way for recognition of the IEA by China and other regional states. We are already seeing signs of a calculated Kabul tilt toward Beijing, primarily because of more pronounced engagement that has thus far resulted in expanded trade, investment, and diplomatic contacts between the two capitals.

On the security and anti-militancy fronts, following discussion in Beijing with PRC officials, Dr. Barney Rubin of New York University wrote: "China's most important security interests in Afghanistan are the related concerns of (1) securing Xinjiang from the three evils (separatism, extremism, and terrorism) and (2) promoting regional security to establish an enabling environment for the Belt and Road Initiative (BRI). These are related, as CPEC (China-Pakistan Economic Corridor) has been attacked by Baloch separatists in Pakistan, and the East Turkestan Islamic Movement (ETIM—the umbrella name used by Beijing for Uyghur militants) has been involved in attacks on Gwadar" (2019). Prior to 2022, China claimed that ETIM enjoyed safe havens in both Afghanistan and Pakistan. With the Taliban's second coming, Beijing seems more confident that the Emirate rulers attach greater importance to bilateral relations and are less inclined to allow anti-China activities to find a base on Afghan territory.

For their part, Taliban leaders have on many occasions reiterated their desire to pursue a balanced foreign policy and avoid entanglements. The IEA tilt to one side or another adheres to a calculated balancing act. Although some countries, institutions and research groups in the West are still skeptical, but Taliban leaders claim that they will stick by their commitments made in the Doha Agreement of 2020, preventing any transnational terror group, especially those threatening Western interests, from using Afghanistan as a base.

Although the Taliban have—with sporadic exceptions—a firm grip on security across the country, however, the record of attacks on Chinese citizens is not entirely without blemish. There have been few instances of terror attacks claimed by ISKP (Islamic State Khorasan province) an offshoot of IS-central, on Chinese citizens working in Afghanistan. In December 2022, Islamic State Khorasan bombed a hotel in Kabul used by Chinese workers just days after the China had asked Taliban officials to increase security around Chinese sites (Yawar 2022). It is estimated that despite Taliban efforts, ISKP may still be trying to recruit fighters among former or current ETIM members.

As far as Beijing is concerned, most meetings with Chinese officials include a call for the Taliban to take action against ETIM supporters in the Af-PAK region. While there were initial reports after the fall of Kabul that the Taliban had moved some groups away from Chinese border areas, the outcome is still not clear. To bolster security and open a new trade route crossing the Northeastern Wakhan corridor part of the Afghan Pamir Mountain range, the Taliban caretaker government is, for the first time, actively funding the construction (Ziwen 2024) of what could become a strategic road linking Badakhshan to the Xinjiang province.

China has a particular focus on Badakhshan in both Afghanistan and Tajikistan, as well as the 596 km long border with Pakistan's Gilgit-Baltistan in the Northern Area along the Karakoram range. However, China has no military bases in Badakhshan. China occasionally participated in a security quadrilateral with

Pakistan, Afghanistan, and Tajikistan, which was led by the military of the four countries. It also participated in a trilateral without Pakistan, which is more effective, as Afghanistan-Pakistan relations usually complicate matters. China was at one point training a unit of Afghan border and security forces under the Republic stationed in Badakhshan, as well as Tajikistan military officers.

Having no military bases in the region, China at times tried to maintain close ties with Kabul for bilateral security cooperation and counterterrorism, primarily to counter ETIM and ISKP. "Through military assistance, China helped Kabul build its military mountain brigade in the Wakhan corridor near Badakhshan province with the primary goal of preventing infiltration by the Islamic State into China. according to Afghan researchers, China provided more than $70 million in military aid to Kabul from 2016 to 2018" (Sun 2020).

According to many analysts, ISKP is now viewed as a growing threat to regional security and stability as it competes with both the Taliban and other more classical groups like al Qaida. ISKP, a Takfiri Sunni extremist group. While trying to draw dissatisfied recruits from other groups, the IS network also poses a serious threat to predominantly Shi'a Iran and Russian interests in the Mideast, Central Asia and the Caucuses.

While the United Nations' multi-national Analytical Support and Sanctions Monitoring Team report to the Security Council in 2023 (UNSMT Letter to UNSC, dated 2 May 2023, S/203/370) claims that "there are indications that Al-Qaida is rebuilding operational capability," the U.S. ODNI, in contrast, in its 2024 report said:" While al-Qaida has reached an operational nadir in Afghanistan and Pakistan," ISIS continues to pause a threat to global security. (*Annual Threat Assessment of the US intelligence Community*—Office of the Director of National Intelligence, Feb. 2024.)

The Taliban, who have been battling ISKP for the last decade, claim to have substantially degraded the group's capabilities inside Afghanistan. However, the more internationalist Islamic State is accusing the more nationalistic Taliban of treason to the greater cause. They claim to attract new recruits from among disgruntled and more radical Islamist elements, many of whom are of Central Asian origin. Although there are, as of this date, still conflicting and unsubstantiated data available on the March 2024 attack on the Crocus City Hall near Moscow, by elements claiming to be affiliated to ISIS, there is proof that the attackers were of Tajikistani origin. Other similar attacks inside Afghanistan and even a deadly in Kerman, Iran, in January 2024, was claimed by the same outfit, included Central Asians, even though there is little evidence presented to point to the real source of the plotting against both Russia and Iran.

3.2 *The Intersection of Geopolitics and Geoeconomics*

Now that the insurgent Taliban have regained power and ended Afghanistan's protracted civil war, the country seems to be almost entirely war-free (with few exceptions of ISKP or ex national security force bombings or opportunistic attacks) after almost 45 years of continuous conflict, which included at different intervals, direct military intervention by the world's two superpowers (the USSR from 1979 to 1989, and the US/NATO from 2001 to 2021). Despite unpopular edicts restricting female education and employment, and Taliban penchant for monopolizing power, the country is for the first time since 1973 under one rule and relatively stable. Most Afghans and regional actors are now of the view that the country sorely needs to remain peaceful and stable, thus emerge as a natural regional trade, transit and investment hub. The net result could well be a win–win outcome for most stakeholders.

In other words, lowering geopolitical heat and rivalries in and around Afghanistan that could lead to proxy warfare, zero-sum games and lose-lose scenarios, can harness the unexplored potential that the country offers overground and underground (as in terms of natural resources and mining) to instigate geoeconomic gains leading to higher wealth generation and productivity along the old Silk Road.

There is a realization that despite the goodwill that existed in some quarters, the spectacular 20-year-long Western investment (to the tune of $2 trillion) in a war economy (Almukhtar and Nordland 2019)—instead of using leverages, resources and opportunities to lay the foundation for a political settlement, reconciliation, assuring fundamental rights and investing in critical infrastructure, good governance, education and other strategic objectives—went astray, fueled corruption or was wasted. Some individuals, groups, corporations, and contractors benefited, however, with the exception of gains in few areas, the net result in terms of infrastructure building, productivity and growth in the country expected after 20 years and $2 trillion is minimal.

It is in such an environment that geopolitical priorities either align, overlap, or clash with geoeconomic realities. They either lead to collaboration, cooperation, subversion, or confrontation. The trajectory of Sino-Afghan relations since the end of the 20-year-long Western presence and expenditures in Afghanistan in August 2021, is in upswing mode. The Chinese, mindful of history, geography, long-haul strategy, overseas militarism, and national priorities, are not only cautious, but also selective by applying due diligence conditions, trying to avoid unrealistic expectations.

A. Old and New Silk Roads

The following areas of infrastructural and transportation initiatives are being considered as the situation in Afghanistan stabilizes:

- The Wakhan Corridor: The IEA has openly embraced the modern-day reincarnation of the old Silk Road and asked for Chinese contributions. Opening the adjoining corridors will need Chinese approval to increase the level of trade between the two countries and become a reliable route for the transit of goods. The ongoing construction of the Wakhan corridor road by the Taliban is such an infrastructural project, even though the Chinese side has yet to fully commit. There are also signs that the Pakistan side may view this particular road as a shorter rival route between China and the Mideast. It is estimated that building such a corridor would be costly and will also need funds to provide security, transportation, and logistical services.

- The BRI-CPEC project worth more than $62 billion could extend to Afghanistan. As of 2013, the focus is on using the Pakistan corridor that includes key infrastructure projects linking the Karakoram Highway to the Gwadar port on the Indian Ocean. China and Afghanistan continue to discuss the connector to the West via existing road and future rail links (Britannica 2024).
- The Jalghoza "Pine nut" corridor: The "pine nut air transport corridor" was launched in 2018 as an air freight route connecting the Afghan supplier to the Chinese market (The Peoples Republic of China 2022). The trade volume reached almost 5,000 tons of Afghan nuts worth more than $200 million in the first year of operations (The Peoples Republic of China 2022). Afghanistan has an estimated annual output of around 20,000 tons of pine nuts, employing more than 100,000 agricultural workers. Since Afghanistan lacks product processing and refinery capabilities, China has replaced Pakistan as the global supplier of Afghan pine nuts, thus benefiting from the value-added process as well (The Peoples Republic of China 2022).
- China-Kyrgyzstan-Uzbekistan-Afghanistan transportation route: Proposed by Uzbekistan in 2022 as a new China-Central Asian route, it now involves railway authorities of all four nations discussing freight and feasibility studies (Lillis 2022).
- The North–South Trans-Afghan railway: In July 2022, China's Special envoy for Afghanistan expressed his country's readiness to support the implementation of a trans-Afghan rail link connecting Uzbekistan via Mazar-i Sharif to Kabul and on to Peshawar, Pakistan (Mishra 2023). The envoy proclaimed that that "Beijing sees Afghanistan as a bridge between Central and South Asia" and added, "these megaprojects will help to a permanent peace on Afghan soil and deepen regional interconnection" (Mishra 2023). If implemented, this link is expected to intersect with an East–West rail route connecting China to Iran via Peshawar and Kabul. To complete the circuit, China and Pakistan are negotiating the building of a direct rail link from Pakistan's Kashgar to China's Xinjiang Province (Mishra 2023).
- The Five-nations Railway: Another major initiative on the drawing board, this project would link China, Iran, Kyrgyzstan, Afghanistan, and Tajikistan, with routes also connecting Afghanistan to Pakistan (Safi and Alizada 2018).
- The Chabahar Alternative: As discussed, CPEC would be the most efficient way to integrate Afghanistan into the BRI, providing Afghanistan the most direct access to warm water ports via Pakistan (Siddiqui 2024). However, to offset any Pakistani attempts at using trade and transit of Afghan goods via its seaports, Kabul has reopened a link through Iran to the Chabahar port on the Gulf (Siddiqui 2024). This initiative has garnered the support of India as well and has recently been revived after tensions flared up between Pakistan and Afghanistan over TTP militant attacks inside Pakistan. Pakistan accuses the Afghan Taliban of providing sanctuary to the TTP, a claim Kabul denies (Siddiqui 2024).
- The Karachi-Mazar Rail Link: Even though Afghan membership in the BRI is pending, there are new attempts at connecting it via the CPEC to Afghan projects.

Earlier in 2023, Pakistan Railway and state-owned China Railway Eryuan Engineering Group issued construction tenders for construction of a rail line between Karachi and Mazar-e-Sharif, the economic hub of northern Afghanistan (Pajhwok 2023).

B. Natural Resources and Mining

Geological surveys show that Afghanistan has approximately 1400 mineral fields, worth between $1–3 trillion underground, including copper, rare earth, coal, gold, iron ore, lithium, zinc, lead, natural gas, oil, semi-precious stones, and others (USGS 2024). However, due to the country's protracted instability and state of conflict between 1978 and 2021, a large portion of such deposits remain untapped. As the new Taliban caretaker government focuses on revenue generation and is courting investment in the mining sector, Chinese businesses have shown particular interest in several areas, including the following two cases:

1. Mes Aynak: On the minerals and other natural resources front, the interest shown by China dates back to 2007 when a state-owned company, China Metallurgic Group (CMG), was awarded the right to develop the Mes Aynak copper mine South of Kabul (estimated to hold copper worth up to $50 billion) (Kullab 2022). Considered the world's largest undeveloped copper field in the world, MCG pledged to invest up to $3 billion over 30 years in addition to investing $500 million for a power plant and more for a rail link (Kullab 2022). Considered as risky given the worsening security environment and other hurdles, the project was put on hold to this day. However, in more recent times, a second Chinese state-owned enterprise, MJAM started discussions on Aynak with the Taliban-led government (Kullab 2022). According to company officials, they face a number of challenges to start the project, including land acquisition issues, and contract revisions that need renegotiating the deal. It has also been decided that an ancient Buddhist archeological site near the mine will be spared and protected (Kullab 2022).
2. Amu Darya Basin: Another major Chinese investment relates to the Amu Darya basin oil and gas extraction bid (Yawar 2023). The region spanning Afghanistan and Tajikistan is estimated to contain almost one billion barrels of crude oil and 52 billion cubic feet of natural gas. The first contract with a private company produced oil in 2012, but operations were suspended in 2016 due to a lack of an export sales contract (Yawar 2023). In January 2022, the de facto government resumed talks and signed a deal with Xinjiang Central Asia Petroleum and Gas Co (CAPEIC) to extract oil from the basin and create an oil reserve in Sar-e Pul province (Yawar 2023). The Chinese pledged to invest up to $150 million annually, increasing to $540 million in three years. The basin is currently producing more than 1350 tons of oil per day (Yawar 2023).

Not in a rush, Beijing is likely to proceed at a slower pace and aim for the long-game in Afghanistan, as "more state-backed projects are likely to move forward as Chinese executives figure out ways to work with the Taliban. Lithium exploration could be a promising area in this regard" (Sun 2020).

Although on the bilateral side, both countries agreed to set up the China-Afghanistan Joint Committee on Economics and Trade (JCET), they have yet to sign a free trade agreement (FTA) or the double taxation avoidance agreements (DTA). However, in December 2022, China renewed a commitment to grant zero-tariff treatment to 98% of taxable trade goods from Afghanistan. On the other hand, Afghan tariffs on raw materials from China is considered as above the norm.

On the geoeconomic front, there are also differing opinions and assessments among experts. Dr. Rubin believes that "Afghanistan's mineral resources are not an important consideration in China's policy. This is so not only because of the difficulties China has encountered in implementing extractive projects in Afghanistan, such as the Mes Aynak copper mine and the Amu Basin petroleum venture, but also because of the transformation of China's own economy: China's reliance on industries using extractive resources is being reduced. It can obtain adequate supplies of the resources it still needs from traditional partners, such as iron ore from Australia and Brazil" (2019).

4 Afghanistan as Space for Great Power Competition Versus the Taliban's Quest for Strategic Equilibrium

4.1 The Bigger Picture

American and Chinese diplomats are adamant that Afghanistan is not part of their strategic competition arena. That may be the case at this moment in time—almost three years after the challenging and badly managed withdrawal of American and NATO forces in 2021—but not many analysts are convinced that the status quo is tenable or durable given the emerging tensions tied to new conflicts erupting in Europe (involving Russia) and in the Mideast. Not only is the Afghan real estate of geostrategic importance because of connectivity and geography, but, as discussed before in this chapter, also because of sought-after untapped resources that remain underground while offering an added-value potential over the next few decades, especially for raw material-importing countries with growing and shifting economies.

Nonetheless, China does not consider itself as a primary party or an instigator to the conflict that just ended in 2021 and would rather prefer to not get entangled with regional or strategic provocations. With the exceptional covert action undertaken during the Soviet occupation period in the 1980s, Afghanistan's reputation as a "graveyard of empires" is a major deterrent for China's leaders looking for economic advantage over geopolitical wranglings. China is also aware of dissatisfaction that exist within its own borders in the Chinese Muslim communities and prefers to crack down on dissent to avoid any external provocation by adversarial forces, whether Islamic or Western. Thus, having a stable and peaceful Afghanistan ruled by a cordial and engaging government—regardless of its social and ideological tendencies—is preferable than to face unrest and turmoil that could destabilize the region and threaten China's own national security.

It is important, however, to have a more accurate reading of Sino-American perceptions of their increasingly complex and multi-layered relationship before projecting it onto other competition theatres. Nicholas Burns, the U.S. Ambassador to Beijing, best described the latest American policy narrative at a recent 2023 Brookings Institution talk: "the US-China relationship is the most consequential relationship that the US has now, but particularly will have for the next couple of decades. We are the two leading economies in the world and will be into the 2030s and 40s; we are the two most powerful militaries in the world, and certainly will be for the next two or three decades; we are the two countries with the widest global reach if you think about the economic, societal, political, strategic breath that the two countries have; and we are vying for global power as well as regional power; I think that we are systemic rivals… in our administration, President Biden has set out a very clear policy, it's called invest, align and compete."

While America says it is ready to compete and does not dismiss "vying for global power as well as regional power," on the PRC side, even though China's views on U.S./NATO have fluctuated since the early days of the Cold War, Beijing's recent statements on the alliance have further sharpened and become bellicose. Both countries do not see eye to eye on Taiwan, which is considered as a historic and national priority issue for Beijing and a red line for the U.S.

Basing the bulk of competition on economic and technology, China claims to want to "connect the world for a shared future" (SCIO, PRC 2023), while the US is more interested in strengthening its own supply chains benefitting American economic and national security priorities. China appears to give the impression of being "outward facing, open and confident," while America over the past decade is viewed by many as "inward facing, exclusionary and fueled by fear." While China is reaching out to the global audience, especially in the global South, America is mostly focused on the domestic audience. Trying to dispel this emerging image, President Joe Biden recently countered and said: the U.S. "never gives up… this moment is the beginning when America won the competition of the 21st century" (C-Span 2021). Some beg to differ, either by saying that it's too soon to pass judgment or by countering the claim and pointing to a gradual decline in American power and prestige.

Although Sino-NATO relations have oscillated between adversary and ally since the 1949 Chinese revolution, Chinese criticism of American "hegemonic" strategies in the Asia–Pacific, Indo-Pacific and Eurasian landmass have mounted in recent years.

The emerging multipolar order is not exclusive to the US and China as others like India, Russia, Brazil, European powers, Japan, Australia, Turkey, Saudi Arabia, Pakistan and others also play a role at varying degrees, either by aligning themselves with one camp or another or by building new alliances or blocs and disrupting traditional lines.

4.2 The United Nations and the Recognition Conundrum

Observers note that Chinese diplomats are actively promoting engagement with the Taliban in international forums now. "Notably, when China took over the rotating presidency of the U.N. Security Council in September 2022, it marked the occasion by sending jars of Afghan pine nuts to other members of the council" (Pantucci 2023).

On their part, Taliban officials have pursued the traditional Afghan position of support for the "One China" policy, which clarifies Kabul's stance on Chinese sovereignty over Taiwan.

Although the security situation has improved since the Taliban takeover, Afghanistan's non-recognition and infrastructural weaknesses have prevented it from becoming a major part of the BRI. A Taliban-led ministerial delegation attended the 3rd Belt and Road Forum held in Beijing in October 2023, expressing a keen interest in joining the multinational mega project.

Meanwhile, several other regional countries i.e., China, Uzbekistan, Kazakhstan, Iran, Turkey, Russia, Qatar, Turkmenistan, and Azerbaijan have lately expressed renewed interest in engaging with Kabul on economic and regional cooperation issues. In January 2024, Kabul's de facto administration hosted the first international meeting since returning to power, entitled "Afghanistan Regional Cooperation Initiative" (Lalzoy 2024). The aim was to address "common challenges" by promoting economic connectivity, cooperation and develop a "region-centric narrative" (Lalzoy 2024). Although some regional attendees prioritized security and counterterrorism (CT) over economic issues, the acting Foreign Minister, Amir Khan Muttaqi, tried to reassure the diplomats that the IEA was committed to peace, stability, and CT (Gul 2024).

This outreach comes on the heels of an independent UN assessment by a Special Coordinator claiming that Afghanistan needs to be fully reintegrated into the international system, whilst admitting that "international engagement is not working," nor does it "serve the humanitarian, economic, political or social needs of the Afghan people" (AAN 2023)—a reference to Taliban restrictions imposed on girls' secondary education, female employment limitations and unwillingness to engage in political talks with other Afghan political movements.

Sensitive to what the Kabul administration deems external interference, the IEA Minister cautioned against "externally imposed" solutions and called for engagement based on "mutual interests" (Gul 2024). The issue of appointing a UN Special Envoy has now become a divisive issue as Kabul, joined by several regional players, prefers a different approach to addressing bilateral and multilateral concerns, and is not comfortable with the appointment of an additional Envoy that does not have full UNSC backing and whose appointment is viewed as redundant. The Taliban claim to be in full control of Afghan territory and government institutions at a time when the international community is divided on priorities and conditions for recognition. A divided UNSC will make it more difficult to reach consensus on key issues such as lifting sanctions, delisting Taliban leaders from terror lists, reconnecting Afghanistan to the international financial and banking grid and bestowing diplomatic recognition to the regime.

On one hand, the current rifts make it more difficult to seek collective solutions to the Afghan case, on the other hand, it offers new opportunities to weaken the sanctions' consensus and pursue different engagement tracks that benefit those countries, like China and Iran, that seek advantage at the expense of others. While some may consider Afghanistan and its vicinity as passé or the last war, others think that the scene is set for another round of great power and economic bloc competition over strategically and geoeconomically sensitive regions, that include countries like Afghanistan, Iran, Central Aisa, the Caucuses and Pakistan.

4.3 *The S.C.O. Option*

Having adopted a somewhat nonaligned stance and pursuing a balancing-act foreign policy as a de facto state, a safer option for the IEA for added regional engagement is through the Shanghai Cooperation Organization (SCO). Kabul signed a protocol in 2005 establishing the SCO-Afghanistan contact group. Seven years later, Afghanistan became an observer state in 2012 but has yet to be awarded full membership, even though others like Pakistan and India gained full membership in 2017.

Besides security and counter terrorism objectives, Afghanistan and the SCO also share common interests in promoting regional economic cooperation, combatting crime and the drug trade. Afghanistan's recent crackdown on opium cultivation, processing and smuggling has been cautiously welcomed by SCO member states and Western countries as well.

During the 2022 SCO foreign ministers' meeting in Uzbekistan, China announced it would not levy tariffs on 98% of goods imported from Afghanistan (Nadin et al. 2022). Within the framework of the SCO, Afghanistan could take advantage of financial and technical support from member states.

4.4 *Afghanistan Prospects in a New World and Regional Order*

For almost 20 years, the U.S. was called the "big elephant" in the Afghan room, not only because of its post-9/11 military footprint and clout, but also because of excessive spending and an attempt at state and nation building that was partly misguided and partly mismanaged.

As the US presence diminishes and security paradigm in Afghanistan takes on new forms, Afghanistan is no longer seen by/the PRC as a territory with the potential to pause a direct threat. Beijing sees a stable Afghanistan as beneficial for its long-term economic plans in the South-Central Asian region. Therefore, China has the political will to push for peace and stability in Afghanistan. Despite a traumatic past and weak structures, Afghanistan's vast amount of minerals and hydrocarbons remain attractive to Chinese investors and companies. In turn, as Kabul actively seeks foreign assistance to help its weak economy, Beijing's economic clout makes it a preferred partner to work with at this stage when most Western countries are seen as unwilling or unable to reduce the impact of sanctions and overcome hurdles attached to human rights and democracy-building principles. Those principles are obviously not on either China's or the Taliban's priority agenda lists.

While some in the West and in the Islamic world may disagree with reports of recent crackdowns against the Uyghur communities, Dr. Yun Sun of the Stimson Center writes: "Chaos in Afghanistan, from Beijing's perspective, stokes Islamic fundamentalism that threatens domestic security in China, particularly in Xinjiang" (2020). She adds: "If anything, China is not a revisionist power in Afghanistan. Given the choice, China would prefer to see an Afghanistan with internal stability and a functional government that is preferably, but not necessarily, neutral among great powers" (Sun 2020). Pointing to the Western disengagement from Afghanistan, she writes: "Having witnessed the quagmire in which Britain, the Soviet Union, and the United States were each entrapped, China has always been convinced that Afghanistan is the 'graveyard of empires.' Traditionally, Beijing believed that it should avoid serious entanglement in Afghan affairs at all costs" (Sun 2020).

While not in a rush and pursuing a long-term strategy towards Afghanistan, China has diplomatically "advised" the Taliban for the past few years to avoid repeating past mistakes, especially the experience of 1996–2001, by imposing strict social policies and exercising a monopoly of power.

5 Conclusion

While America is distracted by wars in Ukraine and the Gaza crisis during an election year, there is little bandwidth left to pay closer attention to other non-vital regional issues, unless an imminent national security threat is perceived. It is also a known secret that US aerial surveillance of Afghanistan using drones and other technologies still continues despite the military pullout in 2021. It is part of America's over-the-horizon (OTH) strategy of monitoring and using any means necessary to target known assets or prevent an attack on US or allied interests.

As a result, balancing China's geoeconomic and geopolitical interests in Afghanistan are made easier, but still a delicate act, fraught with challenges and uncertainties. Whether China can successfully navigate this tightrope walk remains to be seen, but the following factors can tilt the equation to one side or another: 1. Taliban regime durability; 2. Security and stability in Afghanistan and periphery; 3. Anti-terrorism and anti-narcotics cooperation or proxy rivalries; 4. International cohesion and consensus on the way forward; 5. Regional balancing of interests and threat perceptions with countries such as India, Russia, Iran, Pakistan; 6. Conditional engagement based on political inclusivity and human rights standards; 7. Managing the humanitarian and refugee crises; 8. Multilateral and collective partnerships and 9. Aiming for engagement and dialogue as an alternative to renewed fighting or support for armed insurrection. It is within this context that Countries like China and America will calibrate policy and strategy toward countries like Afghanistan and between themselves.

As critical Eurasian trade and investment arteries are being built East–West and North–South, each country situated on these routes will play an important role as a connector as well as a hub and platform for business and new opportunities. While America retrenches from some regions and refocuses on others, China—along with Iran and Russia—may be compelled to pool resources to advance a shared foreign policy and economic strategy that will shape a multipolar world no longer dominated by one side. As competition takes on new forms, geographically sensitive and potentially resource-rich countries like Afghanistan will not be able to stay isolated or disconnected from the strategic dynamics that will shape competition and power plays.

INTERVIEW WITH DR. BARNETT R. RUBIN, AN AMERICAN POLITICAL SCIENTIST AND A LEADING EXPERT ON AFGHANISTAN AND SOUTH ASIA. HE IS CURRENTLY SENIOR FELLOW AND DIRECTOR AT THE CENTER ON INTERNATIONAL COOPERATION AT NEW YORK UNIVERSITY.

SAMAD: What are some of China's vital geopolitical and geoeconomic interests and priorities regarding Afghanistan since the return of the Taliban to power in 2021? To what extent does the BRI/CPEC play a role?

RUBIN: One thing that has changed a great deal is the relationship of the U.S. and China over Afghanistan. The U.S. pulled the troops out of Afghanistan and let the government collapse without even trying to engage China at a senior level. Soon after Biden's inauguration, I was contacted by a top Chinese diplomat. He asked me to arrange a meeting for him with an American counterpart to discuss Afghanistan. He was asking for a meeting to discuss Afghanistan with the U.S. in the context of U.S.-China bilateral relations. Finally, I learned belatedly that they offered him to meet a mid-level diplomat. I think this was shocking to the Chinese. The SRAPs under both Obama and Trump had consistently consulted with China and also had high-level access in the USG themselves. They got a preview of the democracy vs. autocracy frame for foreign policy when the administration held a meeting with NATO partners about the withdrawal and did not consult China. If you think in ethnic stereotypes instead of geography, which no one knows, these two countries have nothing in common. This has to factor into one constant in Chinese discourse on Afghanistan: the U.S. must take responsibility for its failure in Afghanistan. They used to say that the U.S. left Afghanistan "irresponsibly," but later changed it to "hastily." Now, however, the U.S. no longer has an Asian policy but rather an Indo-Pacific one. I think the Chinese like many others are curious what role Afghanistan and Central Asia play in the "Indo-Pacific."

Chinese officials always put terrorism as the number one priority. Of course, there are almost no incidents in China related to "terrorists based in Afghanistan," I don't actually know of a single one, so either the policy is working or there is not much of a threat from ETIM. But as you know the Taliban and Afghans in general, never ever hand people over, from Nazis in the 1940s to Al Qaida in the 1990 and 2000s. Also, if they know they are not going to do something, they just waste your time while you look at your watch. But I think that the bilateral Xinjiang-Badakhshan relations as outlined in my report have become quite important. The Chinese can't get much out of the Taliban, so they are doing it themselves. Taliban have moved ETIM out of Badakhshan... Plus the Chinese have the quadripartite mechanism with Afghanistan, Pakistan, and Tajikistan.

Right now, I do not have the impression that integrating Afghanistan into BRI or CPEC is a high priority. I think that the Chinese have found that making the transition from praising their eternal friendship with Pakistan to actually doing business with Pakistan is quite painful. A Chinese ambassador told me in 2022 that the security situation for Chinese nationals was much worse in Pakistan than in Afghanistan. Also, China is having internal economic problems. The Chinese say that right now they are trying to "stabilize" their relations with the Taliban. Taliban officials have a sacred religious obligation to obey the Amir, but they have not much of a mandate even to listen to anyone else, so it can be frustrating to deal with them.

The Chinese do not want the UN in New York involving itself in Afghanistan. Their consensus with Russia and Iran (also Pakistan, but that investment is not yielding much) is more important to them. They have created a mechanism of meetings among the MFAs of all states bordering on Afghanistan (which does not include the U.S., Russia, or India). They also participate in the Moscow process, but I don't think they want to follow the Russian lead much.

SAMAD: Unrecognized as a de jure state, can the de facto Taliban government maintain equilibrium in its relations with great powers? What could change that formula?

RUBIN: The Chinese say they recognize the state of Afghanistan, as they always have, but that they do not recognize governments anywhere, as that would be interfering in internal affairs. They are willing to upgrade the bilateral relationship to the extent that the Taliban show themselves able and willing to carry out their duties as members of the international community.

The only thing that I can think of would be something that motivated the U.S. to regain interest in Afghanistan. As long as either Biden or Trump are president, that is almost impossible. But you could still have great-middle-power competition in Afghanistan—China-Pakistan-India. I am surprised at how low-key and pragmatic the Indians are being. Perhaps that is Jayshankar's doing. As a grad student at Chicago, I met his father, the late K. Subrahmanian. He was not sentimentally inclined.

Right now, the action is elsewhere. Even the U.S. is not in a hurry to change the situation in Afghanistan, so you can imagine how patient China is.

SAMAD: China and the US say that they do not see Afghanistan as a space for great power competition. Is that practical and what could change that?

RUBIN: What would the U.S. and China compete over in Afghanistan? Who can waste more time and money trying to get precious rare earth minerals out of a country with no infrastructure, where every official has a duty to feed another 50 family members or so and does his job accordingly? Though it turns out that for some purposes the bureaucracy in Afghanistan is easier to deal with than in Pakistan. Afghanistan has a lot of valuable mineral resources. So, I imagine, does the moon. I am not sure which is more difficult to extract and get to market profitably. I heard that Biden issued an instruction that no one should send the White House decision memos on Afghanistan.

Interesting point is that no outside power wants to fund and supply the "resistance." I discussed this with Ahmad Massoud in Paris, He was complaining that nobody was supporting him and trying to convince us that the Taliban are bad. Finally, I said, you don't need to do this. We know the Taliban are bad. The problem is that nobody has any reason to believe that giving you money and weapons will make the situation better. You won't win. At most you could create some chaos that somebody else would take advantage of. If you did win, that would create even more chaos. Russia has only negative interests in Afghanistan: counterterrorism, counter-Americanism.

Thinking about it, I suppose that if somehow a moderate or liberal Talib took over the movement and opened educational institutions for girls, then the U.S. would come up with a small aid program which would still be bigger than anyone else's, and then the intelligence agencies would start monitoring that and then they would start monitoring each other, and if there is no adult supervision anything could happen. But it seems very unlikely. Of course, we have to expect the unexpected, but that's all I can say about that for now.

End.

References

Afghanistan Analysts Network (2023) UN Special Coordinator's Assessment on Afghanistan. Available at: https://www.afghanistan-analysts.org/en/recommended-reading/un-special-coordinators-assessment-on-afghanistan/

Almukhtar S, Nordland R (2019) What Did the U.S. Get for $2 Trillion in Afghanistan? The New York Times. Available at: https://www.nytimes.com/interactive/2019/12/09/world/middleeast/afghanistan-war-cost.html

Brookings Institute (2024) A conversation with U.S. Ambassador to China R. Nicholas Burns on US-China relations. Available at: https://www.youtube.com/watch?v=URRhjgoKRkA

C-Span (2021) 'President Biden Signs Bipartisan Infrastructure Bill.' C-span. Available at: https://www.c-span.org/video/?c4986638/thats-moment-america-began-win-competition-21st-century

China Daily (2023) Full text: a global community of shared future: China's proposals and actions. The state council information office of the People's Republic of China. Available at:https://www.chinadaily.com.cn/a/202309/26/WS65124140a310d2dce4bb7dce.html

Dai SY (1966) China and Afghanistan. China Q 25(1):213–221. Available at: https://doi.org/10.1017/S0305741000027867.ISSN1468-2648.S2CID154014250

Dawi A (2024) China's President Accepts Credentials From Afghan Representative. VOA. Available at: https://www.voanews.com/a/china-s-president-receives-afghan-ambassador-taliban-seek-recognition-from-russia-iran-/7463837.html

Gul A (2024) Afghanistan's Taliban host multilateral huddle to promote regional cooperation. VOA. Available at: https://www.voanews.com/a/afghanistan-s-taliban-host-multilateral-huddle-to-promote-regional-cooperation/7461769.html

Heart of Asia (2024) Available at: https://www.hoa.gov.af/

Hong Z (2013) China's Afghan policy: the forming of the "March West" strategy? J East Asian Aff 27(2):1–29. ISSN 1010-1608

Kullab S (2022) China Eyes Investment in Afghanistan's Mes Aynak Mines Potential wealth from Afghanistan's untapped mining sector has prompted the Taliban's shift in position on the Buddha statues. The Diplomat. Available at: https://thediplomat.com/2022/03/china-eyes-investment-in-afghanistans-mes-aynak-mines/

Lalzoy N (2024) Kabul Hosts 'Afghanistan's Regional Cooperation Initiative.' Tolo News. Available at: https://tolonews.com/afghanistan-187179

Lillis J (2022) Cargo leaves China for Afghanistan on new route via Kyrgyzstan, Uzbekistan. Eurasianet. Available at: https://eurasianet.org/cargo-leaves-china-for-afghanistan-on-new-route-via-kyrgyzstan-uzbekistan

Maizland L (2020) U.S.-Taliban Peace Deal: What to Know. Council on Foreign Relations. Available at: https://www.cfr.org/backgrounder/us-taliban-peace-deal-agreement-afghanistan-war

Mishra A (2023) Is Trans-Afghan Railway linking Central and South Asia a gamechanger or pipe dream? India Narrative. Available at: https://www.indianarrative.com/opinion-news/is-trans-afghan-railway-linking-central-and-south-asia-a-gamechanger-or-pipe-dream-150647.html

Nadin et al (2022) Shanghai cooperation organisation summit 2022: key takeaways. ODI. Available at: https://odi.org/en/insights/shanghai-cooperation-organisation-summit-2022-key-takeaways/

Newby L (2005) The empire and the khanate : a political history of Qing relations with Khoqand c. 1760–1860. Leiden, Brill

OEC (2024) China/Afghanistan. Available at: https://oec.world/en/profile/bilateral-country/chn/partner/afg

Pajhwok (2023) Karachi-Mazar railroad plan seen as boost to trade. Available at: https://pajhwok.com/2023/04/29/karachi-mazar-railroad-plan-seen-as-boost-to-trade/

Pantucci P (2023) China and the Taliban are friends in need of each other. Nikkei Asia. Available at: https://asia.nikkei.com/Opinion/China-and-the-Taliban-are-friends-in-need-of-each-other

Rauf S, Zeidan A (2024) China-Pakistan Economic Corridor international trade project. Britannica. Available at: https://www.britannica.com/topic/China-Pakistan-Economic-Corridor

Rubin BR (2019) Notes from: Visit to Beijing. Center on International Cooperation, NYU

Safi M, Alizada B (2018) Integrating Afghanistan into the belt and road initiative review, analysis and prospects. Friedrich Ebert Stiftung. Available at: https://library.fes.de/pdf-files/bueros/kabul/15587.pdf

Siddiqui S (2024) How significant is $35M Taliban-Iran Chabahar port deal? Al-Monitor. Available at: https://www.al-monitor.com/originals/2024/03/how-significant-35m-taliban-iran-chabahar-port-deal

Singh S (2020) Rebuilding lifelines of its soft underbelly: China engages its Southwest Frontiers. In: The New Great Game [Online]. Redwood City: Stanford University Press, pp 147–170

Starr SF (2004) Xinjiang: China's Muslim borderland. M.E. Sharpe Inc., Armonk, N.Y

Sun Y (2020) China's Strategic Assessment of Afghanistan. Stimson. Available at: https://www.stimson.org/2020/chinas-strategic-assessment-of-afghanistan/

The Peoples Republic of China (2022) Air corridor of pine nuts: a bridge of friendship between China and Afghanistan. The People Republic of China in the Islamic Republic of Afghanistan. Available at: http://af.china-embassy.gov.cn/eng/sgxw/202201/t20220111_10480912.htm

United Nations (2023) Adopting Resolution 2721 (2023), Security Council Requests Secretary-General Appoint Special Envoy for Afghanistan.' 9521ST MEETING (AM); SC/15548, December 29. Available at: https://press.un.org/en/2023/sc15548.doc.htm

United Nations (2024) We all want an Afghanistan at peace, UN chief says in Doha. UN News. Available at: https://news.un.org/en/story/2024/02/1146657

University of British Colombia; Schluessel E (2024) Historical overview of events shaping the politics of Xinjiang. Available at: https://xinjiang.sppga.ubc.ca/timelines/historical-overview/

USGS (2024) USGS Projects in Afghanistan. Available at: https://www.usgs.gov/special-topics/usgs-projects-in-afghanistan

Wiki wand (2024) Afghanistan–China relations. Available at: https://www.wikiwand.com/en/Afghanistan%E2%80%93China%20relations

Wikipedia (2024) Five Nations Railway Corridor. Available at: https://en.wikipedia.org/wiki/Five_Nations_Railway_Corridor

Yawar MY (2022) China 'shocked' at Kabul hotel attack that injured its five citizens. Reuters. Available at: https://www.reuters.com/world/asia-pacific/five-chinese-citizens-badly-injured-kabul-attack-businessman-says-2022-12-13/

Yawar MY (2023) Afghanistan's Taliban administration in oil extraction deal with Chinese company. Reuters. Available at: https://www.reuters.com/business/afghanistans-taliban-administration-oil-extraction-deal-with-chinese-company-2023-01-05/

Zakharov A (2023) The Moscow Format meeting and Russia's Afghanistan policy. Observer Research Foundation. Available at: https://www.orfonline.org/expert-speak/the-moscow-format-meeting-and-russia-s-afghanistan-policy

Ziwen Z (2024) Why China is wary of opening up Afghanistan's new road to Xinjiang. South China Morning Post. Available at: https://www.scmp.com/news/china/diplomacy/article/3249931/why-china-wary-opening-afghanistans-new-road-xinjian

Dragon Over Trident: Lessons for China from Ukraine

Michael O. Slobodchikoff

Abstract China is in a position of strength for the short term and has learned valuable lessons while watching the reaction of the West in Ukraine. However, the time period of having this strength is rapidly fading. Thus, policy makers should expect China to be much more aggressive than it has been in the past. Chinese foreign policy will be one in which impatience will override careful deliberation if China is to realize its strategic goals. Waning interest from Washington coupled with elections in European countries where policy makers were not in favor of continued support to Ukraine has shown Beijing the smartest move is to prepare for short-term pain, which with proper preparation can be overcome. In time, interest in maintaining aid and sanctions will become unpopular. In terms of China's aspirations for Taiwan, it means that Beijing must be prepared for economic sanctions and aid to Taiwan. If China is able to move swiftly enough, interest will fade in helping Taiwan, and Beijing will weather the storm and eventually annex Taiwan. Beijing's greatest problem is that of demography. It is also a falling power, so if it wants to capture Taiwan, its window to do so is closing rapidly.

Keywords China · Taiwan · Ukraine · Russia · Global order · Great power conflict

In 1986, China joined the General Agreement on Tariffs and Trade (GATT), signaling its intent to fully join the liberal international order led by the United States. What followed were fifteen years of intense efforts to meet the conditions of joining the World Trade Organization (WTO). As Zhu Rongji, China's former Prime Minister famously stated, it was long enough to "turn black hair white."[1]

Foreign policy makers in the United States and Western Europe welcomed China's membership and viewed the actions as evidence that as China joined the liberal world order, the Soviet Union would become more isolated, and that it would eventually

[1] As quoted in *The Economist* ("Shades of Grey," 2011).

M. O. Slobodchikoff (✉)
Department of Political Science, Troy University, Troy, AL, USA
e-mail: mslobodchikoff@troy.edu

A. Farhadi et al. (eds.), *The Great Power Competition Volume 6*,
https://doi.org/10.1007/978-3-031-70767-4_15

lead to the defeat of the Soviet Union. Just five years later, the Soviet Union indeed collapsed, leading many observers to declare the end of history and the triumph of the liberal order.[2] Communism had been defeated, showing that there could be only one global order, the liberal order established by the United States.

Prior to the collapse of the Soviet Union, Moscow had pursued a policy of both economic and political reform. While Moscow hadn't embraced the liberal order, it had slowly started moving in that direction through *perestroika*, while at the same time tackling political reform the *glasnost'*. Mikhail Sergeevich Gorbachev, the last president of the Soviet Union, truly believed that the Soviet Union needed massive reform to survive. However, ultimately, these reforms led to the collapse of the Soviet Union.

In contrast to the Soviet Union, China had moved to accept the liberal order much earlier, but carefully chose how to accept the liberal order. China determined that Moscow had made an enormous mistake by taking on both political and economic reforms at the same time. Beijing decided that it would be more practical to reform the economy, but not to undertake any political reforms. Thus, China decided to develop a unique strategy of accepting the liberal economic order, but not undertaking political reforms. The first step in that strategy was to join the WTO.

China's gamble to join the WTO has paid big dividends. China's economy has boomed, and consumers have benefitted from an influx of Chinese goods. China has not only worked within the WTO framework, but has also worked closely with the European Union (EU) through the EU-China Strategic Partnership. Moreover, China has worked to develop bilateral treaties with many states further expanding and cementing its role within the liberal economic order.

In fact, China has been so successful, that some international relations theorists have stated that China's growth will eventually lead to direct conflict with the United States.[3] This group of scholars is made up of mostly realist theorists. In contrast, other international relations scholars have viewed China's economic success and acceptance of the WTO as evidence that China fully accepts the liberal economic and political order, and that the more it is integrated into these international governmental organizations (IGOs), the more likely China will undergo liberal democratic transition.[4] This group of scholars is made up of liberal scholars. Despite China's openness to joining the liberal economic order, China has been very resistant to any

[2] Fukuyama, F., 1989. The End of History? National Affairs, Inc; Ikenberry, G.J., 2008. The Rise of China and the Future of the West: Can the Liberal System Survive? Foreign Aff. 87, 23–37; Legro, J.W., 2007. What China Will Want: The Future Intentions of a Rising Power. Perspectives on Politics 5, 515–534. https://doi.org/10.1017/S1537592707071526.

[3] Glaser, C., 2011. Will China's Rise Lead to War? Why Realism Does Not Mean Pessimism. Foreign Affairs 90, 80–91; Kirshner, J., 2012. The tragedy of offensive realism: Classical realism and the rise of China. European Journal of International Relations 18, 53–75. https://doi.org/10.1177/1354066110373949; Mearsheimer, J.J., 2014. Can China rise peacefully? National Interest 25, 23–37; Mearsheimer, J.J., 2006. China's unpeaceful rise. Current History 105, 160.

[4] Ikenberry, G.J., 2008. The Rise of China and the Future of the West: Can the Liberal System Survive? Foreign Affairs 87, 23–37; Legro, J.W., 2007. What China Will Want: The Future Intentions of a Rising Power. Perspectives on Politics 5, 515–534. https://doi.org/10.1017/S1537592707071526.

organizations requiring political change. The Western liberal order not only values liberal economics, but also stresses liberal freedoms and democracy. For example, the EU's Copenhagen Criteria specifically state which democratic norms must be internalized by candidate states prior to becoming members. Similarly, the International Monetary Fund (IMF) often attaches conditions about liberal values and democracy prior to loaning money. While China has been very open to joining liberal institutions, it has balked at joining institutions that work to spread liberal democratic values. Instead, it has worked to develop its own alternatives to such institutions. For example, in response to the Trans-Pacific Partnership (TPP), China developed the Regional Comprehensive Economic Partnership (RCEP) which includes ASEAN member states, China, Japan, Korea, Australia, and India. In contrast to many of the Western liberal institutions, RCEP will not focus on any politically sensitive topics, instead merely focusing on economic cooperation. Similarly, China has used both the Shanghai Cooperation Organization (SCO) and BRICS organizations to create institutions that foster trade yet do not mix trade and politics. In the SCO, all of the member states agree to cooperate in both security and economics, yet put no pressure on any member states to accept certain political behavioral norms. The BRICS Development Bank is similar to the IMF in many ways except that it provides loans without certain political preconditions having to be met.

While China joined GATT in 1986, it took 15 years for the Chinese government to institute the necessary reforms for China to fully meet the conditions to join the World Trade Organization (WTO). China's GDP growth rate went from 5% in 1981 to close to a 15% growth rate in 1984. Following its agreement to join the GATT in 1986, China spent four years with variable GDP growth rates ranging from approximately a 9% GDP growth rate in 1986 to only a 4% GDP growth rate in 1990. However, following 1990, China's GDP growth rate continued to increase yearly never falling before a 7% growth rate between 1990 and 2015. In 2016, the annual GDP growth rate fell below 7% for the first time since 1990, and hit a low of just above 2% growth rate in both 2020 and 2022.[5] This economic growth allowed China to become the second largest economy in the world, second only to the United States.

According to Power Transition Theory, there will be major conflict between a global hegemon and a rising state if the rising state's power capabilities are relatively equal to that of the global hegemon and if the state is dissatisfied with the global order established by the global hegemon.[6] While China's economy has grown to be second only to that of the United States, its power capabilities are still quite a bit less

---

[5] Data obtained from World Bank Accounts Data.

[6] DiCicco, J.M., Levy, J.S., 1999. Power Shifts and Problem Shifts: The Evolution of the Power Transition Research Program. Journal of Conflict Resolution. 43, 675–704; Kim, W., 2002. Power Parity, Alliance, Dissatisfaction, and Wars in East Asia, 1860–1993. Journal of Conflict Resolution 46, 654–671. https://doi.org/10.1177/002200202236168; Kugler, J., Organski, A.F.K., 1989. The Power Transition: A Retrospective and Prospective Evaluation, in: Midlarsky, M. (Ed.), Handbook of War Studies. University of Michigan Press, Ann Arbor; Levy, J.S., 1998. The Causes of War and the Conditions of Peace. Annual Review of Political Science 1, 139–165. https://doi.org/10.1146/annurev.polisci.1.1.139; Tammen, R.L., Kugler, J., 2006. Power Transition and China-US Conflicts. Chinese Journal of International Politics 1, 35–55. https://doi.org/10.1093/cjip/pol003.

than those of the United States. Nevertheless, China's growth relative to that of the United States made it such that in the not so distant future, China would overtake the United States as the most powerful state in the global system. When that transition will occur, according to power transition theory, war would break out between China and the United States.[7]

One of the major proponents of such a view of China's growth is John Mearsheimer. Mearsheimer strongly believes that China's incredible growth will absolutely lead to conflict with the United States.[8] He believes that while Beijing has been cooperating with Moscow, it will only do so as long as the power remains in favor of Washington. Once Beijing accumulates enough power, then it would be able to challenge the global order established by the United States. Given the vast number of people and resources that China would be able to muster, any war between the United States and China would be devastating. Scholars who fear China's rise believe that the United States has committed several missteps by focusing on the Global War on Terror and the War in Iraq rather than paying attention to China. They argue that China has now effectively gained control over the South China Sea and is close to being able to challenge the United States for regional dominance in Asia. Despite Washington's pivot to Asia, the scholars argue that China has still outplayed the United States, and that more attention needs to be paid to the growth of China. In contrast to international scholars who view China as a future threat to US hegemony, many liberal scholars view the rise of China as an opportunity to integrate China into the current global order and thus bring about internal political change within China itself.

1 Liberal Approach to the Rise of China

In 1959, Seymour Martin Lipset identified that democracy and economic modernization were closely tied together. He argued that economic modernization was a prerequisite to democracy, and that democratization would occur because economic modernization led to a wealthy middle class which would in turn demand political change and a voice in governing their country, which in turn would lead to democracy.[9] While the causal relationship between modernization and democracy has been challenged, nevertheless, economic wealth and democracy are closely intertwined

---

[7] Brzezinski, Z., Mearsheimer, J.J., others, 2005. Clash of the Titans. Foreign Policy 146, 46–49; Glaser, C., 2011. Will China's Rise Lead to War? Why Realism Does Not Mean Pessimism. Foreign Affairs 90, 80–91; Kirshner, J., 2012. The tragedy of offensive realism: Classical realism and the rise of China. European Journal of International Relations 18, 53–75. https://doi.org/10.1177/1354066110373949; Mearsheimer, J.J., 2014. Can China rise peacefully? National Interest 25, 23–37; Mearsheimer, J.J., 2006. China's unpeaceful rise. Current History 105, 160; Mearsheimer, J.J., 2003. The tragedy of great power politics. WW Norton & Company.

[8] Mearsheimer (2003).

[9] Lipset (1959).

concepts.[10] Many scholars believe that the two concepts become reinforcing concepts and that is why most of the wealthy states are democracies.

The end of the Cold War heralded a new era in international relations theory. Theorists were convinced that the end of the Cold War was evidence of the triumph of liberal values. The global liberal order established by the United States was the only one left standing and thus, all other states in the system would fully adopt the liberal order.[11] This was especially true for neoliberal institutionalists who believed that the more interconnected a state became with international governmental organizations, the more likely that state was to have peaceful relations with other states.[12] Further, neoliberal institutionalists argued that as states became interconnected to each other through international governmental organizations that it would have a democratic effect in that it would make a state wealthier, lead to an increased middle class, and thus have a democratizing effect.[13] Finally, it has been noted that membership in international organizations leads to a synchronization of member states' preferences, meaning that as new states join IGOs, they go through a period of socialization as their preferences adapt to those of the other member states within a given organization.[14]

In the case of China, many liberal scholars assumed that China's acceptance of the liberal order and its relationships with international governmental organizations would lead not only to China accepting the liberal economic order, but also eventually lead to the democratization of China itself.[15] The assumption was that China's acceptance of the liberal economic order would create wealth for the Chinese middle class, and that would have a democratizing effect. Further, they expected that Beijing's preferences would become aligned with the other members of the liberal economic order due to interconnectedness through IGOs.[16] Finally, it was expected that China would undergo socialization once it fully accepted the liberal economic order and fully engaged liberal IGOs, which would have a dramatic effect on the domestic preferences of China.[17]

Ultimately, much of the scholarship has focused on Beijing's acceptance of the liberal economic order and the fact that it would be assimilated into the order. This is in

---

[10] Almond, G. A. (1991). Capitalism and democracy. *PS: Political Science & Politics*, *24*(3), 467–474. https://doi.org/10.2307/42009; Arat, Z.F., 1988. Democracy and economic development: Modernization theory revisited. Comparative Politics 21, 21–36; Heo, U., Tan, A.C., 2001. Democracy and economic growth: A causal analysis. Comp. Polit. 463–473.

[11] Fukuyama (1989).

[12] Doyle, M.W., 1986. Liberalism and World Politics. American Political Science Review 80, 1151–1169; Russett, B.R., Oneal, J.R., Davis, D., 1998. The Third Leg of the Kantian Tripod for Peace: International Organizations and Militarized Disputes, 1950–85. International Organizations 52, 441–467. https://doi.org/10.1162/002081898550626

[13] Huntington (1993).

[14] Finnemore (1996).

[15] Lanteigne, M. (2005). *China and international institutions: alternate paths to global power*. Routledge; Volgy, T. J., & Bailin, A. (2003). *International politics & state strength*. Lynne Rienner Publishers.

[16] Abbott and Snidal (1998).

[17] Kent (2002).

stark contrast to the realists who believe that China will directly challenge the global order established by the United States. The problem with both the realist and liberal approach to China's engagement in the liberal economic order is that it examines China's relations from the perspective of the hegemonic power. Both theoretical approaches assume that China will either change or challenge based on the liberal order. At a fundamental level, China's acceptance of the liberal order is a continuing process of negotiation between organizations and Beijing. It is incorrect to assume that as China's power increases, that the liberal order can just assimilate China, but it is also incorrect to assume that China will directly challenge the liberal order. Instead, the liberal order is adapting to China and vice versa. However, to understand this process, it is necessary to understand China's relations with various international governmental organizations and its engagement of the liberal order. I now turn to an examination of first China's relations with global international organizations, and then will examine how Beijing interacts with regional international organizations.

2 Chinese Relations with Global International Governmental Organizations

As China has increasingly become engaged in the liberal economic order, its relations with IGOs has taken on two forms. The first form is an increase in cooperative behavior from Beijing toward these organizations. Beijing has become more open and cooperative in its negotiations and interactions with the IGOs, especially by stressing that the way that cooperation should occur is through multilateralism.[18] Beijing understood that as the world's most populous country, that it needed to become engaged and could add a necessary dynamic to liberal IGOs.[19]

A more open and cooperative policy towards IGOs was not easy for Beijing to achieve. Traditionally, Chinese policy making had not been transparent, and decisions were closely shrouded in secrecy. This was especially true for negotiations, as negotiations were often viewed as more of a zero-sum game, and liberal global IGOs valued a more transparent negotiation process that believed in a positive-sum game. However, as Beijing began to engage with global IGOs, it was forced to adapt to many of the norms of the liberal IGOs. While Beijing's dealings with the International Monetary Fund (IMF) and the World Bank were motivated by self-interest and direct economic benefit, the process of adopting the policies of the World Trade Organization (WTO) prior to gaining membership was not an easy process and required a profound shift in Chinese foreign and domestic policy.[20] For example, the WTO

[18] Ibid.

[19] Kim (1992).

[20] Hempson-Jones (2005a, b).

framework had profound effects on employment in China[21] as well as domestic laws and regulations.[22]

One of the most important changes to the global system and China's relations to the WTO was that any trade dispute would now be handled through the WTO dispute settlement mechanism. Prior to Chinese ascension to the WTO, any disputes were handled through negotiations and threats of trade wars, especially between the United States and China. This was especially true of disputes involving intellectual property rights. For many years prior to joining the WTO, the US and China would argue about intellectual property rights and move to the brink of a trade war before both sides would back down at the eleventh hour to resume negotiations. The WTO ensured that there would be a dispute settlement mechanism and that trade disputes would thus be settled by the WTO instead of between the two states directly.[23] The problem with this mechanism is that neither state might like the outcome of the settlement, however, the dispute settlement would be binding and would try to prevent future conflict between the disputants.

China not only actively engaged the WTO, but also began to engage more within the structure of the United Nations. Prior to the end of the Cold War, China was extremely protective of national sovereignty and was very reluctant to allow any action that might affect another state's sovereignty. This was especially evident in agreeing to UN peacekeeping missions in Cambodia, Somalia and East Timor, in which a "hard-line defense of absolute sovereignty has given way to a more pragmatic stance that sanctions a certain level of interference into another state's affairs."[24]

China's multilateral approach to relations with liberal IGOs has evolved into two distinct categories, economic and security.[25] It has actively engaged global liberal IGOs especially in the economic category. Beijing has become immersed in these global liberal IGOs and has absorbed values such as common security interests and free market economics. It has engaged these global organizations in issues as varied from terrorism to intellectual property rights.

Despite Beijing's openness to engage in relations with liberal global IGOs, Beijing has been very wary of becoming too interdependent. Beijing had very serious concerns that globalization (quanqihua) and modernization (xiandaihua) would create a trap for the its internal politics and lead down to a path of forced democratic changes.[26] Therefore, Beijing developed a strategy to counter such socialization forces. First, Beijing realized that it needed to develop good relations with the liberal IGOs, but it also realized that it was gaining power. By 2012, China had the world's second largest economy, which meant that negotiations with liberal IGOs were begun from a position of strength and not weakness. The other member states

[21] Zhu and Warner (2005).

[22] Potter (2001).

[23] Yu (2005).

[24] Hempson-Jones (2005a, b).

[25] Wang (2000).

[26] Kent (2002).

of the liberal IGOs needed Beijing's membership as much as Beijing needed the liberal IGOs.

Second, Beijing realized that it was mainly interested in engaging with global liberal IGOs on economic issues. While security concerns were of great concern to Beijing, it mainly addressed economic issues multilaterally through relations with global liberal IGOs, and addressed security issues either multilaterally through regional IGOs or bilaterally with individual states. For example, while China actively engaged the European Union (EU)[27] on economic issues, it remained wary of the EU's concerns in regards to human rights protection and activism in climate change.[28] Thus, it negotiated economic treaties with the EU, but moderated those treaties with bilateral treaties with EU member states to ensure that the EU did not become an overly powerful player and force China to adopt the same norms as the EU (Fox and Godement 2009).[29] For example, Fox and Godement (2009) argue that prior to the Czech Republic's ascension into the EU, it tried to reach out to China to cooperate bilaterally. Beijing was not interested at the time. However, once the Czech Republic became a member state, Beijing actively engaged the Czech Republic in bilateral cooperation to ensure that it had some protection and insurance in cooperating with the EU directly.

It should be noted that the Beijing has always held the EU in high regard. Beijing believes strongly in a policy of multilateralism, and believes that the EU is an effective balance against US unilateralism even if the US and EU agree on many different policies. Therefore, Beijing has developed a very close relationship with the EU while still balancing that relationship with bilateral relationships with EU member states.[30] In contrast, the EU has understood the importance of China as a global partner, and has worked to steady their relationship and not to disrupt relations by pushing China too hard on human rights or other issues of importance to the EU.[31] Washington and Brussels understand that they must work to keep China's cooperation to maintain the global order while Beijing believes that the future global order will transform from the current unipolar system to a multilateral system in which the US, EU, and China will be the main players.[32] For this reason, relations between Beijing and Brussels are of strategic importance to maintain in the future.

Despite the fact that China recognizes the power of the US, EU, and China, Beijing also believes that the US is a falling global power. According to Beijing, it believes strongly in multipolarity and shifting away from the unipolar system which became predominant following the end of the Cold War.[33] Further, China has increasingly

[27] While the EU is a regional organization, it has a global reach, and Beijing treats the EU as a global player as opposed to a regional one especially as the EU has no military.

[28] Pan (2010).

[29] Fox and Godement (2009).

[30] Chen (2012).

[31] Pan (2012).

[32] Ross, R., Tunsjø, Ø., & Tuosheng, Z. (Eds.). (2010). *US-China-EU relations: managing the new world order*. Routledge.

[33] Slobodchikoff et al., (2021)

tried to improve its soft power through programs such as the Confucius Institutes[34] and in providing loans to other countries around the world through its Belt and Road Initiative (BRI).[35] What this means is that China is classified as a revisionist power, whereas the United States and its allies are status quo powers.[36]

3 The Challenge to the Global Order

Following World War II, two distinct international orders developed. The United States developed an international order focused on security, trade, international organizations, and cooperation. Organizations such as the World Trade Organization, the World Bank Group, the International Monetary Fund, and NATO strove to protect US hegemony while also developing a culture of cooperation through liberal values. Allies were extremely important, and cooperation among allies was vital to the spread of the US-dominated order. In contrast, the Soviet Union developed an order that was based on Soviet hegemony, creating allies that were completely reliant upon the Soviet Union. There was very little overlap between the US-dominated order and the Soviet order. Most of Western Europe was under the order that the United States established, while Eastern Europe fell under the order established by the Soviet Union.

The period following World War II became a period of intense competition between the United States and the Soviet Union. Both countries were clearly victorious in World War II, and came out as the most powerful countries in the world. However, both countries had decidedly different ideological views of the world. The United States believed in an economy based on capitalism, while the Soviet Union embraced communism. The ideology of the Soviet Union believed that communism would defeat capitalism, and that the two ideologies were in direct confrontation.

Much of the Cold War period was both Washington and Moscow jockeying to try to expand their order over more of the system while limiting the other's ability to extend order over more of the international system. Each region of the world became a battleground state between these two superpowers in their quest to defeat their opponent. In the United States, the government developed a policy of containment to try and prevent the spread of communism around the world. Basically, the policy was established to prevent the Soviet Union from increasing the number of states that fell under its international order.

In 1989, the world's attention was drawn to a monumental event occurring in Berlin. Common German citizens were storming the Berlin Wall, standing on top of it, and symbolically as well as literally destroying one of the most visible symbols of the Cold War. Astounded Western commentators proclaimed this event as a defining moment, which would come to define the end of the Cold War. No longer was the

[34] Davis and Slobodchikoff (2022).

[35] Davis and Slobodchikoff (2023).

[36] Slobodchikoff and Tandon (2022).

Soviet Union and the Warsaw Pact a threat, but rather a new era of peace, cooperation and the triumph of Western values were heralded.

Then Soviet President Mikhail Gorbachev allowed the Eastern Bloc to slip out of the Soviet order, and agreed to accept the reunification of Germany provided that NATO assist to make the reunification smooth. Further, Gorbachev has stated that he had been assured by German Premier Helmut Kohl that NATO would not expand further east than Germany. He was assured that the Soviet Union was no longer considered a threat, and that instead a new era of peace and prosperity for the whole world would ensue.

In just a few short years, not only had the Berlin Wall fallen and much of Eastern Europe replaced Communist governments with Democratic ones, but the Soviet Union itself dissolved. What had once been called the Evil Empire had now broken up into fifteen newly independent states. Gorbachev found himself a leader without a country, and again the West celebrated a final end to the Cold War.

The years following the collapse of the Soviet Union found the Russian Federation in a precarious position. On the one hand, Russian President Boris Yeltsin needed the support of the West and looked to increase Western investment in Russia. On the other hand, Yeltsin found himself under attack from politicians who lamented the reforms that he had undertaken in Russia. Further, Yeltsin felt himself increasingly under siege by an increasingly imperial US foreign policy. First, the US through NATO used force in the Balkan wars. Russia's traditional ally, Serbia, found itself pitted against NATO forces. Yeltsin argued against the use of force against Serbia, instead trying to argue that peacekeeping troops should be used to keep the combatants separated and come to a mutually acceptable solution. However, according to Russian perceptions, NATO did not feel the need to consult with Russia in the Balkans, and conducted operations accordingly. Further exacerbating Moscow's animosity toward Western policy came when NATO ensured the freedom and independence of Kosovo from Serbia. This action caused many nationalist politicians in Russia to determine that the West was not to be trusted, and to believe that Yeltsin was too close to the West to be an effective Russian leader.

In addition to the Balkan Wars, NATO began to expand into what had once been a part of the Soviet bloc. Despite the assurances to Gorbachev at the time of German unification, an organization whose main purpose had been to counter the Soviet threat had now expanded into the former Soviet bloc. Moscow strenuously objected to NATO expansion to no avail. NATO expanded in 1999, accepting the Czech Republic, Hungary, and Poland. While Moscow did not have the power to oppose Washington's wishes, Moscow made clear that it would actively oppose further expansion of NATO, especially if that expansion were to include states within the former Soviet Union (FSU).

Despite these early setbacks in the relationship between Russia and the United States, following the attacks on the World Trade Center in 2001, Russia was one of the first states to offer to cooperate with the United States on counter-terrorism and sharing intelligence that would lead to defeating Al Qaeda and other terrorist groups. Russia also agreed to allow over flights of its territory by American aircraft in the war against the Taliban in Afghanistan. Further, Russian military officers helped to consult with the Pentagon about Afghan tactics, having learned from their own war in Afghanistan.

Cooperation between the United States and Russia was relatively short-lived. US President George W. Bush announced that the United States would withdraw from the Anti Ballistic Missile (ABM) Treaty. The ABM Treaty had been one of the most important cooperative treaties between the Soviet Union and the United States. Signed in 1972, the ABM Treaty did not allow either state to construct missile defenses against long range nuclear weapons. The idea was that if both powers retained the ability to destroy the other side with nuclear weapons, that neither side would be willing to do so because it would be mutually assured destruction.

In June, 2002, the US officially allowed the ABM Treaty to expire without renewal, thus ending an important chapter in the relationship between Russia and the United States. The US began trying to build a missile shield in Poland, insisting that the missile shield was to protect the United States from a missile attack from Iran. Moscow, on the other hand, argued that if the US was seriously trying to protect itself from a missile attack from Iran, that building a missile shield in Poland did not make logical sense. Instead, Moscow believed that the missile shield being built in Poland was designed to protect the US from a Russian missile attack, indicating that Washington still viewed Moscow as a threat, and not a partner and ally.

Relations further soured in the run up to the war in Iraq, where Russia argued that Iraq had nothing to do with the September 11 attacks, and that the proper solution to preventing Iraqi President Saddam Hussein from developing and maintaining weapons of mass destruction was not to attack Iraq, but rather to work within the International Atomic Energy Association (IAEA) monitoring procedures to ensure compliance. The US, however, decided that Saddam Hussein was hiding weapons of mass destruction, and that invading Iraq was the only way to ensure that the weapons would not be used against either Israel or the United States. The events of September 11, 2001 had conditioned Washington to be more aggressive in its foreign policy by trying to prevent a possible crisis through force rather than having to react to a crisis after having been attacked. Bush argued that it was better to invade and take control of the weapons of mass destruction that Saddam Hussein was thought to have possessed than be taken by surprise by an attack by those weapons within the US. Thus, despite international pressure not to invade, the United States and several allies chose to invade Iraq.

The following year, in 2004, NATO again expanded, this time to include the Baltic states, which had once been a part of the Soviet Union. Moscow again raised a significant protest, but again was unable to muster enough power to prevent NATO from expanding. Despite Moscow's protests, NATO began holding talks with Georgia in 2005 on the likelihood of Georgia joining the alliance, and in February, 2005, both states signed a Partnership for Peace Agreement, which began the process of bringing Georgia into the alliance. Further, in 2008, the Bucharest Summit promised Georgia eventual membership in NATO. Also, in 2008, Ukraine officially requested a NATO Membership Action Plan, which is the first formal step to joining NATO.[37]

---

[37] Slobodchikoff et al., (2021)

It should be noted that not all of the NATO member states were in agreement about pursuing rapid expansion of NATO. While the United States and Poland were actively pushing to expand NATO, both France and Germany were worried about Russian opposition and were worried that Russia would treat expansion as a threat to its own security, and that this would cause serious problems in their relations with Russia in the future.

At the Munich Security Conference, in February, 2007, Russian President Vladimir Putin actively scolded the United States for the way that it had handled the unipolar world following the end of the Cold War. He stated that the United States uses an "uncontained hyper use of force—military force—in international relations, force that is plunging the world into an abyss of permanent conflicts."[38] Putin argued that it was time for the countries of the world to discuss the global architecture of the unipolar system established by the United States, and stated that it was time to pursue a multipolar system as opposed to the current system. Further, Putin's speech directly called out NATO expansion and the fact that he believed that the expansion had led to instability, conflict, and was a direct threat to Russian security.

The Munich Security Conference speech symbolized a breaking point of Russia trying to work within the system to reform it, but rather forcefully stated that Moscow was becoming a direct challenger to US foreign policy and the US global order.

Moscow funded opposition parties in many Western European countries, and used a sophisticated tactic of providing disinformation and stoking conspiracy theories to discredit the legitimacy of the US Global order.[39] One of the most successful tactics used by the Russians was called *whataboutism*. Every time Moscow was accused of aggressive behavior and stoking unrest, the Russian response was to point the finger back at Washington and bring up past American actions that hurt international stability. One of the favorite points made by Russians was to talk about the US War in Iraq and to call it an unjust war of choice by the United States.

Moscow's use of whataboutism was very effective in many other countries of the world, especially in the Global South.[40] Many countries in the Global South viewed the United States as a bully, and Russia was seen as standing up against the most powerful country in the world. Moscow began not only to use aggressive rhetoric to oppose the United States, but also began to use more force to challenge the US global order, especially in the countries of the former Soviet Union, which it viewed as within its sphere of influence. For example, in 2008, as a result of Georgian military action in the breakaway republics of South Ossetia and Abkhazia, invaded Georgia. The Georgian War of 2008 reinforced the idea in Washington that Russia was a threat to European security, and that further expansion of NATO was necessary to contain the Russian threat. Discussions began on how best to integrate Georgia into NATO.

---

[38] Speech and the Following Discussion at the Munich Conference on Security Policy • President of Russia (kremlin.ru).

[39] Davis and Slobodchikoff (2018).

[40] Hill, Fiona (2023, September). *Lecture to the Alabama World Affairs Council*. Montgomery, AL.

Further, the US and Poland wanted to include other FSU states such as Moldova and Ukraine into NATO.

It is important to note the importance of NATO expansion. However, NATO was not the only Western organization to expand to the East. The European Union (EU) had followed an expansion policy that was similar to that of NATO. In addition to actually expanding to the East, the EU had also developed closer relations with several FSU states, and worked towards creating articulation agreements with many of these states, with the idea that these states would receive more access to trade, while the EU wanted to be able to spread their normative values to the East while spreading stable democracy.[41]

In 2013, Ukrainian President Yanukovych was at a crossroads. Either he could sign an association agreement with the EU, which would drive Ukraine into the Western sphere of influence, or he could choose not to sign this agreement and remain in Russia's sphere of influence. Both the EU and Russia were negotiating in such a way as to make him choose either side, with little room for compromise. While many Ukrainians were in favor of Ukraine developing closer relations with the West, Russia offered many economic incentives to retain Ukraine in its sphere of influence.[42]

Eventually Yanukovych decided against signing the agreement, which set into motion a huge crisis for Ukraine. Protesters gathered against Yanukovych, and several months chaos ensued. Within a very short time period, Yanukovych lost control of Ukraine, Ukraine lost Crimea to Russia, and civil war ensued in Eastern Ukraine.[43]

The Russian invasion of Crimea and the ensuing civil war in Ukraine was a pivotal moment in relations between Russia and the United States. Ukraine continued to appeal to the West for support, and NATO member states began to train the Ukrainian military forces. Kyiv continued to wage war against the separatist regions of Luhansk and Donetsk, and a period of a stalemate ensued.

During the stalemate, Ukraine continued to press for NATO membership as well as membership in the European Union. However, neither organization was willing to accept Ukraine as a member state. What NATO was willing to do was to help train Ukrainian military and provide defensive weaponry to defend its territory from an invasion from Moscow. For eight years, Ukraine continued to wage operations against the separatist regions of Luhansk and Donetsk while being trained by Western militaries.

During these eight years, Moscow did not remain idle either. The Russian Army received new weaponry while the political narrative remained focused on presenting a liberal institutional culture and ethos from spreading to the former Soviet Union. Putin claimed to be able to provide an alternative to what he referred to as a Godless

[41] Slobodchikoff (2010).

[42] Slobodchikoff (2017).

[43] Slobodchikoff (2015).

West by touting "traditional values."[44] Moscow actively sought to deepen relations with states in the Global South as well as strengthen their ties with China and India.[45]

In early 2022, Putin began amassing forces near the Ukrainian border. Washington claimed that Russia was preparing an elaborate invasion, yet the Ukrainian government remained unconvinced that an invasion was imminent. Moscow claimed that the forces were merely training, but also tried to get NATO to agree that it wouldn't expand into any more former Soviet states. Moscow began negotiating with Washington by presenting a position that NATO should return to its previous membership before the most recent enlargement. Knowing full well that NATO would never agree to that, Moscow was hoping that NATO would agree to withdraw the Bucharest Summit Agreement and walk back its promise to Ukraine and Georgia that they would eventually become member states. Due to NATO's open-door policy, US President Joseph Biden said that it was not even possible to back off that promise. NATO's own rhetoric had painted it into a corner, where if they took away Ukraine and Georgia's eventual membership, they believed that it would undermine NATO's commitment to its member states and deterrence. NATO argued that Putin's actions was a form of blackmail, and that they would not give in to his threats. They believed that negotiating with Putin would only encourage future aggressive and threatening actions.

NATO's stance took Putin by surprise. The Russian invasion of Georgia had convinced him that NATO and the United States were mostly blustering, and that they did not have the resolve to oppose Moscow and would rather negotiate. Instead, Putin was blackballed and was met with silence and refusal to negotiate. Moscow began an invasion of Ukraine in February, 2022, but invading through Belarus and trying to quickly capture Kyiv. Putin believed that Ukrainian president Zelensky was not supported by the people, and that if the Russian military could quickly topple the Ukrainian government, that Moscow could install a new government that would change Ukrainian policy and allow it to become Moscow's puppet state.

The invasion was a disaster at the beginning. The logistics of the attack were not well-planned, and the attack faltered. Russian troops had expected to quickly overwhelm Ukrainian resistance and prevail. Instead, the Ukrainians mounted a very strong and spirited defense, and prevented Moscow from capturing Kyiv. NATO member states declared strong support for Kyiv, and began to supply large amounts of weapons to fight off the Russian invasion.

Moscow adapted its strategy and focused on capturing a land bridge to Crimea and annexed the territories of Donbass and Zaporizhizhiya. Moscow tried unsuccessfully to capture the city of Kharkiv, and Ukrainian troops began to prepare for a counter offensive. Initially the Ukrainian counter offensive gained a fair amount of territory, but eventually Fall made it very difficult to conduct the counter offensive. This pause in the active offensive allowed the Ukrainians time to train with NATO member states to use more modern weaponry, while the Russians used the time to create a no man's land between the two armies by creating formidable defenses. Huge tracks of

---

[44] Davis and Slobodchikoff (2018).

[45] Slobodchikoff and Tandon (2022).

land were mined and defensive trenches were created. An uneasy stalemate settled over the battlefield. That is not to say that there was no fighting. On the contrary, thousands of lives were lost on both sides during this period. However, neither side was close to winning the war.

Initially, the West, especially Washington, tried to portray the invasion of Ukraine not only as a war of choice, but rather a war between democracy and authoritarianism. Ukraine took up the mantle of being the frontline of democracy and was touted as being the last line of defense. Many Ukrainians believed that Ukraine was actively preventing authoritarianism from violently attacking the democracies of Eastern and Central Europe. Zelensky argued that the only way to prevent Putin from directly attacking NATO was to defeat the Russians in Ukraine.

In Washington, the Biden administration despite the rhetoric, was much more cautious in its approach. It wanted to provide the Ukrainians military support to prevent them from being overrun by the Russians, but the administration also wanted to prevent direct confrontation between Washington and Moscow. Biden believed that if direct confrontation were to occur, that we would then face a war between NATO and Moscow, and that World War III would begin. Instead, Biden believed that an incremental approach was more prudent. The US and NATO member states would support Ukraine through weapons, and little by little they began to provide more and more lethal weapons. What ensued was a proxy war between NATO and the United States. Moscow argued that World War III had already begun and that it was only a matter of time before there was a direct confrontation between NATO and Moscow, with a very real outcome of nuclear weapons being used.

While allies of the United States rallied around Ukraine and the United States, much of the rest of the world were much more cautious. Countries like China and India talked about the belief in state sovereignty, but they would not condemn Moscow for its invasion. Instead, they voted to abstain from condemning Russia in United Nations resolutions. Many African countries relied on Russian fertilizers and wheat supplies for their food production. The war in Ukraine and sanctions on Russia exacerbated food insecurity, and led those countries to argue that US policies were forcing African countries and the Global South to bear an unfair burden that the West and the United States did not have to bear.[46] Their economies suffered tremendously, and created more international instability. Sensing that this was an opportunity, both China and Russia began focusing on diplomacy with these countries trying to gain influence, while the United States was singularly focused on the War in Ukraine.[47]

The post-Cold War order established by the United States of America is currently at a crossroads. No longer is the liberal order and United States hegemonic power a given. As General Michael Hayden, former director of both the US Central Intelligence Agency and the National Security Agency stated, "There is now a battle for the formation of the next global order. Russia and China are battling to create a new global order, while the United States is fighting to continue its primacy in the liberal

---

[46] Davis and Slobodchikoff (2023).

[47] Davis and Slobodchikoff (2022).

order."[48] Moscow and Beijing have both begun their challenges to the United States. While long dissatisfied with US hegemony, in February, 2022, Russian President Vladimir Putin ordered the invasion of Ukraine by the Russian military.

The response by the United States and its allies was swift and unprecedented. Wave after wave of sanctions were levied against Moscow, and NATO member states began to provide military support to Kyiv in trying to maintain its independence from Moscow. The first major war on the European continent since World War II was to have a profound global effect on relations with other countries. First, there was a significant demographic impact as migrants left Ukraine either to the West or to Russia. Further, as men were prevented from leaving Ukraine due to mandatory conscription, the refugees were overwhelmingly made up of either old people, or women with children. Russia also was not spared from a huge demographic crisis. Not only has it lost an enormous amount of young men to the war, but it also instituted forced conscription, leading to many men fleeing the country.

Initially Washington described the invasion of Ukraine as a war between democracy and autocracy, with Kyiv being on the front lines of this new war. However, the real situation was much more complicated than a simple framing of the conflict. While Kyiv enjoyed overwhelming support from the United States and its NATO allies, other democracies globally were a lot warier of supporting Ukraine. Countries like India, Turkiye, Hungary, South Africa, Brazil, and other countries began a strategy of hedging. Publicly, they urged caution in the conflict stating that while Moscow possessed legitimate security concerns, that the conflict had to be resolved peacefully.

The hedging strategy employed by much of the world confused Washington, however, the United States found its attention torn between two continents. The war in Ukraine was its main concern, however, a newly resurgent China threatened to begin its own war against Taiwan. Washington could ill afford a diplomatic blitz to force hedging states to support its policies in Kyiv. Moscow, on the other hand, recognized that it stood to gain from Washington's focus. It began to bolster its diplomatic efforts to woo those hedging countries into not aligning with the United States. Further, the alliance between Moscow and Beijing deepened during this period. In other words, a period of tremendous uncertainty about the future of the global order was born.

In 2022, the Biden Administration released its National Security Strategy, and stated that it had to work to manage Russia, but that China had developed into a near peer competitor, in other words, a country that was able to directly challenge the United States economically and militarily. The War in Ukraine has provided China with a very important opportunity to study the response of the United States and its defense of Ukraine. The swift aide that Washington provided Kyiv gave Beijing pause in its efforts to challenge Washington. Despite Washington's strong response to the Russian invasion of Ukraine, Beijing still challenged the United States and its global order at the fringes. For example, China used high altitude spy balloons for surveilling the United States and flying over US air space. In February 2023, one

[48] Hayden, Michael (2018, April). *Lecture to the Alabama World Affairs Council*. Montgomery, AL.

Chinese balloon was spotted flying over the US state of Montana, was tailed by US military aircraft, and finally was shot down over the Atlantic Ocean.[49]

One lesson that China has learned from watching the war in Ukraine play out is that the United States tends to have a short attention span for crises. Initial support for Ukraine was overwhelmingly positive and unprecedented. However, over time, public support for Ukraine has eroded. In the 2024 election cycle, calls have come from conservative lawmakers to cut off aid for Ukraine. Instead, the United States was focused on helping Israel in its war on Hamas. Waning interest from Washington coupled with elections in European countries where policy makers were not in favor of continued support to Ukraine has shown Beijing that the smartest move is to prepare for short-term pain, which with proper preparation can be overcome. In time, though, interest in maintaining aid and sanctions will wane and become unpopular. In terms of China's aspirations for Taiwan, it means that Beijing must be prepared for economic sanctions and aid to Taiwan. However, if China is able to move swiftly enough, that interest will fade in helping Taiwan, and that Beijing will weather the storm and eventually annex Taiwan. Beijing's greatest problem is that of demography. It is also a falling power, so if it wants to capture Taiwan, its window to do so is closing rapidly.

Ultimately, China is in a position of strength for the short term, and has learned valuable lessons while watching the reaction of the West in Ukraine. However, the time period of having this strength is rapidly fading. Thus, policy makers should expect China to be much more aggressive than it has been in the past. Chinese foreign policy will be one in which impatience will override careful deliberation if China is to realize its strategic goals.

References

Abbott KW, Snidal D (1998) Why states act through formal international organizations. J Conflict Resolut 42(1):3–32

Chen Z (2012) Europe as a global player: a view from China. Perspect Rev Int Aff (2):7–29

Davis GD, Slobodchikoff MO (2018) Cultural imperialism and the decline of the liberal order: Russian and Western soft power in Eastern Europe. Rowman and Littlefield

Davis GD, Slobodchikoff MO (2022) Great-power competition and the Russian invasion of Ukraine. J Indo-Pac Aff 5(4):215–226

Davis GD, Slobodchikoff MO (2023) Great power competition following the Ukraine war. In: The great power competition volume 5: the Russian invasion of Ukraine and implications for the central region. Springer Nature Switzerland, Cham, pp 35–45

Finnemore M (1996) National interests in international society. Cornell University Press

Fox J, Godement F (2009) A power audit of EU-China relations. European Council on Foreign Relations, London, pp 20–27

Fukuyama F (1989) The end of history? Natl Interest 16:3–18

Hempson-Jones JS (2005a) The evolution of China's engagement with international governmental organizations: toward a liberal foreign policy? Asian Surv 45(5):702–721

[49] Chinese spy balloon: US officials disclosed new details about the balloon's capabilities. Here's what we know | CNN Politics.

Hempson-Jones JS (2005) The evolution of China's engagement with international governmental organizations: toward a liberal foreign policy? Asian Surv 45(5):702–721, 704
Huntington SP (1993) The clash of civilizations? Foreign Afairs 72(3):22–49
Kent A (2002) China's international socialization: The role of international organizations. Glob Gov 8:343
Kim SS (1992) International organizations in Chinese foreign policy. Ann Am Acad Pol Soc Sci 519(1):140–157
Lipset SM (1959) Some social requisites of democracy: economic development and political legitimacy. Am Polit Sci Rev 53:69–105
Mearsheimer JJ (2003) The tragedy of great power politics. WW Norton & Company
Pan Z (2010) Managing the conceptual gap on sovereignty in China–EU relations. Asia Euro J 8:227–243
Pan Z (2012) Conceptual gaps in China-EU relations: global governance, human rights and strategic partnerships. Springer
Potter PB (2001) The legal implications of China's accession to the WTO. China Q 167:592–609
Slobodchikoff MO (2010) The New European Union: integration as a means of norm diffusion. JEMIE 9:1
Slobodchikoff MO, Davis GD, Stewart B, (Eds.) (2021) The challenge to NATO: global security and the Atlantic alliance. U of Nebraska Press.
Slobodchikoff M, Tandon AA (2022) India as kingmaker: status quo or revisionist power. University of Michigan Press
Slobodchikoff MO (2015) Overcoming barriers to cooperation: territorial and issue neutralization as tools of conflict management. Soviet Post-Soviet Rev 42(2):127–149
Slobodchikoff MO (2017) Challenging US hegemony: The Ukrainian crisis and Russian regional order. Soviet Post-Soviet Rev 44(1):76–95
Wang H (2000) Multilateralism in Chinese foreign policy: the limits of socialization. Asian Surv 40(3):475–491
Yu PK (2005) From pirates to partners (episode II): protecting intellectual property in post-WTO China. Am UL Rev 55:901
Zhu Y, Warner M (2005) Changing Chinese employment relations since WTO accession. Pers Rev 34(3):354–369

African Debt Crises: Causes, the Role of China, and Recommended US Responses

Marcus Snow and Leif Rosenberger

Abstract China's amazing economic growth for four decades inspired African countries to borrow money for their development. But economic growth is easier said than done. Repaying African loans became difficult. Politicians in the US rushed to judgment and accused China of creating a "debt trap" in Africa. Our research reveals multiple reasons for debt crises in Africa, but none were a Chinese-orchestrated debt trap. Two main reasons for debt problems come from transnational external shocks. The global pandemic hurt Central and West African economies. Climate change exacerbated extreme drought and food insecurity in East Africa. The war in Ukraine caused food prices to rise in Africa. US financial actions had unintended consequences in terms of worsening debt crises in Africa. The Fed's repeated interest rate hikes to fight inflation caused the dollar to rise, local African currencies to plunge, and African debt levels to soar. Earlier, a boom-and-bust scenario on Wall Street caused oil prices to collapse and a debt crisis in Angola, China's largest borrower in Africa. Economic issues were exacerbated by internal shocks. Ghana's loose fiscal actions turned the fastest-growing economy in Africa into a country with the world's worst-performing currency and a debt crisis. The analysis delves into Africa's capital demands, the West's difficulty in meeting those demands, China's ability to supply them, and the debt bubble that they generate. What matters now is how to respond to Africa's multiple debt crises. This paper presents specific recommendations that emphasize the critical role of American finance in promoting stability and economic growth in Africa, while also surpassing China's influence on the continent.

Keywords China · Debt · Africa · Liquidity · PRI · Credit-rating

M. Snow (✉) · L. Rosenberger
The Peacekeeping and Stability, Operations Institute (PKSOI), US Army War College, Carlisle, PA, USA
e-mail: Marcus.s.snow.civ@mail.mil

L. Rosenberger
e-mail: leifrosy@yahoo.com

A. Farhadi et al. (eds.), *The Great Power Competition Volume 6*,
https://doi.org/10.1007/978-3-031-70767-4_16

1 Introduction

The US can out-compete near-peer adversaries in Africa by capitalizing on its significant advantage in risk mitigation, capital formation, and debt financing. This analysis delves into Africa's capital demands, the West's difficulty in meeting these demands, China's ability to supply them, and the debt bubble that they generate. This paper puts forth specific recommendations, highlighting the critical role American finance can play in increasing stability and economic growth while surpassing China's influence in Africa.

To understand China's growing success in Africa, it is important to understand the underlying dynamics. It is simplistic and wrong to attribute China's success to nefarious or underhanded back-door dealings. While the high opacity of China's dealings with African nations lends itself to dark speculation, the cold and sobering truth is that China is offering African nations a superior product to meet their development needs. Developing nations in Africa are customers who have a high demand for capital to launch into the 21st century. Current Western capital formation mechanisms are risk-averse and inadequate for the task. China has stepped in and offered the customer what it wants: fast, reliable, and inexpensive infrastructure development projects and financing. China, in short, is offering a better product to the customer.

Many African countries have turned towards China to fulfill their demand for development projects because China proactively leverages its competitive advantage of being fast, cheap, and discreet in construction while addressing the demand signal from Africa. On the other hand, the US does not mobilize its competitive advantage in finance and risk management and is not responding to the demand signal in Africa. This can be fixed by further supporting existing USG initiatives that have successfully leveraged risk mitigation tools that have mobilized significant private capital in Africa.

1.1 Not Debt Trap Diplomacy

To understand China's growing success in Africa, it is important to understand the underlying dynamics. It is simplistic and incorrect to attribute China's success to nefarious or underhanded back-door dealings. While the high opacity of China's dealings with African nations lends itself to dark speculation, the cold and sobering truth is that China is offering African nations a superior product to meet their development needs. Developing nations in Africa are customers who have a high demand for capital to launch into the 21st century. Western capital formation mechanisms are inadequate for the task. China has stepped in and offered the customer what it wants: fast, reliable, and inexpensive infrastructure development projects and financing. China, in short, is offering a better product to the customer.

While China's lending patterns do not appear to be part of a nefarious debt trap diplomacy scheme, they are irresponsible and can lead to something even more dangerous-a debt bubble. This is not to say that China's lending actions are not without significant risk to all parties involved nor lacking in destabilizing subterfuge. While the primary motivation for Chinese infrastructure development does appear to simply be a case of Chinese Supply addressing African Demand.

2 Background

2.1 China's Risk Mitigation Strategy

China has a strategy to reduce risks associated with its infrastructure projects in Africa. The aim is to ensure that the projects yield positive returns on investment. China can do this by using its expertise in construction, state power, and corruption. Through this, China can turn potentially unprofitable projects into lucrative ones by reducing the costs and risks that deter Western investments. Traditionally, Western lenders have been discouraged from investing in Africa because of the risks involved such as low creditworthiness, uncertain legal systems, and high transaction costs. However, China minimizes labor and input risks by using an Chinese construction process. Also, direct bilateral financial arrangements and clandestine dealings with government officials help to alleviate transaction and legal risks, respectively. Furthermore, China mitigates credit risks by sidestepping conventional credit seniority and placing pledged assets under Chinese control. This enhances the feasibility and profitability of its ventures in Africa.

2.2 China's Balance Sheet "Three Card Monty"

Even if the projects were unprofitable or only marginally profitable, they still might be worth the effort for China because there is a larger benefit to be gained: Cooking China's financial books. Businesses and financial institutions must be solvent to avoid bankruptcy. That is, their assets and liabilities must balance and their cashflows must be able to finance any acquired debt. Balancing the books is a basic task of any business. But what happens when you can't? An easy way out is to acquire assets at a discount that can be posted at full value (par). This is easy with thinly traded assets like real estate or art. What it is something worth? Whatever someone is willing to pay for it or whatever a financial institution or government is willing to state its value is.

2.3 How It Works

When lenders issue loans, they lose cash but gain assets. If you are a bank and you issue a loan, you have lost cash but gained an asset (the loan) as well as an income stream from interest payments. The value of the loan is based on stringent criteria to evaluate risk such that higher-risk loans are worth less (they are discounted) and lower-risk loans are worth more (premium). We see this in bond markets where, for the same coupon, junk bonds sell below their face value and high-quality bonds above their face value. The same is true for loan valuation, even if the loan is never sold or transferred. High-risk borrowers are compelled to pay a higher rate of interest to compensate for the risk. Banks tightly curate their loan portfolios to ensure that they have a reliable and accurate assessment of risk. Excessive defaults would reduce the value of their assets and potentially drive the institution into bankruptcy. Banks then, avoid risk and uncertainty.

If you are a struggling Chinese bank, there is a way around this. If you lend to a high-risk borrower in Africa, you gain an asset and income stream as before. Were you to try and sell that loan on the open market however, you'd find it to be heavily discounted and worth less than the cash you lent out. Not to worry, since you don't intend to sell it to Western financial institutions, it is worth whatever the government assesses it to be. If the Chinese government, which is non-transparent[1] and disinclined to see its state-affiliated banks become insolvent, assesses the loan at a premium, who is to say they are wrong? Instantly, the struggling bank has a robust balance sheet with a high-value loan portfolio.

2.4 The Real Risk-Bursting Bubbles

In such a scenario, asset value was fabricated by the fiat of the Chinese state. The Chinese government has assumed the risk of the loan to a certain extent. Keep in mind, that the risk didn't go away, it was simply transferred and swept under the Government's skirt. This is a familiar story to anyone who experienced the 2008 financial crisis and will have the same result.

Previously, the US was concerned that China's economy growing at a double-digit clip of 10% and higher was growing too fast. The risk was that a strong Chinese economy could drive a strong Chinese military threat. However, there is now a new US security risk—that China's economy is not growing fast enough.[2] During Covid, China had restrictions that kept economy weak. After the pandemic, US assumed China's economy would bounce back That did not happen. China's GDP is now only growing at 4–5%.

Why so slow? Long-term headwinds to China's growth are making themselves felt. China's population has peaked. China has fewer workers to generate growth, faces the familiar "middle income trap" and is struggling to shift from poor-to-middle-income economic growth. At home, Chinese borrowers used empty skyscrapers as collateral. The bill for these bad real estate investments is coming due. The result is a huge mortgage crisis. Overseas, China loaned too much money to Africa and other countries for economic development. That contributed to a debt crisis.

Thus, the real risk isn't that China is exercising Sun-Tzu level debt diplomacy in a great power competition, but rather, they are creating a debt bubble for themselves and creating an untenable debt burden in Africa. It is destabilizing in both directions. For the United States, the harm China is doing to itself is the greater threat as, like it or not, we have a vested interest in a prosperous China and will suffer mightily if it collapses.

2.5 How We Respond

It is telling that rather than attempt to provide a better product to compete, the collective West has largely chosen to complain about China and the customer states buying their development services. In business, complaining about the competition and customer preferences is not a recipe for success- understanding customer needs and meeting them better than the competition is. This is a useful mentality to adopt for US policymakers seeking to compete with China.

3 Africa's Experience with Debt

For almost five decades, Africa's economic experience has included a cycle of booms and busts—and debt crises have punctuated many of the busts. The last 20 years of the 20th Century were considered two "lost decades" in Africa. That convinced many African specialists that Africa was a "hopeless continent." But as it turned out, the first decade of the new millennium was not so hopeless after all.

Near the start of the second decade of the 21st Century, African specialists talked about "Africa Rising" like Asian specialists talked about the 'Rise of China." Near the end of the second decade global powers were frantically involved in a "new scramble for Africa." But that new scramble for Africa didn't last long either. In the 2020s, a bust has replaced the boom.[3]

In this regard, there were two periods when debt levels accelerated. The first was the rising borrowing in Africa in the 1980s that led most of Africa into a two-decade-long debt crisis. The second was after debt relief created a clean slate. Debt levels rose again after 2010. A large group of lenders including China enabled most African countries to boost their borrowing.[4]

It's important to understand that debt is not necessarily bad. Debt can be a benefit, not a burden. If the African governments borrow money wisely, good things happen, and economic conditions improve. But if the African governments mismanage the money, bad things happen. Debt levels become unsustainable, and the debt crisis causes the economy to decline.[5]

African governments borrow capital for many reasons. The most important reason is for economic development. Gregory Smith, a former World Bank economist, notes that "(UN) sustainable development goals (SDGs) will fail without complementary private capital. African governments have long lists of investment ideas ... but all fall short of the financing required to meet investment needs.[6]

Sometimes people ask why African governments and companies just can't go to their domestic markets for their financing. The problem is the pool of domestic financial savings is generally too small in African countries to finance the scale of their investment needs. However, since all African investment needs are not the same, African economies can be broken down into three groups: emerging market economies, frontier economies, and poorer and low-income economies.

- Emerging market economies include countries South Africa, Mauritius, and Morocco. This group has had access to global capital markets for decades and can access its own sizable capital in domestic markets.
- Frontier economies (including Ivory Coast, Ghana, Kenya, Rwanda, and Senegal. This group has varying access to global capital markets and issues Eurobonds—a debt instrument denominated in a currency other than the home currency in which it is issued). In addition, this group experienced a large increase in borrowing from emerging lenders, like China, while still accessing lending from traditional official sources.
- The poorer countries remain aid-dependent and have had little access to private finance. This group includes countries such as Burundi, Sierra Leone, and Mali.[7]

3.1 *Many Reasons for Concern*

Starting in 2017, concerns rose over African debt levels for the first time since the large debt relief initiatives of the early 2000s. In 2018, Vera Songwe, the Executive Director of the Economic Commission for Africa, claimed that there are many reasons to be worried about African debt. Debt risks rose and included defaults by Mozambique and Congo Brazzaville. In addition, several other countries such as Zambia were ignoring risks and heading into a debt crisis. A few others such as Sudan and Somalia were still stuck in past debt crises.[8]

3.2 *Exacerbating Concerns About Debt Crises*

In addition, several external and internal shocks began to exacerbate high African debt levels and confirm Vera Songwe's concerns about African debt crises emerging. US financial actions had unintended consequences in terms of worsening debt crises in Africa. The Fed's repeated interest rate hikes to fight inflation caused the dollar to rise, local African currencies to plunge, and African debt levels to soar. Earlier, a boom-and-bust scenario on Wall Street caused oil prices to collapse and a debt crisis in Angola, China's largest borrower in Africa.[9]

There were also internal shocks. For instance, Ghana's loose fiscal policy turned the fastest-growing economy in Africa into a country with the world's worst-performing currency and a debt crisis. In addition, debt problems came from transnational external shocks. Climate change exacerbated extreme drought and food insecurity in East Africa. The war in Ukraine also caused food prices to rise in Africa. Finally, the global pandemic hurt Central and West African economies.[10]

But while the rising risks of a debt crisis needed to be urgently addressed, it should not be done in a way that indiscriminately kills off private financing for economic development in Africa. In November 2020, Cyril Ramaphosa, the president of South Africa and then chair of the African Union, underscored this point when he wrote: "For three decades, Africa has worked strenuously towards a more sustainable future. The world cannot allow short-term debt dynamics to derail its march towards a green, digitally enabled and globally connected future."[11]

3.3 UN SDGs

Most analysts agree that the 17 UN Sustainable Development Goals (SDGs) are an excellent blueprint for economic development for 2030. That said, it will arguably be impossible to achieve without adequate investment capital from the private sector. While governments need to lead the SDGs, public financing is far below what is needed, and complementary private financing is essential.[12]

Unfortunately, after debt relief for covid, the private capital pendulum swung away from economic development in Africa. This is a big mistake. The idea that private finance is needed for the SDGs to be successful was suddenly forgotten. Little thought was given to the risk of throwing away hard-won African access to private financial markets.[13]

3.4 Solving the Debt Crisis

Too often, lenders take a short view after debt crises. They point to those African countries that mismanage debt. So, lenders are once burnt, twice shy about lending again. They tell borrowers not to borrow so much. But lenders should not low-ball debt. Low debt will keep Africans poor.

Instead, lenders should improve the quality of their financial management. The best kind of debt sustainability is robust, sustainable GDP growth rising faster than the size of the debt. To make this happen, lenders need to do their due diligence ... but also take the long view. Rome wasn't built in a day.

Sound private investment will drive larger and more productive economies. Education and health care investments will lead to a healthier, more resilient, and more skilled workforce. Building industrial parks, railways, electricity, and more functional cities in Africa will eventually generate more revenues and exports. The big concern is that a debt crisis will interrupt this process, put off capital flows, and undermine creditworthiness.[14]

As we look back over almost a half-century of reoccurring debt crises and boom and bust economics, one takeaway seems clear. Standard fixes to Africa's debt problems look outdated. Debt fixes during the pandemic highlighted how the systems in place to deal with debt crises are in urgent need of reform.

What are some of important lessons learned from Africa's debt crisis?

- Getting creditors together is difficult.
- Debt transparency remained too weak.
- Not enough safeguards were available for emerging and frontier African borrowers who needed to preserve access to global capital markets.
- More innovative solutions are needed to achieve the ultimate goal of adequate and sustainable long-term capital.[15]

In addition, efforts are needed to keep African economic development efforts on track. When a country's debt is unsustainable, it makes sense to default and restructure the debt. Otherwise, they will not be able to solve their debt crisis.[16]

But where debts are sustainable, lender efforts should be made to:

- Increase access to proven risk mitigation measures such as political risk insurance.
- Ensure the inputs to measuring investment risk are well and widely understood in Africa.
- Expand regional and global market access for African debt and equity securities.

4 Africa's High Demand for Capital

Africa is growing rapidly and has a large need for capital to fund its development. Many African nations are experiencing rapid economic and population growth and have a significant need to expand their infrastructure, and urban areas, and develop their natural resources. According to the U.N. Economic Commission for Africa, "the continent must invest $1.3 trillion every year to attain the Sustainable Development Goals by 2030. … Africa's annual SDG financing gap is estimated at $200 billion."[17] Obtaining that capital is paramount for Africa's future.

As population and economic activity expand, the demand for transportation, energy, water supply, and telecommunication infrastructure accelerates and outpaces the supply. Investments in infrastructure are crucial for sustained economic development and improved living standards. With this expansion comes urbanization as migrants leave the less developed rural areas for the more developed urban areas. This places a greater demand for infrastructure build-out. Africa's wealth of natural resources requires significant infrastructure and capital outlays to fully develop. The extraction, processing, and transportation of commodities play a large role in African economies, and significant infrastructure improvements are needed to realize their economic potential. This infrastructure requires enormous capital to build and maintain.

4.1 Developing Countries Need Debt to Grow

Debt instruments play a crucial role in the development of emerging economies by serving as vital resources for financing a wide range of development initiatives, such as infrastructure, education, healthcare, and technology. They also help bridge funding gaps, ensuring uninterrupted operations and investments when faced with budget deficits or inadequate revenue.

Debt acts as a catalyst for economic growth by fueling investments in industries and initiatives aimed at job creation, thereby stimulating overall economic expansion. Additionally, during times of crises such as economic downturns or natural disasters, debt provides quick access to funds, facilitating urgent humanitarian assistance and infrastructure rehabilitation efforts.

Moreover, debt helps in bolstering institutional frameworks and regulatory systems, essential for sustainable development and attracting investment capital. Finally, it channels resources into human capital development, through investments in education, healthcare, and social welfare programs, ultimately enhancing productivity and laying the groundwork for long-term economic prosperity.

4.2 Africa Struggles to Get Financing

John Adams famously said, "There are two ways to conquer and enslave a country. One is by the sword. The other is by debt". What Adams left out is that the lack of debt access can condemn a nation to poverty. Debt is the cash needed to build roads, bridges, dams, and virtually any major construction project- in short: debt is essential for development.

Developing countries often face higher borrowing costs compared to their developed counterparts. This is due to a combination of factors. Poor market access limits their ability to tap into diverse funding sources, which constrains competition among lenders and drives up borrowing rates. Credit rating biases, influenced by perceptions of political instability or economic volatility, further exacerbate borrowing costs by deterring investors and lenders. Moreover, the higher perceived risks associated with economic, political, and social instability in developing nations increase lenders' demands for risk premiums, elevating borrowing expenses. Additionally, the lower liquidity of debt instruments in these markets intensifies borrowing costs as investors demand higher returns to compensate for the lack of liquidity.

Collectively, these factors contribute to the financial challenges faced by developing countries. Therefore, it is crucial to foster market access, improve creditworthiness, and implement robust risk management strategies to mitigate borrowing costs and promote sustainable economic development.

Figure 7: Developing countries pay much more for their borrowing
Bond yields (2022-2023)

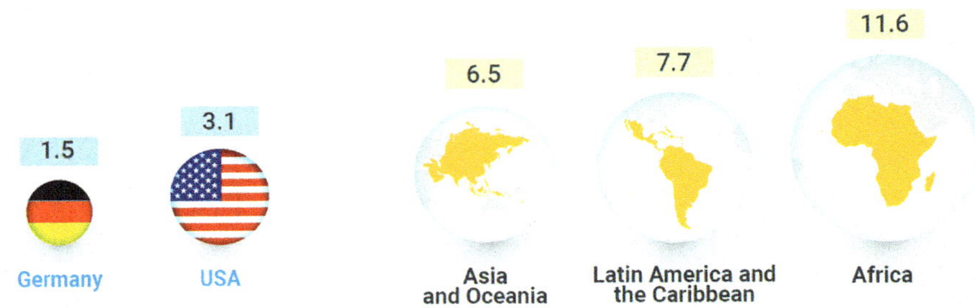

Source: UN Global Crisis Response Group calculations based on IMF World Economic Outlook (April 2023).
Note: Illustrative comparison of the average JPM EMBI Global Diversified USD bond yields per region with the 10-year bond yields of Germany, and the United States from January 2022 to May 2023.

Developing countries' debt trends have caused a rapid increase in total public interest payments relative to the size of their economies and government revenues. 4 Currently, half of developing countries devote more than 1.5% of their GDP and 6.9% of their government revenues to interest payments, a sharp increase over the last decade. The rise of interest payments is a widespread problem. The number of countries where interest spending represents 10% or more of public revenues increased from 29 in 2010 to 55 in 2020.[18]

4.3 *Public Debt is Growing Faster in the Developing World Than in the Developed World*

Public debt is increasing at a faster rate in developing countries than in developed ones. This trend highlights the fiscal challenges faced by emerging economies as they try to meet their developmental needs with limited resources. Expenses such as infrastructure investments, social welfare programs, and economic stimulus measures contribute to the mounting debt burden in these regions. The consequences of this accelerated debt accumulation include increased vulnerability to economic shocks, reduced fiscal space for future investments, and higher debt servicing costs. These factors could potentially hinder long-term sustainable growth and development efforts.

Crucially, growing debt is not necessarily bad if it leads to even more GDP growth. That is if the debt went to capital improvements that create a surplus beyond the cost of financing the debt. Africa has a hit-or-miss story with this. While developing economies and sub-Saharan Africa in particular are expected to post very healthy growth rates (over 4% for 2024 and 2025 projected),[19] the African share of external debt to GDP has been climbing over the last decade. The chart below illustrates the growth of African external debt as a share of GDP since 2009. This largely coincides with the increased pace of Chinese lending to Africa. Of note, the ratios have moderated in the or guarantees can also secure favorable interest rates and loan terms at 22.8% for Africa collectively.[20]

African external debt as a share of GDP has been increasing

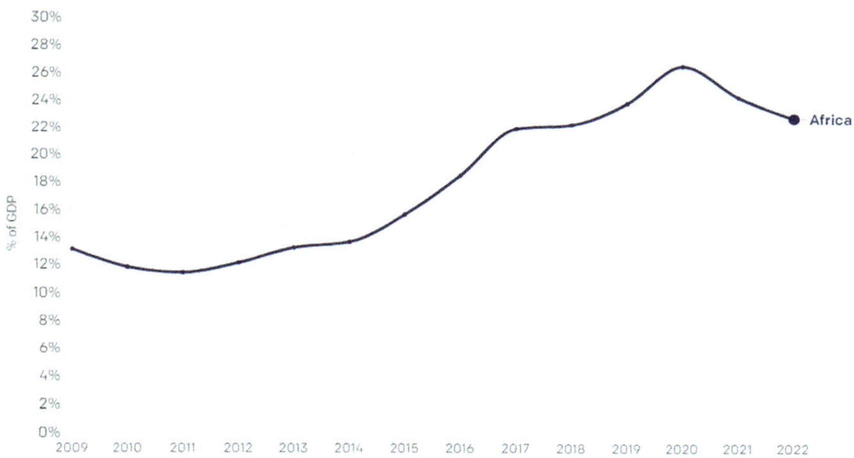

Sources International Debt Statistics (IDS) database

Debt trends in developing countries have led to a significant surge in public interest payments. Over half of these countries allocate more than 1.5% of their GDP and 6.9% of their government revenues for interest payments. This has risen over the past decade and is a growing concern, as interest spending constitutes 10% or more of public revenues in 55 countries, up from 29 in 2010.[21]

Figure 8: Developing countries use more resources to pay interest on their public debt
Net interest payments of developing countries as a share of GDP and government revenues

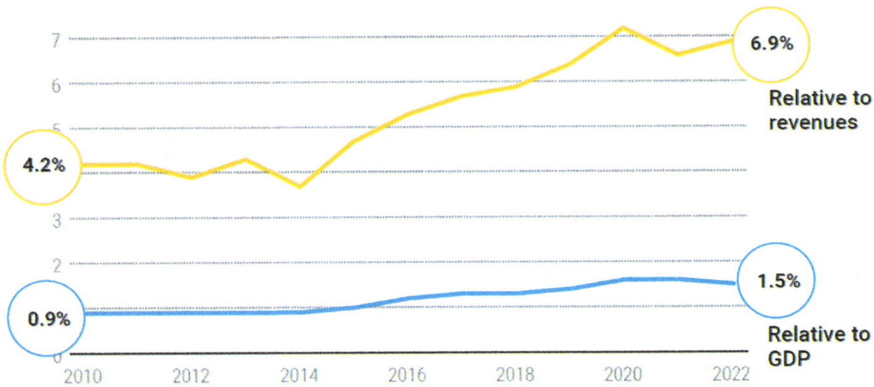

Sources: UN Global Crisis Response Group calculations, on IMF World Economic Outlook (April 2023).
Note: Median shares across developing countries.

5 Supplying the Demand: The Rise of Chinese Lending

At first, Africa's demand for investment capital was much larger than the shortfall of supply from traditional Western lenders. Thankfully, China stepped up to fill the gap in unmet demand. By 2019, China funded "one out of every five projects (20.4%) in Africa, just behind the share of projects funded by African governments (22.8%)." [22] This process didn't happen overnight. It was many decades in the making and emerged from China's own inspiring experiences as a rapidly developing nation.[23]

China experienced rapid growth in the 1990s. This surge in the demand for commodities spurred China's interest in Africa. Henceforth access to resources became a crucial factor driving deeper engagement with Africa. China also pursued various other objectives. These included expanding trade relations, encouraging private foreign direct investment (FDI), providing military support, facilitating commercial lending, extending official lending, and offering foreign aid. Following the launch of their "Go Out" strategy in 1999, China continued to scale up its lending to Africa in the early 2000s. The result was economic interdependence along the pathway to shared prosperity.[24]

China's unprecedented economic growth stands as a monumental achievement in the annals of history. Its rapid transformation from poverty to wealth served as an inspiring narrative, positioning China as a role model for nations aspiring to replicate its success in economic development. The Chinese example has notably inspired African countries, igniting aspirations to emulate China's trajectory. Moreover, China has emerged as a more credible source of economic development advice than traditional donors, offering valuable insights and strategies garnered from its remarkable journey of growth and development.

China's rapid export-led growth strategy successfully delivered a substantial trade surplus. As a result, US dollars poured into the country. To further leverage this economic momentum and expand its global presence, the "Go Out Strategy," urged Chinese companies to seek business opportunities abroad. Toward this end, Beijing encouraged Chinese companies to finance infrastructure projects across Africa. In the process, China showed its.

Enthusiasm and increasing involvement in international markets and its commitment to supporting development initiatives in Africa and across the world.[25]

The start of the Forum on China-Africa Cooperation (FOCAC) conferences in 2006 marked a pivotal moment in solidifying the relations between China and Africa. A significant aspect of China's rising partnership with Africa was China's approach to lending. China's non-interference policy was distinctly different than the Western approach. Unlike other lenders, China's lending practices were notably less intrusive, largely devoid of critiques on recipient governance systems and devoid of long, imposing lists of policy conditions. Most importantly, China's rapid infrastructure lending added to its appeal. Not surprisingly, there was a high demand for Chinese loans from nearly all African governments. This collegial Chinese approach stood in stark contrast to the more selective approach often adopted by Western aid agencies.[26]

There was also a stark contrast between what Western aid providers offered and more tangible product lines China offered. The West talked about Millenium Development Goals and financing for social sectors like education and health care, China focused on infrastructure. African governments were free to choose the type of infrastructure they wanted. This included sports stadiums, government buildings, hydroelectric dams, roads, and railways. Africans like to say it boiled down to the following: "We had a choice: China offered us fast infrastructure; the West offered us long lectures."[27]

Thanks to the China Africa Research Initiative at Johns Hopkins University, we can appreciate the large scale of Chinese lending. There was a remarkable increase in Chinese lending to Africa over the past two decades. China's lending surged from $133 million in 2000 to a staggering $6.4 billion in 2010. However, this growth accelerated even further after 2011. Following a brief slowdown due to the global financial crisis, China's economy quickly rebounded China's economy was fueled by a significant domestic stimulus package. That in turn drove a period of rapid expansion in lending activities, not only in Africa but also worldwide.[28]

By 2013, China's annual lending to Africa had soared to $17.6 billion China's lending reached a peak of $28.3 billion in 2016 before gradually moderating to $12.7 billion in 2017 and $7 billion in 2019. From 2010 to 2019 alone, China extended a total of $127 billion in loans to African nations, a stark comparison to the $27 billion lent between 2000 and 2009. Most of these loans were for infrastructure projects Transportation projects received around 30% of the funding while power projects accounted for about 25%.[29]

The significant surge in Chinese lending to Africa during the period spanning 2010 to 2016 aligns with China's broader trend of expanding its lending activities worldwide during that time frame. A substantial portion of the heightened lending observed after 2013 can be attributed to the Belt and Road Initiative (BRI), a comprehensive infrastructure and economic development strategy spearheaded by China. This initiative, as outlined earlier, encompasses various regions and countries, with several nations along Africa's eastern coast notably positioned along the maritime routes integral to the BRI's objectives.[30]

6 Supplying the Demand: The Chinese Advantage

China holds a significant advantage in its unparalleled capacity to execute large-scale infrastructure projects swiftly and affordably, surpassing its global counterparts. This advantage stems from China's proficient construction capabilities, which it combines with accessible and unobtrusive lending packages. Notably, all construction and financial assistance are exclusively provided by Chinese firms. However, this advantage comes with its caveats; these firms are frequently marred by corruption, often resorting to bribery with local authorities. Despite concerns over construction quality, deemed satisfactory albeit mediocre for a rapidly growing economy, China abstains from imposing political or intrusive prerequisites, unlike Western lenders.

Chinese construction companies have a very real competitive advantage in construction. They are highly skilled, experienced, and a well-oiled machine. Chinese projects are finished faster and cheaper than their competitors. China has a very real and significant competitive advantage in the ability to complete projects quickly with a large, experienced workforce, project management skills, and the use of prefabrication techniques. Chinese construction firms are perceived as being able to offer cost-effective solutions with low labor costs, economies of scale, and efficient procurement strategies. Finally, Chinese companies have experience in large-scale infrastructure development, as China itself has undergone significant infrastructure expansion in recent decades.[31] This experience can contribute to their ability to efficiently execute projects.

6.1 China's Favorable Interest Rates

China has been able to keep interest rates below market rates through collateralization and in-kind payments. In some cases, natural resource-producing countries agreed to pay back Chinese loans with commodities. Angola, Congo-Brazzaville, Ghana, Niger, and South Sudan have made such commitments with the oil they produce. The DRC made similar commitments with gold, cobalt, and copper. Ghana and Guinea made agreements for infrastructure loans in exchange for bauxite between 2017 and 2019. Zimbabwe also put up platinum. Although details are opaque, not transparent. Chinese resource-backed loans to African countries had interest rates as low as 0.25%.[32]

Risk Appetite China has shown a willingness to accept lower than market standard risk/return ratios due to its comprehensive Return on Investment (ROI) framework. This framework includes strategic benefits such as access to essential resources and the maintenance of strategic construction capacity. It also maintains a perceived balance in the accounting of Chinese banks, which helps to avoid defaults. Additionally, the CCP benefits from providing employment and revenue to firms operated by it directly, which contributes to China's overall ROI. These projects also present opportunities for insider corruption within the CCP, which further enhances its gains.[33]

Chinese Value Proposition: Fast, Cheap, and Good Enough A study by Neil Gerard from "Construction Briefing", a construction industry journal excellently summarizes what China brings to the table. Following is an excerpt from his 14 March 2023 article, "Why have Chinese construction firms become so influential in Africa?" which neatly captures China's construction advantage[34]:

> *"There are various reasons for this success. One is the fact that these contractors are able to deliver complex projects with excellent technical specifications in short timeframes, something that African governments and their elites have appreciated. The massive construction boom in China since the 1980s constituted a major advantage in terms of gaining technical experience, scaling up construction activities and learning to manage complex projects,"*
>
> ***What is driving Chinese contractors' success?***
> ***1) Scale of lending:*** *Frank Kehlenbach, director of European International Contractors explains: "China offers package deals, which are attractive to African clients because they are organised more or less as a one-stop shop. After the financing agreement between the Chinese and the African government the loan is provided quickly by a Chinese policy bank and then the contract is awarded to a Chinese state-owned contractor. Sometimes, the African country pays back the loan with natural resources in a barter deal. Such one-stop project delivery is not possible for a country like the UK or Germany. Development policy processes in Europe are time-consuming and much more complex."*
> ***2) Competitive pricing:*** *Chinese construction companies have often proven able to offer more competitive tenders than their western or African counterparts. According to European International Contractors (EIC), Chinese contractors are under-cutting market-dependent businesses from elsewhere in the world by up to 20% in open competitions.*
> *Because some Chinese finance deals are tied to the use of Chinese contractors, the companies already benefit from having supply chains and labour in place in Africa. That helps them to bring down cost.*
> ***3) Speed of construction:*** *CSCEC took just 808 days to build the entire concrete core for the Iconic tower in Egypt. It built the steel framework at a rate of one floor every three days. That reflects the speed at which Chinese firms are able to build, sometimes working off pre-existing designs for structures they have already built elsewhere. That speed appeals to local leaders, anxious to see their projects completed before the next election.*
> ***4) Western companies' attitude to risk:***
> *Construction work in Africa can carry risks including political instability and escalations in conflict. Some western companies also hold concerns about business practices. Meanwhile, beefed-up anti-corruption laws in places like America and Britain mean that paying bribes can land them in serious trouble.*

6.2 On Project Quality

Criticism of Chinese Corruption and quality concerns around large infrastructure projects in Africa are similar to criticism of the construction of the US rail system in the 19th century. At that time, European nations critiqued the quick building and lower-quality construction of American railroad projects. This wasn't corner-cutting on the American part but rather a deliberate strategy to capture the economic benefits of rail lines in the near term to fund further improvements and development later on. Timber was used when the Europeans would have used stone. The Americans deduced the lifespan of projects was 10–20 years and timber was adequate. Who knew if the rail or project would be viable after 20 years? It is only semi-permanent. Europeans used stone because they conceived of projects as permanent. Timber was faster and cheaper and got the rails moving now rather than later. The American concept was technically more efficient and amortized the construction costs over the expected useful life of the project. The cost of later repair or renovation work was more than paid for with the added revenues provided by the initial investment. It is not surprising then that developing nations are making similar choices.

7 The Problem with the Chinese Way

China's ambitious projects, which are brokered and built at below-market costs, may pose significant financial risks for the country. In addition to the cost of corruption, China is assuming the risk of defaults. China may not have accurately evaluated the non-monetary strategic value of these projects if there is significant graft on either side of the contracts. In such cases, China may end up underwriting bad loans without gaining any strategic access, goodwill, or partners. All of this puts China at risk.

Napoleon once quipped "Never interrupt your enemy when he is making a mistake." In the case of China, however, the US has a vested interest in the stability of both the Chinese and developing world's debt profile. After all, the US 2008 sub-prime implosion had a globally negative impact. Similar poor lending practices between China and developing nations may do the same when the bubble bursts.

7.1 China Brings Risk and Instability

China's lending practices can create significant risks and instability for borrowing nations. This is because many countries tend to borrow too much, which weakens their financial resilience. The consequences of this practice can be severe, as financially vulnerable countries are more susceptible to catastrophic outcomes in the face of various shocks. These shocks could range from shifts in interest rates to crises in food, energy, and commodity prices, or disruptions in the supply chain. Recent incidents, such as those involving the Gulf of Aden and the Houthis, highlight the potential impact of these shocks. Strong financial positions can help nations absorb such shocks, ensuring stability even in tough times. Conversely, those without such resilience face increased instability, which can affect not only global supply chains but also cause humanitarian crises.

In addition, how Chinese loans are negotiated—and their lack of transparency—raises significant concerns from development, governance, and financial perspectives. For instance, any Chinese loan contracts reportedly have provisions that can lead to "hidden debt." This hidden debt can result in unexpected public liabilities when Chinese-funded projects fail to generate strong economic returns. Additional provisions ensure that in the event of financial distress, Chinese banks are the first to be paid off.[35]

In the face of the challenges brought on by the COVID-19 pandemic and Russia's full-scale invasion of Ukraine, rising interest rates have brought many African borrowers to economic distress. In short, China's hidden debt complicates an African borrower's ability to negotiate with lenders and achieve long-term economic stability.[36]

8 The Way Ahead: What the USG Can Do to Win

Winners focus on winning, losers focus on winners—Eric Thomas

The path to winning in Africa lies in meeting the African demand for financing better than the Chinese do. Supplying the demand for capital will lead to the long-term governance and transparency reforms needed to release the pressure on the debt bubble China has created.

To supply the demand for capital, it is important to understand why the private market does not supply enough capital (either debt or equity) on its own. The short answer is that capital is a coward and avoids risk and the unknown. African development projects are perceived as high risk, largely due to political and legal concerns, and the lack of experience investing in Africa makes it an unfamiliar landscape with hidden and unknown risks. Africa is competing for scarce financing dollars with nations that have proven and well-known risk profiles and where the 'unknowns have been reduced or accounted for. Finally, investors want to know that they can get their money back or trade a debt or investment with relative ease.

Luckily, the US Government is well positioned to mitigate risks and mobilize private capital to address Africa's demand for financing while simultaneously mitigating the consequences of poor Chinese lending practices and reasserting US influence in Africa that may have been lost to China's easy money policies. The US must reinforce its efforts along the following three winning lines of effort:

- Mitigate political risk through insurance and guarantees for investment projects.
- Remove the unknowns by optimizing debt credit ratings to more accurately reflect investment risks in Africa.
- Increase marketability and liquidity of debt instruments in Africa by supporting regional African debt facilities.

Collectively, addressing these issues will mitigate political risk, lead to more accurate and transparent risk assessments, and provide attractive market access for private capital.

9 Recommendation 1: Mitigate Risk Through Insurance

Dramatically increase the size of the US Development Finance Corporations (DFC) Political Risk Insurance (PRI) program and target African nations where loose Chinese lending has created untenable levels of debt service

The US Development Finance Corporation has led the way in mitigating risk in fragile states in Africa. They have achieved this through the use of Political Risk Insurance (PRI), a non-invasive and efficient way to reduce the high-risk premiums that would make speculative investments otherwise untenable to private capital. The DFC in effect, insures investments to some extent removing significant risk from the table for private investors while at the same time reducing the interest rates that borrowers have to pay. An analogy might be VA home loans or subsidized disaster insurance for coastal homeowners in the US. The DFC's use of PRI is an inexpensive way to attract billions in additional private FDI that would not otherwise occur.

9.1 The Value of PRI

Political risk insurance (PRI) is a valuable tool for investors to manage country-specific risks and increase the financial viability of their projects. It allows investment in countries with high levels of risk that would otherwise be considered unfeasible. In addition to its primary function of risk allocation, PRI has a broader impact on developing nations, as it helps to foster positive externalities such as boosting investor confidence, enhancing credit conditions, and increasing the value of investment ventures.[37]

9.2 PRI Makes Borrowing More Affordable

Political Risk Insurance (PRI) can help to improve the creditworthiness of debt securities in emerging markets. Additionally, PRI effectively lowers country risk premiums, which increases the value of investments within these markets. As a result, this enhances the internal rate of return for developmental projects.[38] As shown in the chart below, in the case of Ghana, PRI reduced the risk premium of debt instruments to one-third of the non-insured rate.[39]

9.3 PRI Lengthens Credit Terms

According to research carried out by the Islamic Development Bank, guarantees from The International Bank for Reconstruction and Development (IBRD) have a significant impact on the duration and cost of financial agreements in 13 countries. The study found that the presence of insurance or guarantees can extend credit terms by 10–15 years. However, these same insurance or guarantees can also secure favorable interest rates and loan terms[40]:

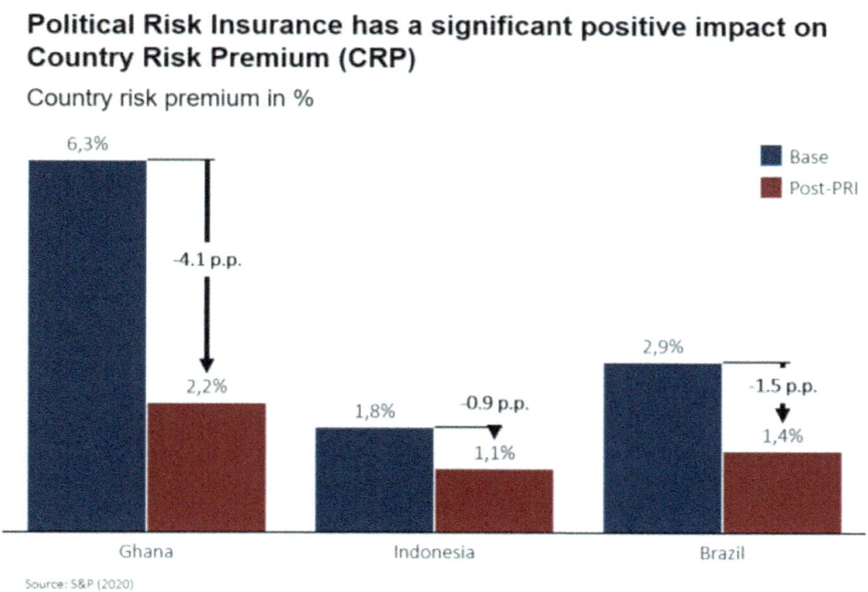

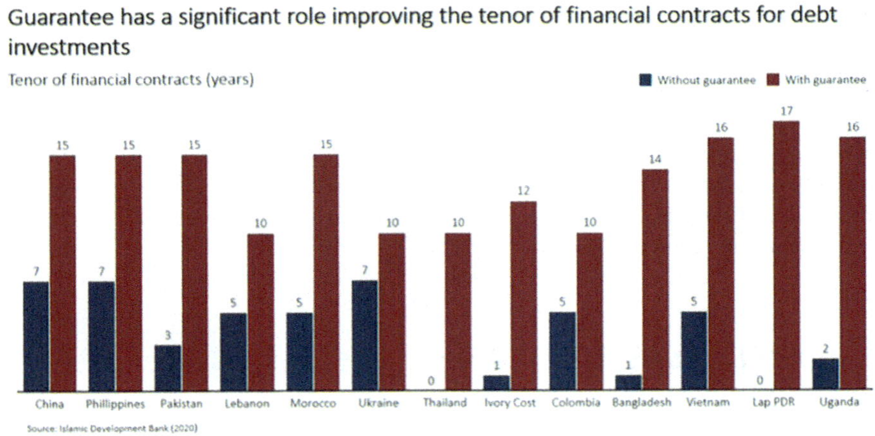

9.4 DFC's PRI Drives Private Capital Mobilization (PCM)

A McKinsey study of the effects of DFC's use of PRI in 14 projects in Africa found that it drove the mobilization of private capital through both first and second-order effects.

The first-order effect of DFC's PRI was to attract private capital for the projects by reducing or reallocating risks for private investors. This lowered the investment's risk profile and enhanced expected returns. As a result, it boosted investor confidence and mobilized capital. The McKinsey report found that private investments would not take place without PRI coverage. DFC's PRI activities have a direct impact on capital mobilization by improving expected returns, securing investor commitment based on reduced risk, enhancing credit and lending terms, and decreasing the likelihood of political risk events through diplomatic engagement.[41]

The second-order effects of PRI pertained to a broader perspective on capital mobilization beyond individual transactions and encompass the risk mitigation effect that comes with familiarity and experience. In short, private investors attracted by the PRI were subsequently more likely to invest future dollars based on their increased understanding of the market realities and conditions in the developing countries.[42]

9.5 DFC Has Room to Grow Using PRI to Attract Private Capital

A recent report on the impact of the DFC's use of PRI found that the DFC demonstrates competitiveness in pricing, tenor, and exposure limits comparable to leading public insurers and solid prospects for expansion.[43] When surveyed, commercial and institutional lenders indicated that additional PRI would likely lead to increased private capital in lower-income and high-risk states. The report also found that DFC had a competitive advantage in serving large-scale projects that could serve as an opportunity for growth.[44]

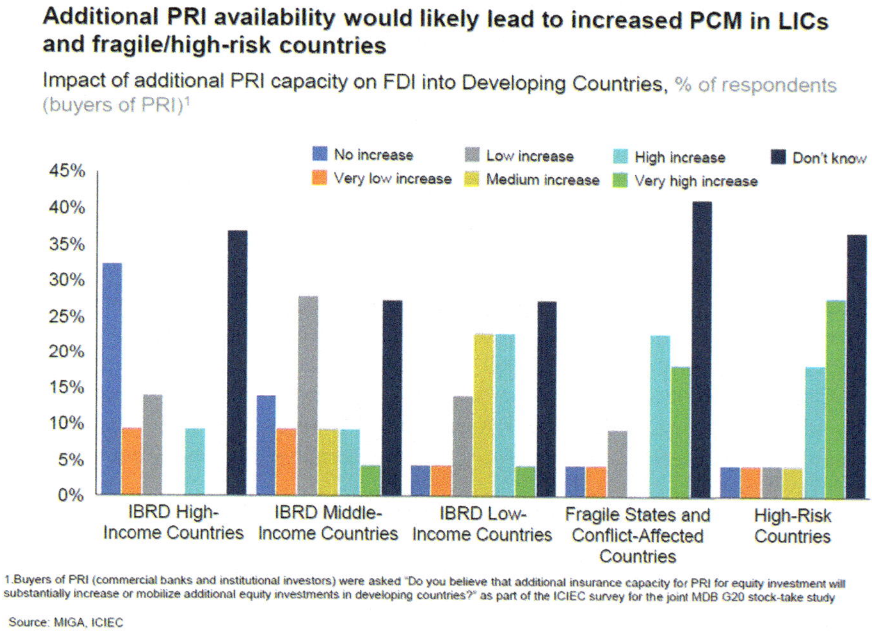

Private capital mobilization refers to the amount of private capital that is invested in a project, sector, or country as a result of some type of DFI investment. In the case of the DFC PRI, private capital mobilization refers to the amount of private capital that is invested because PRI was provided to the projects. Two main drivers of PRI enable first-order private capital mobilization. First, PRI attracts investments into high-risk countries by managing unacceptable levels of risk. Political risk insurance can help secure investor buy-in that is conditional on reducing risk to an acceptable level and can also help secure an affordable debt by improving the credit ratings of project debt. Second, PRI, and especially those coverages provided by public entities, reduce the likelihood of political risk events because of political and diplomatic engagement.[45]

DFC has a competitive edge in the market due to its underwriting process being perceived as faster and smoother than its competitors, such as the World Bank's Multilateral Investment Guarantee Agency (MIGA). That said, the McKinsey report found DFC's underwriting process to be lengthy. Furthermore, DFC was viewed as a passive entrant to the marketplace with little to no business development presence. If DFC plans to expand its PRI activities, it will need to increase its PRI underwriting capacity and embrace proactive marketing measures to maintain its reputation as an efficient and private-sector-friendly underwriter.[46]

10 Recommendation 2: Optimize Credit Ratings

Increase USG technical assistance to at risk African nations to optimize their sovereign debt rating. This is achieved through close engagement with the major international Credit Rating Agencies (CRAs) (Moody's, Fitches, S&P) and individual African states and Regional Economic Communities (RECs)

According to the UNDP, the CRAs do not accurately rate African debt.[47] The US can work with the CRAs and African states to review their credit ratings, understand where the scores are derived, and work to increase them.

Critics argue that external credit rating agencies lack a comprehensive understanding of African economies, cultures, and contexts. This perceived lack of understanding may lead to biases and inaccuracies in their assessments. Some argue that credit rating agencies often use standards and criteria that are more applicable to developed economies, potentially leading to misjudgments when applied to the diverse economic structures of African nations. Furthermore, African economies, especially those with smaller market sizes, may receive less attention from major international credit rating agencies. This limited coverage can result in less accurate or outdated assessments. Finally, CRAs may be influenced by global economic conditions and geopolitical factors, which might not fully capture the unique circumstances of African nations.

To what extent these critiques are valid is difficult to say. Given that the CRAs are unlikely to create an Africa-specific benchmark, however, the most productive step is to engage closely with the CRAs to understand the criteria, feed the analyst with timely data, and ensure that each nation is getting the credit score they deserve and nothing less.

African countries are paying a premium in borrowing costs of 2.9% points more than what the macroeconomic fundamentals and current credit ratings might suggest. The UNDP analysts argue that higher coupon payments can only be explained by a penalty on the borrowing nations resulting from investors' biased perceptions.[48] There is certainly merit to this assertion in that the major credit analysts have relatively thin coverage of African sovereign debt and may be missing a robust perspective of African sovereign debt.

According to a report from the United Nations Development Program (UNDP), the credit ratings for African countries have some idiosyncrasies due to a lack of relevant data and methodological subjectivity. These idiosyncrasies development financing in Africa. To estimate the impact, the ratings of the Big Three CRA's were compared with those provided by an independent and more objective algorithm from trading economics. The report shows that 16 African countries had to pay an additional $28 billion in interest payments due to these peculiarities and lost $46 billion in potential flows. The total cost is higher than what is needed to reduce malaria in Africa by 90%[49]:

Figure 5: Development Costs of Credit Ratings Idiosyncracies

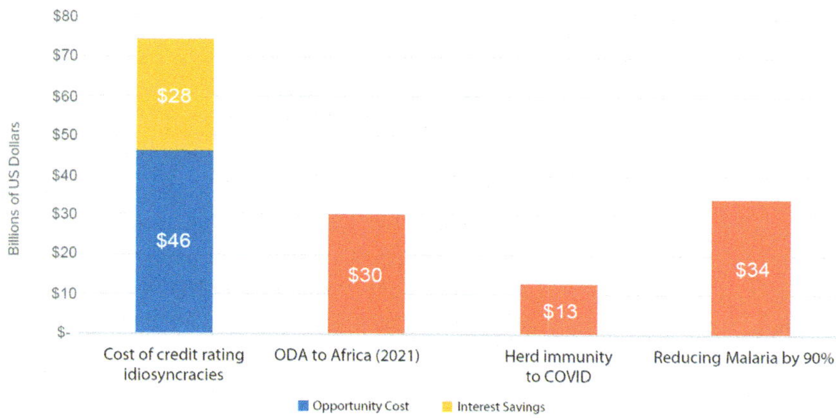

Reducing interest rates paid by African countries on both domestic and foreign debt could greatly decrease the debt service burden they face. This, in turn, would enable them to repay the principal faster and free up funds for more investments in development.[50]

10.1 Solutions

To minimize credit rating differences, it is important to increase transparency in credit rating methodologies, ensuring that these methodologies are well understood by the nations being rated. Ratings should be in line with the macroeconomic fundamentals of each nation, and credit analysts and rated nations should use the same set of accurate economic data.[51] It is important to engage with rating agency advisors and understand how ratings are determined and how economic policies will impact credit scores. A nation cannot receive a better rating than it deserves, but it should aim to receive at least what it deserves.[52] By following these guidelines, African countries can reduce their spending on debt service and allocate more resources towards other priorities.

The USG can play an outsized role in reducing credit rating idiosyncrasies. USAID and DFC already offer technical assistance to both African countries and US firms to understand and mitigate investment risks. This should be extended to a formal program of technical assistance to for sovereign debt issuers to work with CRA's to minimize credit idiosyncrasies. A technical assistance program that pursued the recommendations of the 2023 UNDP study and worked bi-laterally with select African nations would be able to eliminate tens of billions lost to inaccurate risk analysis.

11 Recommendation 3: Increase Liquidity of Debt Instruments

Increase marketability and liquidity of debt instruments in Africa through USG support to regional African sovereign debt facilities. Increasing liquidity will decrease interest rates and significantly enhance debt sustainability

In 2020, Ghana issued a 29-year US €1 billion bond at a coupon rate of 8.95%. Greece, with the same credit rating, issued a 30-year bond at a coupon rate of 1.88%. Why the difference?[53] It has to do with liquidity. Greek debt paper is cash and can be bought at a discount by the European Investment Bank (EIB). It is liquid. When liquidity goes up, financing costs go down because spreads are not just credit, they're also liquidity. Global markets are competitive, and money is fungible, it goes to highly liquid markets.[54] If African nations had access to a liquidity facility for sovereign debt, it would help reduce interest rates.

Nations with similar credit ratings will have lower interest rates if their notes are highly marketable. Increasing the marketability of African sovereign debt. Liquidity facilities serve to repurchase (or "repo") debt issues so that they then can be further traded. This allows thinly traded debt instruments to be traded more efficiently and makes them more marketable.

The United Nations launched the Liquidity and Sustainability Facility (LSF) in 2021 with the intent to support the liquidity of African Sovereigns Eurobonds and incentivize SDG-related investments such as SDG and green bonds on the African continent. It aims to improve African Sovereign debt sustainability and enhance the liquidity in the market.[55] Per the LSF: *"Developed countries have long enjoyed the existence of large repo markets for their government bonds, facilitating the creation of stable and additional funding sources."*[56] The creation of the LSF provides a large repo market to Africa.

Smaller liquidity facilities, such as the African Development Banks Emergency Liquidity Facility[57] and the Regional Liquidity Support Facility (RLSF), for sub-Saharan African Independent Power Producers[58] play more niche but important rolls.

In part due to poor liquidity for sovereign debt securities, African governments face high costs of borrowing. Debt facilities reduced transaction costs and improve global market access which has the potential to save African nations an estimated USD 11 billion on borrowing costs over the next five years.[59] This supports the debt sustainability of African nations.

USG support to the LSF and other niche liquidity facilities can have a tremendous impact on creating stability in Africa and counter the negative effects of poor Chinese lending practices. This should occur in three areas: Direct financial support to strengthen the LSF's ability to provide liquidity; technical assistance and staffing support to the LSF; and bi-lateral technical support to sovereign debt issuers to maximize the benefits of the LSF.

12 Conclusion

The United States has a significant advantage in risk mitigation, capital formation, and debt financing that can be leveraged to out-compete near-peer adversaries in Africa. This analysis focuses on the capital demands in Africa, the challenges faced by the West in meeting this demand, China's ability to supply it, and the debt bubble that it creates. This paper presents specific recommendations that emphasize the critical role of American finance in promoting stability and economic growth in Africa, while also surpassing China's influence on the continent.

To comprehend China's increasing success in Africa, it is essential to consider the underlying dynamics. It is incorrect and oversimplified to attribute China's achievements to shady or deceitful backdoor dealings. While China's dealings with African nations may seem opaque and lend themselves to dark speculation, the reality is that China is providing African nations with a superior product that meets their development needs. Developing nations in Africa require a significant amount of capital to progress into the 21st century, and the current Western capital formation mechanisms are inadequate and risk-averse to accomplish this task. China has taken the initiative and offered African nations what they require: fast, dependable, and economical infrastructure development projects and financing. In essence, China is providing the customer with a superior product.

Many African countries have looked to China for development projects because China is known to deliver construction projects quickly, cheaply, and discreetly while responding to the demand signal from Africa. In contrast, the US has not effectively utilized its competitive advantage in finance and risk management and has not been responsive to this demand signal in Africa. However, this situation can be improved by providing further support to existing US government initiatives that have successfully employed risk mitigation tools, thereby mobilizing significant private capital in Africa.

The USG has a significant role to play in mitigating the adverse effects of Chinese lending while simultaneously accelerating the development of strategic partners in Africa and showing a development path that is open and transparent. A strong reinforcement of DFC's successful PRI initiative will mobilize billions of private investment dollars and equip US firms with the experience to knowledgeably invest in some of the world's fastest-growing economies. An effort to bridge the information gap between the major CRAs and the issuers of sovereign debt will ensure that risk assessments are more accurate and fairer, saving billions lost to market inefficiencies. Finally, the USG has a role to play in supporting the growth and integration of African securities markets to provide desperately needed liquidity and capital access for African debt instruments.

Notes

1. Written Statement of Janean Davis, Deputy Assistant Administrator, Bureau for Africa, before the House Foreign Affairs Subcommittee on Africa. USAID https://www.usaid.gov/news-information/congressional-testimony/apr-18-2023-great-power-competition-implications-africa-chinese-communist-party, April 18, 2023, accessed 28 March 2024.
2. Martin Sandbu, "The other economic security threat from China", FT, May 25, 2023.
3. Gregory Smith, Introduction, "Where Credit is Due", Oxford, New York, 2021.
4. Ibid.
5. Ibid.
6. Ibid.
7. Ibid, 3–4.
8. See Vera Songwe, "African Countries Should Demand Loans Are Made in Local currencies", Financial Times, (FT), 18 November 2018.
9. For an in-depth discussion regarding how the boom-and-bust scenario on Wall Street caused oil prices to collapse, see Leif Rosenberger, The Energy War: How Europe Turned the Tables on Russia, pp. 258–260 in Adib Farhadi et al. in The Great Power Competition, Volume 5: The Russian Invasion of Ukraine and the Implications for the Central Region, Springer, Switzerland, 2023.
10. For an in depth whole of government discussion of how the pandemic hurt Africa's comprehensive security, see Leif Rosenberger, Root Causes of Violent Extremism in Africa: Understanding Multi-Disciplinary Factors, The Street, 19 January 2021.
11. Smith, op.cit, 6–7.
12. Ibid.
13. Ibid.
14. Ibid.
15. Ibid.
16. Ibid.
17. Eziakonwa, Ahunna, "The Impact of Credit Ratings on the Costs of Development in Africa", The Brookings Institution Seminar Transcript, Washington, April 14, 2023.
18. UN Global Crisis Response Group, "A World of Debt: A Growing Burden to Global Prosperity", July 2023.
19. Data from IMF World Economic Outlook, Oct 2023, https://www.imf.org/external/datamapper/NGDP_RPCH@WEO/OEMDC/ADVEC/WEOWORLD/SSA, accessed 28 March 2024.
20. Data One, African Debt, https://data.one.org/topics/african-debt/, accessed 28 March 2024.
21. UNDP report.
22. See World Bank chart on China's official lending exposure to African countries, 2005–2019, Smith, op. cit, 52.
23. Smith, Ibid. 52.
24. For a discussion of Chinese lending before 2009, see Deborah Brautigam, "The Dragon's Gift: The Real Story of China in Africa", Oxford: Oxford University Press, 2009.
25. Smith, op. cit., 48.
26. Smith, op. cit, 49–50.
27. Ibid, 49.
28. Ibid, 50.
29. See Sebastian Horn, Carmen M. Reinhart, and Christoph Trebesch, China's Overseas Lending, Working Paper, 26,050, National Bureau of Economic Research, July 2019.
30. See Storm Brews over China Deal, Zimbabwe Independent, 9 July 2009.
31. Gerard, Neil, **"Why have Chinese construction firms become so influential in Africa?", Construction Briefing, 14 Mar 2023,** https://www.constructionbriefing.com/news/why-have-chinese-construction-firms-become-so-influential-in-africa-/8026251.article, accessed 28 March 2024.

32. Smith, op. cit., 55.
33. 2020 Report to Congress of the US China Economic and Security Review Commission: https://www.uscc.gov/sites/default/files/2020-12/2020_Annual_Report_to_Congress.pdf
34. Gerard.
35. See Aid Data, a research lab at William and Mary, for more information on China's hidden debt.
36. Davis, House Foreign Affairs Subcommittee on Africa.
37. McKinsey & Co. "Capital Mobilization Impacts Resulting from DFC's Political Risk Insurance Product: Impact Assessment Report", May 2023, 7.
38. Ibid, 8.
39. Ibid.
40. Ibid, 9.
41. Ibid, 11.
42. Ibid.
43. Ibid, 12.
44. Ibid, 23.
45. Ibid, 27.
46. Ibid, 34.
47. UNDP Strategy, Analysis and Research Team, "Reducing the Cost of Finance For Africa: The Role of Sovereign Credit Ratings, April 2023, 37.
48. UNDP, 37.
49. Eziakonwa, The Brookings Institution Seminar Transcript, Washington, April 14, 2023.
50. UNDP, 39.
51. Sembenem Daouda, Managing Director, Africatalyst, The Brookings Institution Seminar Transcript, Washington, April 14, 2023.
52. Mahesh Thatcher, The Brookings Institution Seminar Transcript, Washington, April 14, 2023.
53. Ken, Ofori-Atta, Republic of Ghana Minister of Finance, The Brookings Institution Seminar Transcript, Washington, April 14, 2023.
54. Thatcher, Ibid.
55. The Liquidity & Sustainability Facility, https://lsfacility.org/, Accessed 31 March, 2024.
56. Ibid.
57. African Development Bank, https://www.afdb.org/en/topics-and-sectors/initiatives-partnerships/emergency-liquidity-facility-elf, Accessed 31 March, 2024.
58. RLSF, https://safe.menlosecurity.com/doc/docview/viewer/docNAC9C0B955E8F84e6a2dd4000a6c17d00309fe68385b412ae0c6b43c015aeb035ed0296295281. Accessed 31 March, 2024.
59. LSF website.

The manufacturer's authorised representative in the EU is Springer Nature Customer Service Centre GmbH, Europaplatz 3, 69115 Heidelberg, Germany. If you have any concerns regarding our products, please contact ProductSafety@springernature.com

Printed and bound by CPI Group (UK) Ltd, Croydon, CR0 4YY
30/09/2026
02225586-0013